Advanced Field-Effect Transistors

Advanced Field-Effect Transistors: Theory and Applications offers a fresh perspective on the design and analysis of advanced field-effect transistor (FET) devices and their applications. The text emphasizes both fundamental and new paradigms that are essential for upcoming advancement in the field of transistors beyond complementary metal–oxide–semiconductors (CMOS). This book uses lucid, intuitive language to gradually increase the comprehension of readers about the key concepts of FETs, including their theory and applications.

In order to improve readers' learning opportunities, *Advanced Field-Effect Transistors: Theory and Applications* presents a wide range of crucial topics:

- Design and challenges in tunneling FETs
- Various modeling approaches for FETs
- Study of organic thin-film transistors
- Biosensing applications of FETs
- Implementation of memory and logic gates with FETs

The advent of low-power semiconductor devices and related implications for upcoming technology nodes provide valuable insight into low-power devices and their applicability in wireless, biosensing, and circuit aspects. As a result, researchers are constantly looking for new semiconductor devices to meet consumer demand. This book gives more details about all aspects of the low-power technology, including ongoing and prospective circumstances with fundamentals of FET devices as well as sophisticated low-power applications.

Advanced Field-Effect Transistors
Theory and Applications

Edited by
Dharmendra Singh Yadav
Shiromani Balmukund Rahi
Sukeshni Tirkey

CRC Press
Taylor & Francis Group
Boca Raton London New York

CRC Press is an imprint of the
Taylor & Francis Group, an **informa** business

Designed cover image © Shutterstock

First edition published 2024
by CRC Press
6000 Broken Sound Parkway NW, Suite 300, Boca Raton, FL 33487-2742

and by CRC Press
4 Park Square, Milton Park, Abingdon, Oxon, OX14 4RN

CRC Press is an imprint of Taylor & Francis Group, LLC

ISBN: 9781032493800 (hbk)
ISBN: 9781032493879 (pbk)
ISBN: 9781003393542 (ebk)

DOI: 10.1201/9781003393542

Typeset in Times LT Std
by KnowledgeWorks Global Ltd.

Dedication

I would like to dedicate this book and express my hearty gratitude toward my respected parents, uncle, aunty, younger brothers, and sisters for their affection and persistent efforts in my education. Also dedicated to my wife and our loving son for their everlasting supports, encouragements, and understanding. This work is dedicated to my family; my teacher, the late Dr. Dheeraj Sharma Sir; my dear student, Prabhat Singh; and others, who have always been a source of my continued efforts for academic excellence.

—Dr. Dharmendra Singh Yadav

This textbook is dedicated to "almighty GOD and our family members," for their affection, untiring efforts, and patience.

—Dr. Shiromani Balmukund Rahi

To my revered parents, for their affection and untiring efforts in my upbringing. Also dedicated to my husband, Ashish, and our loving son, Avyan, for their precious time and patience.

—Dr. Sukeshni Tirkey

Contents

Preface

Complementary metal–oxide–semiconductor (CMOS) technology has evolved as the dominant fabrication process in the past few decades, and it has become the only option for semiconductor industries, but it's not best suited for low-power applications. The uses of developing semiconductor devices in electronic circuits are explored throughout this book. It addresses systematic and comprehensive engineering problems as well as applications of sophisticated low-power devices. The advent of low-power semiconductor devices and related implications for upcoming technology nodes provide valuable insight into low-power devices and their applicability in wireless, biosensing, and circuit aspects. The book gives more details about all aspects of the low-power technology, including ongoing and prospective circumstances. The book also covers the fundamentals of field-effect transistor (FET) devices and highlights new and upcoming FET technologies. An evaluation of the use of FET devices in diverse fields, such as biosensing, wireless, and cryogenics applications, is also included. In the digital and analog realms, the chapters additionally look at device–circuit co-design difficulties. The information is provided in an easy-to-understand approach, making it excellent for those who are new to the subject. For the reader's reference, we have included many device architectures presented by various researchers. This will help readers understand the strategies for improving the properties of FETs. Furthermore, these strategies will inspire readers to create their own device optimization techniques and research FETs to make them appropriate for forthcoming circuit applications. Finally, the book covers how to perform numerical simulations of FETs using the technology computer-aided design (TCAD) tools Silvaco ATLAS and Sentaurus to assist new researchers in the subject of FETs. In addition, this book endeavors to cover all recent scholarly articles on FETs to ensure that it also covers the state of the art. We have emphasized the qualitative qualities of the devices throughout this book. This was done so that readers could obtain a spontaneous grasp of the devices rather than being obstructed by complex equations. This book is projected to accomplish this prerequisite for device-to-circuit-level research work.

SHORT EXPLANATION OF THE CHAPTERS

Effective communication among research scholars working on device processes, device physics and modeling, circuit designs, sensing applications with advanced materials, and quantum mechanics is a critical facilitator for hatching and advancing innovations from investigation to implementation. Chapters in this book cover:

Chapter 1: This chapter predicts that the CMOS transistor must be able to scale to at least the 3 nm node by the year 2021 to 2022. CMOS might finally run out of steam, prompting the creation of a new switch technology. In order to continue improving the power and performance of computing, it is crucial for the semiconductor industry to find a new integrated circuit technology that can take us beyond the CMOS era. Devices based on electron spin (spintronics) and nano-magnetism are examples of fields that are currently being explored, as are quantum electronic devices such as the tunneling field-effect transistor (TFET).

Chapter 2: This chapter describes tunneling briefly. The tunneling phenomenon and the methodology for calculating the tunneling rate are also offered. By describing the current–voltage statistics, TFET functioning is explained. Ambipolarity and low ON-state current are two concerns that affect TFET effectiveness; methods to address them, such as gate–drain under- and overlap, hetero gate dielectrics structures, and area-scaled TFET topologies, are detailed. It is shown that III-V TFETs can overcome the diminished I_{ON} constraint. A TFET-based circuit design methodology is also reported. Ways to deal with these issues, such as prolonged saturation, increased Miller capacitance, and unidirectional conductivity, are discussed.

Chapter 3: Modeling a FET requires a detailed understanding of its physical operation and an accurate representation of its behavior. This work shows different approaches to FET device modeling, including analytical models, semi-empirical models, and physics-based models. Each approach has its own advantages and disadvantages.

Chapter 4: Organic transistors can be fabricated on flexible substrates, resulting in flexible electronic applications that cannot be achieved with conventional silicon-based transistors. Along with this, trap states and mathematical modeling are explained in detail throughout the organic material structure as deep and shallow traps. The current–voltage equation has been reviewed to discuss the effect of trap parameters on the performance metrics of the device. The trap states affect the device capacitance and add a brief hump of short duration due to trap capacitance.

Chapter 5: This chapter discusses the basic TFET architecture and its operation using the concepts of energy band diagrams. ON- and OFF-states of the device are explained by adding their band diagrams to help readers grasp an understanding of tunneling junction and barrier width. Various simulation, fabrication, and noise-related challenges and issues are addressed in this chapter. A detailed analysis of a TFET-based biosensor is presented.

Chapter 6: This chapter includes an investigation of hetero buried oxide (HBOX) doped-pocket gate-engineered TFET structures. The low-K and high-K oxide layers that are present in the buried oxide layer at the drain and source terminals of TFETs boost the ON-state current and cause the ambipolar current to recede by varying different parameters such as HBOX and silicon thickness. HBOX-TFET performance is analyzed for direct current (DC) and analog/radiofrequency (RF) performance. The exceptional parameters extracted by the proposed HBOX-TFET prove its applicability for low-power DC and analog/RF applications.

Chapter 7: This chapter consists of a brief history of transistors and how they evolved, and it introduces ferroelectric materials and their properties. The addition of a ferroelectric substance to a traditional L-shaped TFET (L-TFET) and its impact are analyzed through the characteristics of the proposed device. The proposed device, a negative-capacitance L-TFET (NC-L-TFET), is compared with L-TFETs in terms of various parameters.

Chapter 8: In this chapter, a Si-doped MoS_2-based step-structure double-gate TFET (MoS_2-SS-DG-TFET) with a wide variety of properties is examined. To enhance device performance, this device uses a low dielectric thickness at the source–channel junction. The MoS_2 material has a special property: Its energy gap varies with the number of material layers. Using this varied number of layers causes a small energy gap at the source side that improves ON-current and a large energy gap at the drain side that minimizes OFF-current. The suggested device can be employed for high-temperature applications, as evidenced by the fluctuation in DC and analog characteristics. In the TCAD simulator, temperatures between 250 K and 500 K have been examined to measure the device's performance.

Chapter 9: In this chapter, compact implementation of logic gates for digital applications using a step-channel TFET (SCTFET) is presented. OR, AND, NOR, and NAND logic gates are implemented by optimizing the gate work function and utilizing the gate-to-source overlap approach. The gate-to-source overlap method is used to implement the AND and NOR logic functions.

Chapter 10: In order to design an effective static random-access memory (SRAM) cell, this chapter provides in-depth research on several SRAM characteristics, including read delay, write delay, read stability, and write stability, of the circuit that is to be designed. Along with these parameters are others like delays, average power dissipation, and stability via changing the cell ratio, the pullup ratio, and different supply voltages. For this, 7T to 13T SRAM (odd numbers of transistors) was investigated to accomplish the improved static noise margin.

Chapter 11: A detailed overview of the Synopsys TCAD software suite (Sentaurus Structure Editor, SDevice, SVisual, and Inspect) is briefly discussed in this chapter. In later sections of the chapter, a novel, vertically stacked, gate-all-around nanosheet field-effect transistor (GAA-NSFET) is designed and simulated. The GAA-NSFET structure enables us to continue metal–oxide–semiconductor field-effect transistor (MOSFET) scaling beyond 10 nm and proves to be a better alternative for the substitution of fin field-effect transistors (FinFETs). The GAA-NSFET structure consists of several stacked channels, thus effectively improving the I_{ON} (ON-state current) and output characteristics of a MOSFET.

Chapter 12: This chapter provides a detailed explanation of basic simulation steps of TCAD tools. This includes basic information on device structure definition, mesh generation, material property assignment, simulation setup, simulation execution, postprocessing, and visualization. Finally, the simulation results are compared to experimental data to validate the accuracy of the simulation. A conventional silicon MOSFET has been designed using a Cogenda TCAD simulator tool to understand the TCAD simulation process.

We believe that learners, researchers, and training engineers in the industry will find this book beneficial.

THANKS TO THE CHAPTER AUTHORS

We extend our heartfelt thanks to all the chapter authors for their contributions and diligent work in this book. When the chance arose to begin writing this book on TFET technology, all writers agreed that now was the opportune moment to look back on the previous decade's research on TFETs as well as compile the most recent findings. With this in mind, the respective authors dedicated their time to produce a thorough, comprehensive, and informative new book for sharing and distributing information.

Editors Biography

Dharmendra Singh Yadav received his PhD in electronics and communication engineering from the PDPM-Indian Institute of Information Technology, Design and Manufacturing, Jabalpur, India. He is currently working as assistant professor (Grade-I) at the National Institute of Technology (NIT), Kurukshetra, Haryana, India. He has more than 60 international publications in reputed journals and four book chapters. His current research interests include

- Very large-scale integration (VLSI) design: nanoelectronics devices, thin-film transistors, semiconductor devices, negative capacitance, nanosheet FETS, and circuits.
- Device modeling: MOS device modeling and numerical simulation, analysis of semiconductor devices, and electrical characterization of semiconductor devices in MHz and THz frequency ranges.
- Circuit design: ultra-low-power SRAM/DRAM/RRAM-based memory circuit design from devices to array architecture using CMOS and advanced CMOS device technologies
- Machine learning in semiconductor device/circuit-based applications in research.

Shiromani Balmukund Rahi received his PhD from the Indian Institute of Technology Kanpur, Uttar Pradesh, and did his postdoctoral research work at the Electronics Department, University Mostefa Benboulaid of Batna, Algeria and Korea Military Academy Seoul, Republic of Korea. He is working at University School of Information and Communication Technology Gautam Buddha University Greater Noida, Uttar Pradesh, India. He has published 25 journal articles, 18 book chapters and 2 proceedings. He has edited successfully 7 books. He is also associated for advanced research work at the Indian Institute of Technology Kanpur and the Electronics Department of the University Mostefa Benboulaid for the development of ultra-low-power devices such as TFETs, negative-capacitance (NC) TFETs, and nanosheet FETs.

Sukeshni Tirkey received her PhD in electronics and communication engineering from the PDPM-Indian Institute of Information Technology, Design and Manufacturing, Jabalpur, India. She is currently working at the Maulana Azad National Institute of Technology (MANIT), Bhopal, Madhya Pradesh, India. She has more than 20 international publications in reputed journals and conference proceedings. Her current research interests include VLSI design, nanoelectronics devices, thin-film transistors, semiconductor devices, negative capacitance, nanosheet FETS, and circuits.

Contributors

Farkhanda Ana
Baba Ghulam Shah Badshah University
Rajouri, Jammu and Kashmir, India

R.K. Baghel
Department of Electronics and
    Communication Engineering
Maulana Azad National Institute of
    Technology (MANIT)
Bhopal, Madhya Pradesh, India

Varun Bharti
National Institute of Technology
Hamirpur, Himachal Pradesh, India

Brinda Bhowmick
National Institute of Technology
Silchar, Assam, India

Bharat Singh Choudhary
Department of Electronics and
    Communication Engineering
Maulana Azad National Institute of
    Technology (MANIT)
Bhopal, Madhya Pradesh, India

Vibhash Choudhary
Department of Electronics and
    Communication Engineering
National Institute of Technology
Hamirpur, Himachal Pradesh, India

Priyanka Goma
Department of Electronics and
    Communication Engineering
National Institute of Technology
Hamirpur, Himachal Pradesh, India

Yuvraj Kadale
National Institute of Technology
Hamirpur, Himachal Pradesh, India

Priya Kaushal
Department of Electronics and
    Communication Engineering
National Institute of Technology
Hamirpur, Himachal Pradesh, India

Gargi Khanna
Department of Electronics and
    Communication Engineering
National Institute of Technology
Hamirpur, Himachal Pradesh, India

Mamta Khosla
National Institute of Technology
Jalandhar, Punjab, India

Haider Mehraj
Baba Ghulam Shah Badshah University
Rajouri, Jammu and Kashmir, India

Sirisha Meriga
National Institute of Technology
Silchar, Assam, India

Archana Pandey
Jaypee Institute of Information
    Technology
Noida, Uttar Pradesh, India

G. Boopathi Raja
Velalar College of Engineering and
    Technology
Erode, Tamil Nadu, India

Shiromani Balmukund Rahi
University School of Information and
    Communication Technology
Gautam Buddha University
Greater Noida, Uttar Pradesh India

Ashish Raman
National Institute of Technology
Jalandhar, Punjab, India

Ashwani K. Rana
Department of Electronics and
  Communication Engineering
National Institute of Technology
Hamirpur, Himachal Pradesh, India

Soumya Sen
National Institute of Technology
Jalandhar, Punjab, India

Abhay Pratap Singh
Department of Electronics and
  Communication Engineering
Maulana Azad National Institute of
  Technology (MANIT)
Bhopal, Madhya Pradesh, India

Anushka Singh
Jaypee Institute of Information
  Technology
Noida, Uttar Pradesh, India

Prabhat Singh
Department of Electronics and
  Communication Engineering
National Institute of Technology
Hamirpur, Hamirpur, India

Sukeshni Tirkey
Department of Electronics and
  Communication Engineering
Maulana Azad National Institute of
  Technology (MANIT)
Bhopal, Madhya Pradesh, India

Najeeb-Ud-Din
National Institute of Technology
Srinagar, Jammu and Kashmir, India

Dharmendra Singh Yadav
Department of Electronics and
  Communication Engineering
National Institute of Technology
Kurukshetra, Haryana, India

1 Future Prospective Beyond-CMOS Technology

From Silicon-Based Devices to Alternate Devices

G. Boopathi Raja

1.1 INTRODUCTION

Manufacturing integrated circuits often involves the use of complementary metal–oxide–semiconductor (CMOS) technology as the primary technology. The CMOS technique is used in a wide variety of electronic components, including microcontrollers, batteries, and digital sensors, because of the many significant benefits that it offers. In order to implement a wide variety of logic operations, this technology makes use of both the N and P MOSFET channels (NMOS and PMOS, respectively), which are both intended to have matching qualities when they are developed.

Before CMOS logic came along, PMOS and NMOS logic were the most common ways to build logic gates. Eventually, the NMOS technology, which had been the industry standard for making integrated circuits, took over from the PMOS technology. CMOS started off being a slower and more costly option than NMOS. The primary benefits of NMOS technology are its straightforward physical process, high functional density, lightning-fast processing speed, and cost-effective production. Electrical asymmetry and the loss of static power are two of the most significant drawbacks associated with NMOS technology. Utilizing CMOS technology helps to reduce the impact of these limitations. The fact that power loss is only ever experienced during the switching of circuits is the primary benefit offered by CMOS technology. This results in significantly improved performance, since it makes it possible to integrate a greater number of CMOS gates into an integrated circuit.

Based on the 2001 edition of *The International Technology Roadmap for Semiconductors*(ITRS), which showed how quickly metal–oxide–semiconductor field-effect transistor (MOSFET) technology was getting better, several new technologies have been developed to expand CMOS into nanoscale MOSFET architectures. In 2023, the semiconductor industry will concentrate on lithography and advanced process nodes. Major semiconductor companies might introduce 3nm and 5nm nodes to increase the density and efficacy of transistors. These developments will facilitate the development of processing units, memory modules, and SoCs.

DOI: 10.1201/9781003393542-1

The packaging innovations allow for miniaturization and integration, resulting in more efficient and compact devices. There is a growing hope that, with the help of these new technologies, it will be possible to expand MOSFETs to the 22-nm node (which corresponds to a 9-nm physical gate length) by the year 2016, if not before. There will almost certainly be some new, innovative materials in these new devices. New MOSFET topologies that do not need bulk materials have been skillfully crafted using these materials. They will have a ravenous desire for power and be exceedingly quick and dense at the same time. There is a possibility that intrinsic device veloci-ties may be more than 1 THz, and there is also a possibility that integration densities could be greater than 1 billion transistors per square centimeter. However, due to their high power consumption, these high-performance devices ought to be used only in critical pathways where their enhanced performance is required. The road-map for semiconductors will emphasize energy management and efficiency. Modern electronics contend with heat production and energy usage. Dynamic frequency and adaptive voltage scaling increase energy efficiency and battery lifespan. It is possible to minimize the total power consumption of the device by using two or even three extra MOSFETs that have lesser performance but greater power efficiency. These MOSFETs might be utilized to carry out functions that are less performance-critical.

In addition to CMOS, fundamentally new technologies as well as architectural designs are now being developed for the processing and storage of data in the future. Rather than attempting to "replace" CMOS, a few of these new ideas, when paired with a CMOS platform, could make it possible to employ microelectronics in meth-ods that are not feasible with CMOS on its own. A new technique of processing infor-mation that is successful will almost surely need an alternative platform technology. It comprises a fabric of connected primitive logic cells, maybe in three dimensions. This is because the old method of processing information was inefficient. This new logic paradigm might also recommend an updated information-processing architec-ture that is consistent with the logic fabric and that makes the most of the capabilities of the logic fabric. The Information Society Technology Project of the European Commission developed the *Technological Roadmap for Nanoelectronics*. It gives a good description of nanoelectronics devices (emerging technologies).

"Beyond CMOS" refers to improvements in digital logic that are expected to come in the future and go beyond the scaling limits set by CMOS. These limitations are intended to minimize heat impacts by restricting the speed and intensity of the device. Digital logic is necessary for the creation of electronic and logical devices. It enables us to perform activities critical to the success of a system, such as design-ing circuits and testing computer chips. CMOS is the acronym for complementary metal–oxide–semiconductors, which are the standard on-and-off switches in mod-ern semiconductor products [1].

In other words, "beyond CMOS" refers to new technologies in digital logic or emerging technologies in general that are used to explain the events and signals in a digital circuit. It is expected that these technologies would get around the scaling problems caused by CMOS [2, 3]. It has already been surpassed by an order of mag-nitude for feature size and by two orders of magnitude for speed [4, 5].

Moore's law will be partially met by new materials, architectures, devices, and topologies that will be developed over time and referred to as "beyondCMOS." One illustration of this is the speculation that the 32-nm CMOS integrated circuit will be developed by the year 2020. Its primary purpose is to operate as a supplement to the CMOS technology that is now in use, enabling the production of circuits with a greater switching capacity and an improvement in the information storage capacity of such devices. This is necessary since the CMOS technology that is currently in use will not be able to surpass certain capacity constraints.

Experts say that beyond-CMOS technology will eventually replace silicon because the electrical properties of graphene are better than silicon's physical limits. BeyondCMOS is looking into new technologies like nanostructures made of carbon nanotubes (CNTs) and spintronics, which send information using the charge and spin of electrons.

The following are the primary motivations for launching beyond-CMOS research and development:

- When the amount of power used goes up, the speed at which the circuit works doesn't go up enough to keep up.
- Because of the increase in power, there will also be a rise in the temperature of the gadget, which will harm its functionality.
- An increase in the number of faults at both the lithography and design levels. This is because, when dealing with such tiny scales, it is quite simple to make mistakes in either the printing or the circuit design.

The following is the chapter's organizational structure: Section 2 discusses the various alternatives to beyond-CMOS technology that are based on previous semiconductor technologies. Section 3 elaborates on the role of industries and researchers in the advancement of beyond-CMOS technology. Sections 4 and 5 discuss applications of beyond-CMOS technology and alternatives to CMOS electronics. Section 6 explains futuristic gadgets that use beyond-CMOS technology. Section 7 summarizes, in the conclusion, the important key features of technology beyond CMOS.

1.2   BEYOND-CMOS TECHNOLOGY FROM EXISTING AND PAST SEMICONDUCTOR TECHNOLOGIES

"Beyond CMOS" refers to the computer logic techniques. It may be created in the future, and it will be able to scale beyond the CMOS scaling constraints. It will restrict the density and speed of devices since they heat up. Beyond-CMOS technologies will be capable to scale beyond the CMOS scaling limits.

The International Technology Roadmap for Semiconductors (ITRS) 2.0 was indeed a significant document in the semiconductor industry, published in 2013 as a continuation of the original ITRS roadmap. It aimed to provide guidance and predictions for the future development of semiconductor technology. However, ITRS 2.0 was succeeded by the International Roadmap for Devices and Systems (IRDS) in 2017. Within ITRS 2.0 (2013) and its successor, the *International Roadmap for Devices and Systems*(IRDS), there are a total of seven focus groups, and one of them is called Beyond CMOS. It is focused on exploring and defining technologies and devices that go beyond conventional Complementary Metal-Oxide-Semiconductor

(CMOS) technology. CMOS technology has been the basis of semiconductor manufacturing for several decades, but as it approaches its physical limits, the industry is researching and developing new technologies to continue advancing the field.

The Beyond CMOS focus group within IRDS is tasked with assessing and guiding the development of emerging technologies that can potentially replace or complement CMOS technology as it becomes increasingly challenging to scale down. This includes exploring novel materials, devices, and architectures that could pave the way for the next generation of semiconductor technology.

Central processing unit (CPU) clock scaling: Computers with CMOS processors (like the 12 MHz Intel 80386) were first sold to the public in 1986. The clock speeds of CMOS transistors increased in tandem with their dimensions being reduced. The clock speeds of CMOS CPUs have remained stable at approximately 3.5 GHz since approximately 2004.

A comparison of the efficiency gains that could be made with "more Moore" (i.e., more improvements to the current technology) and "beyond CMOS": The following is an excerpt from the IRDS. CMOS device sizes are continuing to shrink; for examples, see Intel Tick-Tock and ITRS: Ivy Bridge at 22 nm was released in 2012, and the first processors at 14 nm were distributed in the fourth quarter of 2014.

Samsung Electronics displayed a 300-mm wafer of 10-nm fin field-effect transistor (FinFET) chips in May 2015. Beyond-CMOS research and development focuses on the extension of integrated circuit technology to truly innovative methodologies in order to construct the optimal paths for technological breakthroughs far beyond the conclusion of CMOS dimensional scaling. Even though semiconductor technology has changed a lot over the years, other technologies that are based on research have also changed a lot.

The primary objective of this chapter is to survey, evaluate, and classify promising new technologies in terms of their long-term viability and technological maturity. This chapter also talks about the scientific and technical problems that keep the semiconductor industry from using them. These problems are seen as acceptable risks for future development. Researchers are developing both nonvolatile as well as volatile memory technologies in an attempt to replace static random-access memory (SRAM) and FLASH in the appropriate applications. Due to present scaling limitations, it is essential to obtain electrically accessible memories that are:

- Embeddable
- Low power
- High density
- High speed
- Possibly nonvolatile

The sizes of modern memory systems vary substantially. Some, especially mobile systems, might be sufficiently small to be gigabyte-based, whereas others might need some extra bytes of storage. Regardless of system size, the issue is the same: The majority of computer systems seldom run at peak demand. Even with modest consumption rates, modern data servers still need a great deal of power.

These kinds of difficulties highlight the potential usefulness of permanent memory. If this capability did not need to be renewed regularly, the operating strain on logic devices could be greatly reduced. Refresh power may use as much as one-third of the energy required by a large computer system; if this power were made available, customers might use it to operate devices that are far faster and more powerful.

1.2.1 Moore's Law

Gordon E. Moore, the cofounder of Intel, hypothesized in 1965 that the total amount of transistors that could be crammed into a given volume would approximately double every two years. This prediction was based on the fact that transistors are used in electronic circuits.

Gordon E. Moore did not refer to his observation as "Moore's law," and neither did he set out to produce a "law" when he remarked. Moore based his comment on observations he made at Fairchild Semiconductor, where he worked, on future trends in the manufacture of chips. Moore's observation was eventually turned into a prediction, which led to the idea that became known as Moore's law [5].

Moore's law states that the number of transistors on a microchip may double every two years. The law ensures that computer speed and capacity will continue to improve at a rate of 100% every two years, while their costs will fall. The second rule of Moore's law states that this growth is happening at an exponential rate. Gordon Moore, who helped build Intel and was once the CEO of the company, is credited with coming up with the idea for the law. Figure 1.1 shows the evolution of Moore's law.

1.2.2 Opinion of the IRDS™ on Beyond CMOS

The goal of the IRDS™ is to find beyond-CMOS devices that will help make possible new ways of using computers that go far beyond what is possible with traditional CMOS designs and technology. The Beyond CMOS focus group of the IRDS™ was made with the specific goal of finding research opportunities, evaluating existing devices, and mapping future devices [6].

The IRDS™ Beyond CMOS section has several pillars on which its projects are built. New gadgets for cognitive processing, logic, and storage receive considerable attention in this chapter. The IRDS™ community has a strong interest in many emerging application domains. Elements of big data analytics, device security, and intelligent computing are discussed in this chapter.

The Beyond CMOS study found that the circuit, architect, and device groups are now trying to reach more than one goal. Coordinating the creation of new devices is very important if they want to make big steps forward in the field of computing.

While Moore's law and typical CMOS scaling are expected to continue for the next few years, the IRDS™ recognizes that we are fast nearing the limits of conventional scaling. Scaling as we know it will likely stop due to issues such as increasing pricing and fundamental physical effects. The roadmap admits that challenges such

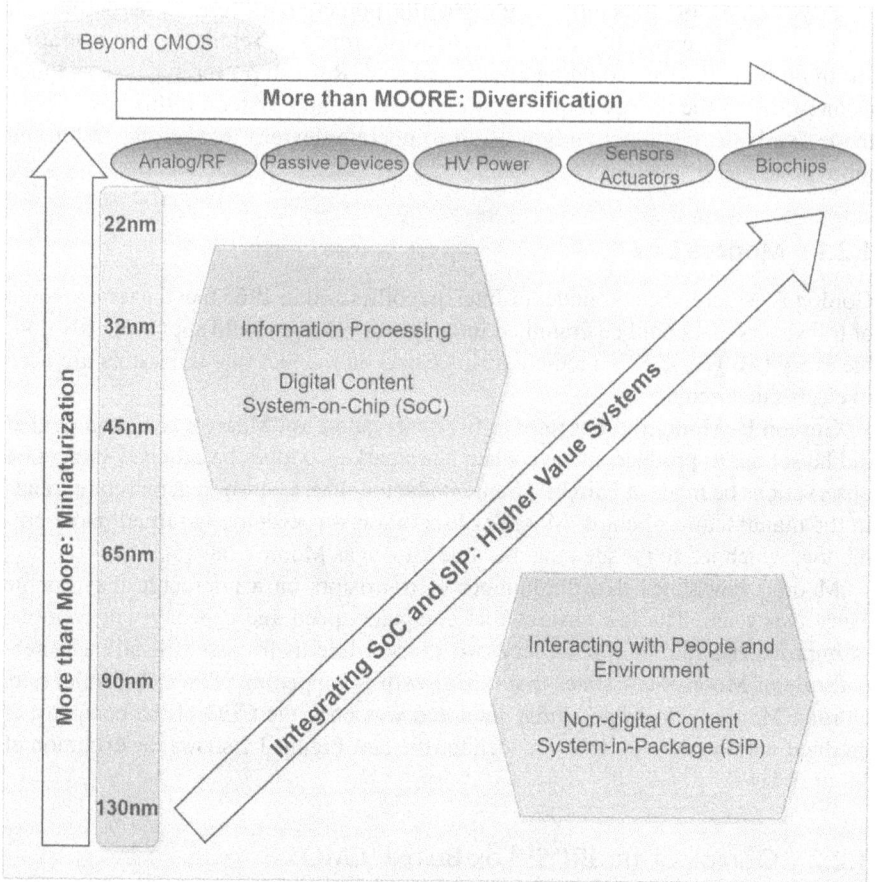

FIGURE 1.1 Evolution of Moore's law.

as gate length (outside system connectivity) are one illustration of how statistical distributions and critical dimensions may impede CMOS scaling attempts.

The IRDS™ is designed to provide maximum performance with the lowest energy usage. As a consequence of recent advances and improvements in the technological sector, the emphasis has shifted toward new computing paradigms, capabilities, and applications. Several guiding principles for present and future beyond-CMOS research are stated in the roadmap.

1. Nanoscale thermal management
2. Nonthermal equilibrium systems
3. Computational state variable(s) other than solely electron charge
4. Sub-lithographic manufacturing process
5. Novel energy transfer interactions
6. Alternative architectures

1.2.3 EVOLUTION OF CMOS AND BEYOND CMOS

Even though the performance of integrated circuits has been getting better and better over the past 30 years, CMOS technology has been around for much longer. Frank Wanlass came up with the method in 1963, but it didn't get much use until the 256 Kb CMOS dynamic random-access memory (DRAM) came out in 1984. Since then, CMOS technology has been a key part of the development of many digital circuits, including electronic parts, microcontrollers, embedded systems, sensors, memory, and many more.

Since the unfortunate discovery (made by Gordon Moore in 1965) that integrated circuit transistors quadruple every 18 months, a lot has changed. Researchers have put in a lot of work to find ways to deal with the problems that standard CMOS causes. High gate leakage currents, gate stack reliability, source-to-drain leakage, and channel mobility degradation have all played a big part in how CMOS has changed over time.

A new CMOS technology node (i.e., the advancement of CMOS technology at 32 nm and upwards) has developed about every 24 months during the past 15 years. Each node has its changes, which usually means that density goes down by 2% and performance goes up by 35% per node. Because of limits on power and the ability of chips to use more power, CMOS scaling has become a technology that focuses more on density. Figure 1.2 shows how the number of transistors used in CPUs and memory-integrated circuits has increased over time [7].

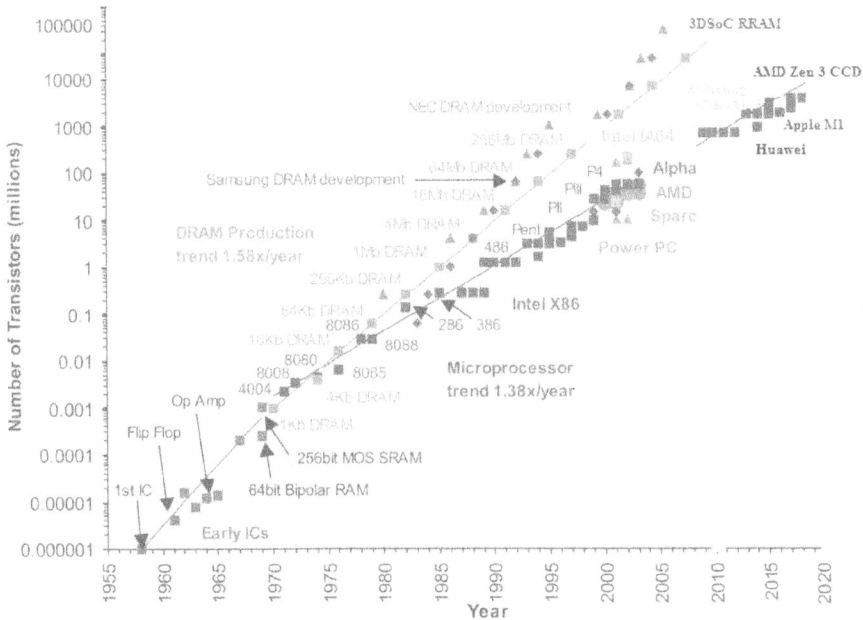

FIGURE 1.2 Illustration of how the number of transistors in central processing units (CPUs) and memory-integrated circuits has grown over time [7].

Nonclassical CMOS structures include improved MOSFETs. Utilizing novel transistor structural designs provides the way for scaling CMOS within the timeframe of the 2001 *Roadmap*. This may be accomplished via the use of nonclassical CMOS structures. Nonclassical submissions include an ultrathin-body silicon-on-insulator (SOI), a band-designed transistor, and three double-gate architectures. These structural alterations are contrasted in Table 1.1, and also their implications are described.

1.2.4 What Makes Beyond-CMOS Technology Feasible Now?

It is important to keep in mind that a lot of beyond-CMOS technology can't be used in the real world right now. Commercially, specialized products are available, although beyond-CMOS capabilities are seldom used by the general population. It is now possible to work toward these capabilities [8].

This study on beyond CMOS is possible right now because of several factors, such as the need for it, external technological innovation and progress, and a very high level of adaptability [9, 10]. Scaling CMOS technology is rapidly losing its potential. Academicians and researchers from across the globe are driven to investigate these technologies due to the lack of alternatives [11–13].

The huge amount of work that has gone into making computers better over the past few decades has also made a big difference in how good they are now. Professional research organizations have a keen interest in resistive-switching electronics, magnetic devices, and a range of other technologies and capabilities that have been studied for a considerable amount of time.

Because the possibilities of beyond-CMOS technologies are so diverse, researchers have a lot of leeway in deciding where to focus their attention and efforts. This means that people who know a lot about circuits and designs, for example, can focus on them and give detailed feedback on development, while other groups can work on building nano- and microelectromechanical systems (NEMS and MEMS, respectively) or improving magnetic logic [14, 15].

Together, these factors, as well as the efforts of forward-thinking individuals and organizations that recognized the need for this technology, made it possible for us to begin developing beyond-CMOS technology. Research and development might stop moving forward if there weren't so many ways to go, so many specialties to choose from, and new ideas coming from outside the field.

"Beyond-CMOS technology" refers to the new materials, structures, devices, and architectures that will be made in the far future (when CMOS technology reaches its physical limits) and should be in used semiconductor industry by 2020. The main goal of this innovation is to work as a supplement to CMOS technology so that it is easier to make circuits with more space to store information.

Table 1.2 provides a review of the documented research efforts that have been made on numerous different memory technologies. As may be observed, current research initiatives are investigating a wide array of fundamental memory systems. These processes consist of charge retention by Coulomb blockade potential, charge retention by magnetic phenomena, surrounding dielectrics, chemical phenomena, and phase alterations in the material. Shortly, the majority of these memory options will likely be integrated into a CMOS-based technology platform. In the context of

TABLE 1.1
Nonclassical CMOS

Device	Methodology	Application/Driver	Advantages	Scaling Issues	Design Challenges
Ultrathin-Body SOI	Fully depleted SOI	• Higher performance • Higher transistor density • Lower power dissipation	• Improved subthreshold slope • Vt controllability	• Si film thickness • Gate stack • Worse short-channel effect than bulk CMOS	• Device characterization • Compact model and parameter extraction
Band-Engineered Transistor	SiGe or strained Si channel Bulk Si or SOI		• Higher drive current • Compatible with bulk and SOI CMOS	• High-mobility film thickness, in case of SOI • Gate stack • Integration	• Device characterization
Vertical Transistor	Double-gate or surround-gate structure (no specific temporal sequence for these three structures is intended)		• Higher drive current • Lithography-independent Lg	• Si film thickness • Gate stack • Integrability • Process • Accurate TCAD complexity, including QM effect	• Device characterization • PD versus FD • Compact model and parameter extraction
FinFET			• Higher drive current • Improved subthreshold slope • Improved short-channel effect • Stacked NAND gate	• Si film thickness • Gate stack • Integrability • Process • Accurate TCAD complexity, including QM effect	• Applicability to mixed-signal applications
Double-Gate Transistor				• Gate alignment • Gate stack • Integrability • Process • Accurate TCAD complexity, including QM effect	

Note: CMOS: Complementary metal–oxide–semiconductor; FD: Fully Depleted; FinFet: Fin field-effect transistor; PD: Partially Depleted; QM: Quantum Mechanics; SOI: Silicon-on-insulator; TCAD: Technology CAD; Vt: Threshold Voltage.

TABLE 1.2
Emerging Research Memory Devices

Storage Mechanism	Device Types	Availability	General Advantages	Challenges	Maturity
Baseline 2002 Technologies	DRAM NOR FLASH	2002	• Density • Economy • Nonvolatile	• Scaling	Production
Magnetic RAM	Pseudo-spin valve Magnetic tunnel junction	2004	• Nonvolatile • Long-lasting • Rapid reading and writing • Hard radiation • NDRO	• Integration problems • Material quality • Magnetic property management for write operations	Development
Phase-Change Memory	OUM	2004	• Low power • Nonvolatile • NDRO • Hard radiation	• New materials and integration	Development
Nano Floating-Gate Memory	Barrier-engineered tunnel nanocrystals	2005	• Nonvolatile • Fast read and write	• Material quality	Demonstrated
Single/few Electron Memories	SET	2007	• Density • Power	• Dimensional control (room temperature operation) • Background charge	Demonstrated
Molecular Memories	Bistable switch molecular devices	2010	• Density • Energy • Equivalent switches • Larger I/O disparity • Possibilities for 3D • Less difficult to communicate • Tolerance-based circuitry	Volatile thermal stability	Demonstrated

Note: DRAM: Dynamic random-access memory; I/O: Input/output; NDRO: Non-Destructive Read Out; NOR: NOR gate; OUM: Ovonic Unified Memory; SET: Single-Electron Transistor.

CMOS platform technology, fabrication may be thought of as a sort of modification or addition. The following technologies are used as benchmarks: DRAM and FLASH NOR. Table 1.2 compares the properties of developing research memory devices with those of present DRAM and FLASH NOR technologies. DRAM and FLASH NOR are the two types of memories that create the most volume at the moment. Magnetic random-access memory (MRAM) and phase-change random-access memory (PCRAM) are two of the new memory technologies that may become available in the near future by 2030. Nonvolatility is one of the primary motivating factors for the development of both of these technologies.

BeyondCMOS is being studied in a lot of different ways. Some of the things that are being looked at are magnetic devices, MEMS, response electronics, and devices that use two-dimensional (2D) materials. The following are some of the most noteworthy investigations:

- Circuits and architectures include both circuits that only use new technology and hybrid circuits that use both new and old technology. Research is being done on massively parallel architectures, processors for applications like image processing and pattern recognition, and circuits for radiofrequency (RF) systems.
- *Magnetic logic*: The logic that is built with new magnetic devices is called "mLogic," and it is being looked at as a possible technology for developing electronic systems in a world where energy is a limited resource. Because it is made without semiconductors and only uses magnetic devices, each logic gate can act as its own independent nonvolatile storage element. The mLogic systems can be powered with as little as 100 mV of voltage.
- MEMS and NEMS are embedded and specialized miniature systems that consist of one or more components or micromachined structures that function as sensors or actuators to enable functions at a higher level inside the structure of a more complex system. Embedded and specialized miniature systems can be broken down into two categories: generalized and specific. Whereas MEMS technology operates on a micrometric scale, NEMS technology operates on a nanometric scale.
- *Resistive switching electronics*: The materials described here can be used to add new features to classic CMOS, like memory cells, nonlinear two-terminal selection devices, RF signal switches, oscillator relaxation, and surge protection devices. These materials might be as simple as metal oxides or as complex as phase-transition chalcogenides.
- *Devices based on 2D materials*: Graphene and other materials have unique electrical and optical properties that can be used to make brand-new electronic devices. These could be used in areas like photonics and neural networks.

In contrast to memory technologies, which are thought to have a wide range of potential applications, beyond-CMOS solutions for logic devices are projected to have a greater emphasis on application specificity. One further key difference between beyond-CMOS memory logic is that, although it is believed that memory alternatives may be incorporated into a CMOS technological platform, it may be more difficult to do so for logic devices. This distinction is one of the most important differences between the two. Table 1.4 provides an overview of published research on key developing logic technologies.

In an approach similar to that of memory devices, contemporary logical technologies need to satisfy some basic needs. In the context of beyond CMOS technology, akin to memory devices, modern logical technologies must fulfill fundamental requirements and exhibit compelling characteristics to justify the substantial investments needed for the development of new infrastructure. This rationale is crucial, as the establishment of fresh infrastructure entails significant resource allocation. Every newly developed information-processing technology needs to first and foremost be capable of satisfying the following requirements:

- Functionally scalable considerably beyond (>100x) CMOS
- A remarkable pace of data processing as well as throughput
- Minimum amount of energy required for each operational function
- The lowest possible scalable cost for each functional activity

Spintronics is the technique that has been used the most in the creation of beyondCMOS. It is used to fix the problem of power loss that can happen in integrated circuits with a lot of transistors. Quantum mechanics, which shows how the spin of an electron can be used, backs up this idea. It is possible to think of it as a brief magnetic moment that is experienced by an electron. Having control over the spin's polarity is a huge benefit. This makes it possible to take into account the amount of energy that is lost in a transistor.

1.3   ROLE OF INDUSTRY IN THE ADVANCEMENT OF BEYOND-CMOS TECHNOLOGY

Beyond CMOS would not be where it is now without the enormous contributions made by Intel. When combined with ultralow power while sleeping, the business's unique magneto-electric spin-orbit (MESO) logic circuit can cut voltage by five times and energy use by 10–30 times [16]. The technology was created and presented by the company. Modern semiconductors based on complementary metal oxides are incapable of such feats.

The term "beyond CMOS" refers to digital logic technologies that might be developed in the future. These will be capable of scaling beyond the current CMOS scaling constraints. These scaling limits restrict the density and speed of devices because they heat up; hence, "beyond CMOS" refers to digital logic technologies that will be able to scale beyond these limits [17].

Intel is doing secret research on possible ideas for the time after CMOS, and several other companies and scientists have also helped. The Nanoelectronics Research Initiative (NRI) is going on a journey to evaluate these potential alternatives to one another as semiconductor research organizations throughout the globe scramble to develop a viable successor to CMOS.UCLA's California NanoSystems Institute (CNSI) has pioneered several research programs in the field, recognizing the global

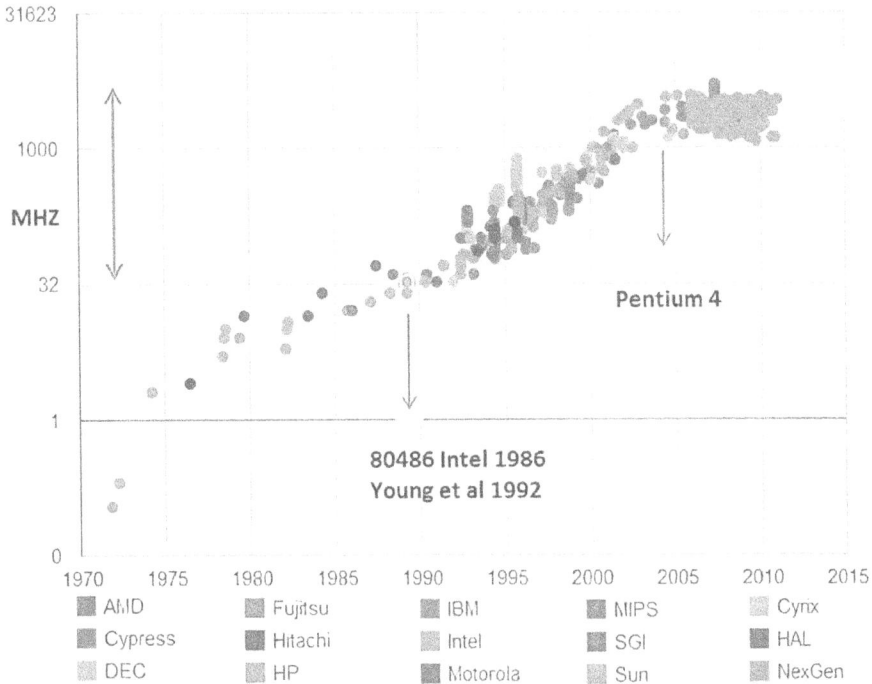

FIGURE 1.3   Advancements in CPU clock scaling.

need for beyond-CMOS technology. CNSI researchers are devoted to producing adaptive, accessible electrical gadgets that enable us to imagine a future without CMOS.

As mentioned earlier, "Beyond CMOS" is the name of one of the seven focus groups of ITRS 2.0 (2013) and its successor, the IRDS. Beginning in 1986, there was a proliferation of CMOS-based CPUs (e.g., the 12 MHz Intel 80386). Figure 1.3 shows the advancements in CPU clock scaling over the years.

1.3.1   THE IMPACT OF BEYOND-CMOS TECHNOLOGY DEVELOPMENTS ON OTHER RESEARCH FIELDS

Most likely, the easiest way to find related study disciplines is to look at the IRDS emphasis areas again. Beyond CMOS is a big and important part of the goals that have been set, but there are more important parts for research and development [6, 18, 19]. Several roadmap International Focus Teams (IFTs) were made to see where things stood and how much progress could be made in the following areas:

- Systems and architectures
- Lithography

- Yield enhancement
- Application benchmarking
- More Moore (a term from Moore's law that describes how technology has gotten better over time)
- Making standard integrated circuits
- Outside system connectivity
- Emerging research materials

Figure 1.4 presents a performance comparison between "more Moore," which refers to the practice of producing more enhancements to the now-available technology, and "beyond CMOS," which refers to the practice of making significant alterations to how technology operates.

The demand for the creation of new system designs is being driven by the development of innovative technologies for wafer bonding. This is urging both research toward the three-dimensional (3D) integration of silicon devices as well as the

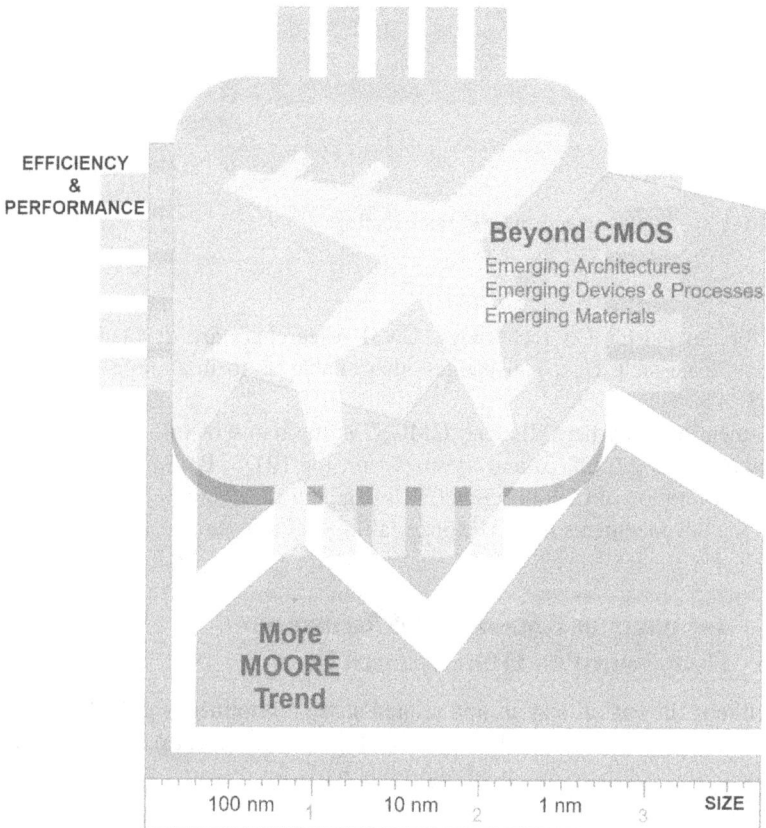

FIGURE 1.4 A comparison of the potential productivity increases that may be achieved with "more Moore" (that is, more enhancements to the existing technology) versus "beyond CMOS."

construction of new system architectures. Similarly, the creation of optical input/output with the potential for high bandwidth could influence the design of a new architecture that might take advantage of this capacity. This possibility exists because of the progress of optical input/output. As a consequence of this, the development of new processes might be seen as the possibility of new architectures that were not previously realizable. The following research architectures that are on the horizon are outlined in Table 1.3.

TABLE 1.3
Emerging Research Architectures

Architectures	Device Implementation	Challenges	Advantages	Maturity
3D Integration	CMOS employing a variety of different material systems	• Heat elimination • Absence of design tools • Challenging test and measurement	• Less interconnect delay • Enables mixed technology solutions	Demonstration
Quantum Cellular Automata	Arrays of quantum dots	• Limited fan output • Dimensional control (low-temperature operation) • Background charge sensitive	• Excellent functional density • No connections in signal path	Demonstration
Defect-Tolerant Architecture	Intelligently assembles nanodevices	• Requires pre-computing test	• Helps hardware that has defect densities that are greater than 50%	Demonstration
Molecular Architecture	Molecular switches and memories	• Limited functionality	• Allows for computing that is done in memory	Concept
Cellular Nonlinear Networks	Single-electron array architectures	• Subject to background noise • Tight tolerances	• Makes it possible to use single-electron devices at room temperature	Demonstration
Quantum Computing	• Spin resonance transistors • NMR devices • Single-flux quantum devices	• Extreme application limitation • Extreme technology	• Enables unbreakable cryptography and exponential performance scaling	Concept

Note: CMOS: Complementary metal–oxide–semiconductor; NMR: Nuclear magnetic resonance.

1.4 APPLICATIONS OF BEYOND-CMOS TECHNOLOGY

1.4.1 EXISTING BEYOND-CMOS TECHNOLOGY APPLICATIONS

Beyond-CMOS process technology is being used in a variety of specialized applications. For instance, Josephson junctions are essential for the deployment of qubits as well as the management systems that govern them. Due to the temperature sensitivity of qubits, Josephson junctions are the optimal option due to their orders-of-magnitude lower power dissipation per computation than CMOS.

Aside from CMOS, technology has almost always played a significant role in the medical industry. Superconducting quantum interference devices (SQUIDs), which are sensitive superconducting devices, have made noninvasive evaluations of electrophysiological activity feasible. Magnetoencephalography (MEG)is a magnetic-source imaging technique that has a price tag of over $500 million USD by the year 2020. Aside from CMOS technology, medical professionals can use magnetocardiography to detect major issues, such as abnormalities in the fetal heart rhythm.

1.5 ALTERNATIVES FOR CMOS ELECTRONICS

In information technology, the performance of microprocessors has gotten a lot better over the last 20 years. Moore's law states that every two years, the device complexity and hence the performance of microprocessors will double. For this performance boost to happen, the size of CMOS transistors, which are the brains of microprocessors, must be greatly reduced. Currently, CMOS electronics are composed of structures of around 100 nm. Consequently, the functionally critical components of the transistor now have just a few dimensions on the atomic-layer scale. Soon, CMOS component designs will approach the sub-100-nm level by the year 2030. Extreme ultraviolet (EUV), X-ray, and electron-beam lithography, as well as scanning probe methods, nanoimprinting, and self-organization processes, are now being investigated for the fabrication of structures smaller than 100 nm.

Conventional CMOS technology suffers from physical restrictions. CMOS will ultimately have a structural width limit of 20–30 nm due to the properties of electron waves. Miniature wires require a reduction in current density to address the electromigration issue. Further circuit miniaturization is also hampered from a financial aspect, since it is projected that manufacturing costs for such microchips would increase faster than their market return. Several technologies, including spintronics, molecular electronics, and quantum information processing, are proposed as possible future alternatives for CMOS technology.

All of these ideas incorporate genuine, practical nanotechnology components. The market could shift in the future to 3D devices that are monolithic or stackable. Surprisingly, 3D electronics, graphene, and CNTs did not meet the requirements according to the most recent performance benchmarks from Intel. The top five technologies on Intel's most promising devices are the spin-majority gate, spin-wave devices, III–V tunnel field-effect transistors (TFETs), heterojunction TFETs, and graphene nanoribbon (GNR) TFETs. GNR TFETs are a new type of TFET that uses GNRs as the tunneling material. Decisions need to be taken in the next 10 years to find workable CMOS replacements by the year 2025. Benchmarking techniques and measurements are being utilized to organize and direct the study of materials, devices, and circuits.

1.5.1 MOLECULAR ELECTRONICS

The main goal of the field of molecular electronics is to use organic and/or biological molecules to make electronic functions and/or parts. Some of the main problems that this field of study tries to solve are how to change switching processes, how fast switching happens, how to make large molecular circuits, how to design good processors, and how molecular circuits interact with the larger world [20, 21]. Approaches to self-organization are being studied, specifically for the development of molecular circuits. These methods should make it possible to produce these circuits at a lower cost. Since fundamental research is still in its infancy, molecule electronics are not yet commercially viable. Due to their unique electrical properties, CNTs are a promising class of materials for molecular nanoelectronics. This includes components for transistors and logic circuits as well as the small wires that connect them.

1.5.2 SPINTRONICS

Magnetoelectronics is frequently viewed as the natural progression that leads to spintronics. For data processing, spintronics uses both the magnetic moment and charge of a potential electron. Existing projections indicate that components that just alter the spin of electrons will process information far more quickly than those that depend on electrical charge for very rapid data processing. In addition, switching would cost less energy compared to an equivalent charge transfer. Due to the magnetic moment providing an extra degree of freedom for the electron, spintronics may indeed be developed alongside charge-based data processing. Quantum computers might be developed more quickly in the long run if they make use of nuclear spin in addition to electron spin. The first component in the field of data storage to make use of electron spin has been successfully manufactured in mass quantities. The most recent generation of the thin-film read heads for hard disc drives makes use of a read head known as a spin valve. It takes advantage of the giant magnetoresistance (GMR)effect. MRAM is a highly promising candidate for the development of future spintronic data storage components. MRAM is an alternative that could be used in place of DRAM or FLASH memory [22–25].

1.5.3 QUANTUM COMPUTING

Quantum effects appear as disconcerting nanoscale effects that decrease the performance of conventional components. Quantum information processing, on the other hand, is based on the exact use of quantum effects for a brand-new sort of massively parallel data processing. The fundamental unit of information in a quantum computer, known as a "qubit" or "quantum bit," is more quaternary than binary. A unique sort of "quantum parallelism" that enables the "simultaneous" exploration of an exponentially large number of computing routes on a single device looks to be realizable by using the superposition and entanglement properties wisely. Since its inception, quantum information processing has made several significant strides forward, such as the creation of two- and three-qubit quantum computers that can carry out fundamental arithmetic operations and sort data. The development of a quantum computer that could compete with the most advanced digital computer available today is hampered by a few potentially major obstacles. Error correction, decoherence, and hardware design are three of the areas that have the potential to be the most challenging.

1.5.4 LOGICS WITH TUNNELING COMPONENTS

Tunneling components, such as resonant tunneling diodes, take advantage of the very fast quantum mechanical tunneling effect. This provides a substantial gain in speed as compared to components that are more traditionally used. Photodetectors, optoelectronic switches, and high-frequency oscillators in the terahertz region are all examples of applications for resistive temperature detectors (RTDs) manufactured from III–V semiconductors. There are applications for digital electronics with ultra-fast, energy-efficient processors in satellite communication systems. Unfortunately, there are now so few RTD-based transistor designs. Nevertheless, the earliest logical circuits were already in existence. Because the qualities of the components depend greatly on the geometry of the component, a rigorous manufacturing process is very difficult.

RTDs will continue to be a specialist application if there is no more research and development done in this field. Si/SiGe RTDs are thought to have more potential because they can be built into standard silicon circuits. However, there are still a lot of technical problems to solve. Their sensitivity to radiation should be taken into account when thinking about how RTDs could be used in space. This research was done in part by working together. Radiation defects significantly reduce the tunneling current, according to early tests on the radiation sensitivity of RTDs.

Table 1.4 shows the performance analysis of different memory technologies such as SRAM, DRAM, FLASH, and so on. The comparison of various memory cells was done based on several parameters, including read time, write time, write power, and full scalability.

A comparative performance analysis of 6T, 7T, 8T, 9T, and 10T SRAM cells is shown in Table 1.5.

The efficiency of graphene nanoribbon field-effect transistor (GNRFET)-based SRAM cells is compared with that of CMOS-, FinFET-, and carbon nanotube field-effect transistor (CNTFET)-based SRAM cells in Table 1.6.

1.6 FUTURISTIC GADGETS USING BEYOND-CMOS TECHNOLOGY

In the near future, several types of novel devices will be crucial for the development of security applications. Since hardware security turns into a greater aspect of the design, it is anticipated that beyond-CMOS technologies will assume the passive function that CMOS technology currently performs in security [17, 27]. Certain characteristics may be useful for simplifying circuit structures to improve protection. Figure 1.5 shows that, around the turn of the 20th century, an era in which silicon technology was scaled geometrically to ever smaller dimensions came to an end [14, 28–30].

TABLE 1.4
Performance Comparison of Different Memory Technologies

Parameters	Typical Memory Technology			New Memory Technology				
	DRAM	SRAM	Flash	FeRAM	ReRAM	PCRAM	STT-MRAM	SOT-MRAM
Cell Size	6–10	50–120	5	15–34	6–10	4–19	6–20	6–20
Nonvolatility	No	No	Yes	Yes	Yes	Yes	Yes	Yes
Write Time (ns)	50	≤2	10^6	10	50	10^2	10	≤10
Read Time (ns)	30	≤2	10^3	5	1–20	2	1–20	≤10
Write Power	Low	Low	High	Low	Medium	Low	Low	Low
Future Scalability	Limited	Good	Limited	Limited	Medium	Limited	Good	Good

Note: DRAM: Dynamic random-access memory; FeRAM: Ferroelectric random-access memory; PCRAM: Phase-change random-access memory; ReRAM: Resistive random-access memory; SOT-MRAM: Spin–orbit torque magnetic random-access memory; SRAM: Static random-access memory; STT-MRAM: Spin-transfer torque magnetic random-access memory.

TABLE 1.5
Performance Comparison of 6T, 7T, 8T, 9T, and 10T SRAM Cells

SRAM Memory Cell	SNM	Write Margin	Dynamic Power	Delay
6T SRAM	202	340	10.05	5.9
7T SRAM	223	360	4.87	3.6
8T SRAM	397	379	12.46	6.5
9T SRAM	410	330	16.55	7.6
10T SRAM	432	475	17.58	8.0

Note: SNM: Static noise margin; SRAM: Static random-access memory.
Source: [12]

TABLE 1.6
Performance Comparison of a Standard 6T SRAM Cell at a 32 nm Technology Node [12], [26]

Parameters Used	CMOS-Based Design	FinFET-Based Design	GNRFET-Based Design	CNTFET-Based Design
Average Power Consumption	16.6 nW	10.23 nW	8.23 nW	6.21 nW
Total Voltage Source Power Dissipation	5.51 nW	34 nW	0.40 nW	32.9 pW
Average Delay	0.29 μs	4.65 ns	3.75 ns	2.50 ns

Note: CMOS: Complementary metal–oxide–semiconductor; CNTFET: Carbon nanotube field-effect transistor; FinFET: Fin field-effect transistor; GNRFET: Graphene nanoribbon field-effect transistor.

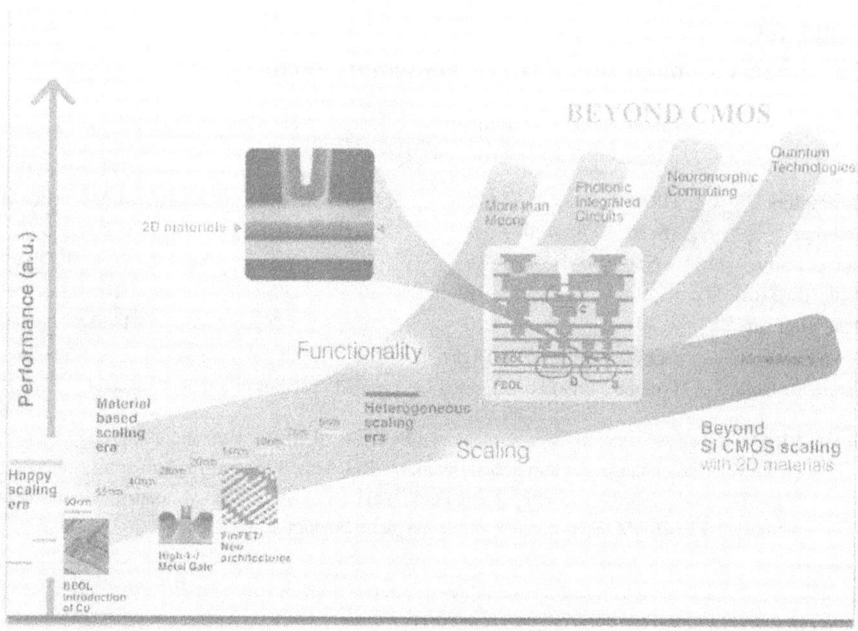

FIGURE 1.5   Silicon technology reached the end of geometric scaling around 1900.

Several topologies, for instance, have been created using magnetic tunnel connections. Some researchers have even used varying write times to obtain unique responses in phase-change arrays. These new devices are well-suited to random number generation because many of them include an inherent element of unpredictability [9, 16, 31].

Digital and quantum computing will likely use these and other technologies in the future. There are already quantum annealing processors on the market. These computers, which depend on superconducting qubits, are currently a costly and unusual solution to very specific problems. At this moment, none of them have shown the adaptability of current technology. However, these technologies will not be required for at least 10 to 15 years.

1.7   CONCLUSION

As researchers work to move beyond CMOS technology, more powerful tools like these are likely to become available. For example, memory technology and microprocessors are now undergoing development. Even though they are not yet appropriate for public or commercial use, their prototypes demonstrate that beyond-CMOS technology can be employed in much more devices than previously believed. Even though it is impossible to know how the market for digital superconductor computing will change in the future, experts and researchers are getting ready for more and more people to be interested in this technology. These new technologies need time, money, and skill to be refined; for the time being, the focus is on developing small-scale systems and finding growing markets. As the level of technology goes up, very-large-scale integration integrated circuit (VLSI IC) designers will have a lot of new options to choose from. Designing effective and reliable processors requires a fundamental understanding of technological advancements and specialized applications. Designing complex integrated circuits requires dealing with several intriguing and difficult obstacles. If the semiconductor industry is capable of maintaining its phenomenal historical growth and proceeds to adhere to Moore's law, it requires advancements on all fronts. These advancements will include front-end process and lithography, as well as the development of innovative high-performance processor architectures and system-on-chip (SoC) solutions. The purpose of the roadmap is to bring together experts in each of these sectors to determine the challenges that need to be addressed and, if possible, to identify potential solutions.

REFERENCES

1. https://www.texaspowerfulsmart.com/space-applications/alternatives-for-cmos-electronics.html.
2. Raja, G. B., & Madheswaran, M. (2013). Logic fault detection and correction in SRAM based memory applications. 2013 International Conference on Communication and Signal Processing. https://doi.org/10.1109/iccsp.2013.6577046.
3. Boopathi Raja, G., & Madheswaran, M. (2013, July). "Design of improved majority logic fault detector/corrector based on efficient LDPC codes" ijareeie International Journal of Advanced Research in Electrical, Electronics and Instrumentation Engineering Vol. 2, Issue 7.
4. (2007). International Technology Roadmap for Semiconductors (ITRS). San Jose, CA: Semiconductor Industry Association. https://www.semiconductors.org/resources/2007-international-technology-roadmap-for-semiconductors-itrs/.
5. https://www.eetimes.com/moores-law-dead-by-2022-expert-says/.
6. Beyond CMOS: the future of semiconductors. https://irds.ieee.org/home/what-is-beyond-cmos.
7. Stojčev, M. K., Tokić, T. I., & Milentijević, I. Z. (2004). The limits of semiconductor technology and oncoming challenges in computer micro architectures and architectures. Facta universitatis-series. *Electronics and Energetics, 17*(3), 285–312.
8. Raja, G. B., & Madheswaran, M. (2013). Design and performance comparison of 6-T SRAM cell in 32nm CMOS, FinFET and CNTFET technologies. *International Journal of Computer Applications, 70*(21), 1–6.
9. Manipatruni, S., Nikonov, D. E., & Young, I. A. (2018). Beyond CMOS computing with spin and polarization. *Nature Physics, 14*(4), 338–343.
10. Hutchby, J. A., Bourianoff, G. I., Zhirnov, V. V., & Brewer, J. E. (2002). Extending the road beyond CMOS. *IEEE Circuits and Devices Magazine, 18*(2), 28–41.
11. Raja, G. B., & Madheswaran, M. (2016). Performance comparison of GNRFET based 6T SRAM cell with CMOS FINFET and CNTFET technology. *International Journal of Innovative Research in Science and Engineering, 2*(05), 197–204.
12. Raja, G. B. (2022). Performance review of static memory cells based on CMOS, FinFET, CNTFET and GNRFET design. *Nanoscale Semiconductors*, 123–140. https://doi.org/10.1201/9781003311379-6.
13. Nikonov, D. E., & Young, I. A. (2013). Overview of beyond-CMOS devices and a uniform methodology for their benchmarking. *Proceedings of the IEEE, 101*(12), 2498–2533.

14. Zhao, Y., Gobbi, M., Hueso, L. E., & Samorì, P. (2021). Molecular approach to engineer two-dimensional devices for CMOS and beyond-CMOS applications. *Chemical Reviews*, *122*(1), 50–131.
15. Nikonov, D. E., & Young, I. A. (2015). Benchmarking of beyond-CMOS exploratory devices for logic integrated circuits. *IEEE Journal on Exploratory Solid-State Computational Devices and Circuits*, *1*, 3–11.
16. Banerjee, S. K., Register, L. F., Tutuc, E., Basu, D., Kim, S., Reddy, D., & MacDonald, A. H. (2010). Graphene for CMOS and beyond CMOS applications. *Proceedings of the IEEE*, *98*(12), 2032–2046.
17. Topaloglu, R. O., & Wong, H. S. P. (Eds.). (2019). *Beyond-CMOS technologies for next generation computer design*. Berlin/Heidelberg, Germany: Springer.
18. Hao, Z., Yan, Y., Shi, Y., & Li, Y. (2022). Emerging logic devices beyond CMOS. *The Journal of Physical Chemistry Letters*, *13*(8), 1914–1924.
19. Lemme, M. C., Akinwande, D., Huyghebaert, C. *et al.* (2022). 2D materials for future heterogeneous electronics. *Nature Communications*, *13*, 1392. https://doi.org/10.1038/s41467-022-29001-4.
20. Tyagi, P., Riso, C., & Friebe, E. (2019). Magnetic tunnel junction based molecular spintronics devices exhibiting current suppression at room temperature. *Organic Electronics*, *64*, 188–194.
21. Tyagi, P., & Riso, C. (2019). Magnetic force microscopy revealing long range molecule impact on magnetic tunnel junction based molecular spintronics devices. *Organic Electronics*, *75*, 105421.
22. Psaroudaki, C., & Panagopoulos, C. (2021). Skyrmion qubits: A new class of quantum logic elements based on nanoscale magnetization. *Physical Review Letters*, *127*(6), 067201.
23. Jeon, J. C. (2021). Designing nanotechnology QCA–multiplexer using majority function-based NAND for quantum computing. *The Journal of Supercomputing*, *77*(2), 1562–1578.
24. Chae, S., Choi, W. J., Fotev, I., Bittrich, E., Uhlmann, P., Schubert, M., & Fery, A. (2021). Stretchable thin film mechanical-strain-gated switches and logic gate functions based on a soft tunneling barrier. *Advanced Materials*, *33*(41), 2104769.
25. Moradinezhad Maryan, M., Amini-Valashani, M., & Azhari, S. J. (2021). A new circuit-level technique for leakage and short-circuit power reduction of static logic gates in 22-nm CMOS technology. *Circuits, Systems, and Signal Processing*, *40*(7), 3536–3560.
26. Raja, G. B., & Madheswaran, M. (2013). Design and analysis of 5-T SRAM cell in 32nm CMOS and CNTFET technologies. *International Journal of Electronics and Electrical Engineering*, 256–261. https://doi.org/10.12720/ijeee.1.4.256-261.
27. Knechtel, J. (2020, March). Hardware security for and beyond CMOS technology: an overview on fundamentals, applications, and challenges. In *Proceedings of the 2020 International Symposium on Physical Design* (pp. 75–86).
28. Bernstein, K., Cavin, R. K., Porod, W., Seabaugh, A., & Welser, J. (2010). Device and architecture outlook for beyond CMOS switches. *Proceedings of the IEEE*, *98*(12), 2169–2184.
29. Chen, A., Datta, S., Hu, X. S., Niemier, M. T., Rosing, T., & Yang, J. J. (2019). A survey on architecture advances enabled by emerging beyond-CMOS technologies. *IEEE Design & Test*, *36*(3), 46–68.
30. Knechtel, J. (2021, March). Hardware security for and beyond CMOS technology. In *Proceedings of the 2021 International Symposium on Physical Design* (pp. 115–126).
31. Nikonov, D. E., & Young, I. A. (2012, December). Uniform methodology for benchmarking beyond-CMOS logic devices. In *2012 international electron devices meeting* (pp. 25–4). IEEE.

2 Design and Challenges in TFET

Soumya Sen, Mamta Khosla, and Ashish Raman

2.1 INTRODUCTION

The tunnel field-effect transistor (TFET) functions via quantum mechanical and band-to-band tunneling, which have been shown to exhibit outstanding switching qualities that go beyond what a regular transistor is theoretically capable of. As every new technology is faced with a significant number of challenges, the same is true for the TFET structure, which is greeted with an inferior ON-current, the issue of ambipolarity, and also meager radiofrequency (RF) behavior. This chapter focuses on the detailed concepts of TFETs and the deep-rooted physics behind their work. Readers will get the complete idea, from the evolution of TFET structures outsmarting the shortcomings of metal–oxide–semiconductor field-effect transistors (MOSFETs) to the different challenges faced by the novel technology and the remedies to restrict them. The conception of the various TFET structures will also be highlighted, and readers will get an impression of modern-day TFET exploration. Section 2.2 introduces the TFET device, the reason behind its existence in the semiconductor industry, and outclassing MOSFET. Readers will get a complete description of properties as well as a functioning analysis of the two devices, including better steepness in the subthreshold slope. Section 2.3 discusses the basic physics of tunneling, which is the major driving force for TFETs. The detailed Fermi-level analysis at various temperature conditions will be analyzed here, as well as its contribution to the working of TFETs. Section 2.2 introduces the readers to a basic TFET structure, its working, I_d-V_d characteristics, and the I_{ON}-to-I_{OFF} ratio. Section 2.5 will focus on the different recently researched TFET structures and their characteristics: the physics behind their functionality, including the heterostructures and how they are superior to the homogeneous TFETs. Section 2.6 provides particulars about different types of TFETs and short analysis of the same with figures. Section 2.7 throws light upon the different challenges faced by TFETs during simulation, which may lead to hindrances at the time of fabrication and structural analysis; a discussion and review of the few structural modifications done to eradicate them are also shown, along with a proper explanation.

2.2 TFET TECHNOLOGY: AN EVOLUTION

The major quest of the semiconductor industry these days is to boost the drive current and cutoff frequency, resulting in brisk switching. This makes complementary metal–oxide–semiconductor (CMOS) technology desirable to them. In order to be on track with Moore's Law, given in Figure 2.1 (i.e., the number of devices on a

DOI: 10.1201/9781003393542-2

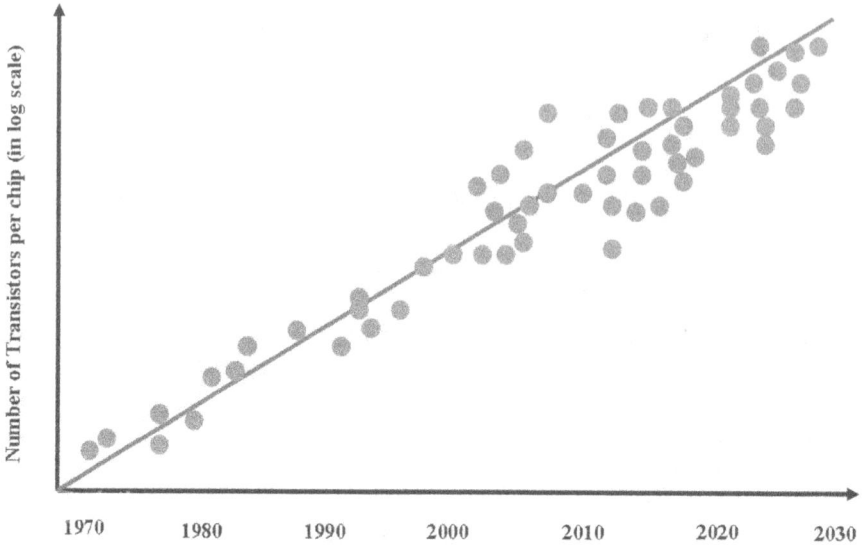

FIGURE 2.1 Moore's theory, illustrating the number of transistors on a chip versus the year; every two years, the transistor's quantity fabricated on a chip multiplies. This graph depicts the best probable situation through 2030 [1].

single chip will double every two years), semiconductor devices based on CMOS technology are exposed to scaling. Now, as the dimensions of the devices are scaled down to nanoscale limits, the modern era is faced with the challenge of handling power and lithographic processes.

The invariable nature of the subthreshold slope in the I_D-versus-V_{GS} graph in logarithmic scale for the wonted MOSFET is a significant downside. Camouflaging the mentioned shortcomings of the traditional MOSFET, a novel device recognized as the TFET was developed. The following are a few of the most notable characteristics of TFETs:

1. Leakage current diminishing is done by the prodigy of band-to-band tunneling.
2. TFETs have a steeper subthreshold slope of less than 60 mV/decade.
3. They are appropriate for low-power tasks, with greater control for short-channel effects.
4. They possess an exceedingly strong I_{ON}/I_{OFF} current ratio.

Essentially, designers seek a device similar to the field-effect transistor (FET) with enhanced features. As in TFETs, with the variation of the gate voltage, it is essential for the tunnel current to stream throughout the device in the activated state following the band-to-band tunneling phenomenon. The primary objective is actualizing a low quantity of current in the OFF-state and enhancing the magnitude of subthreshold swing to greater than 60 mV/decade. The steeper the slope, the better will be the ON-current and the healthier the device performance, which is a mess in

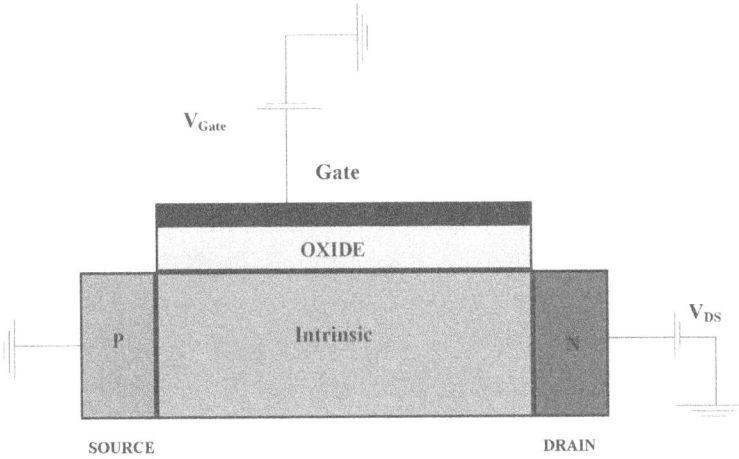

FIGURE 2.2 P-I *(Intrinsic)*-N structure of a tunnel field-effect transistor (TFET).

an orthodox MOSFET. The lack of flexibility for the subthreshold slope in a typical MOSFET debars it from use in any low-power device with a curtailed gate voltage (V_g).

The TFET, as its name implies, works in accordance with the notion of band-to-band tunneling. Quantum tunneling gets light of focus, because of its reliance on tiny particles. The uniqueness of the doping of the source and drain regions makes the device's nature asymmetric, which is the primary distinction between the TFET and its MOSFET equivalent. The P-I-N nature of the device portrayed in Figure 2.2 features gate electrode controlling, as well as intrinsic electrostatic potential causing aggression due to deployment of biasing, and with proper control there is band-to-band tunneling.

Energy band alignment with each other causes electron tunneling via valence to the conduction band p-region, resulting in current transmission [1–8]. Drop-in gate voltage (V_g) leads to distortion of the bands, and the current flow is insufficient. Both the ON- and OFF-states are depicted in Figure 2.3(a) and (b), respectively.

The subthreshold slope given in Equation (2.1), lowering down the technology nodes, increases the steepness of the curve, thus reducing the OFF-current and making the I_{ON}/I_{OFF} ratio healthier as given in Figure 2.4, which in turn makes switching faster.

$$SS\ (mV) = 60(1 + (C_d/C_{ox}) \tag{2.1}$$

where *SS* is the subthreshold slope, C_d is drain capacitance, and C_{ox} is oxide capacitance [9, 10].

2.3 PHYSICS OF TUNNELING

The charged particle's tendency to seep across a potential hindrance is a quantum process arising at minuscule scales because of their wave nature; this phenomenon is collectively called "tunneling." Whenever any charged particle is faced with a

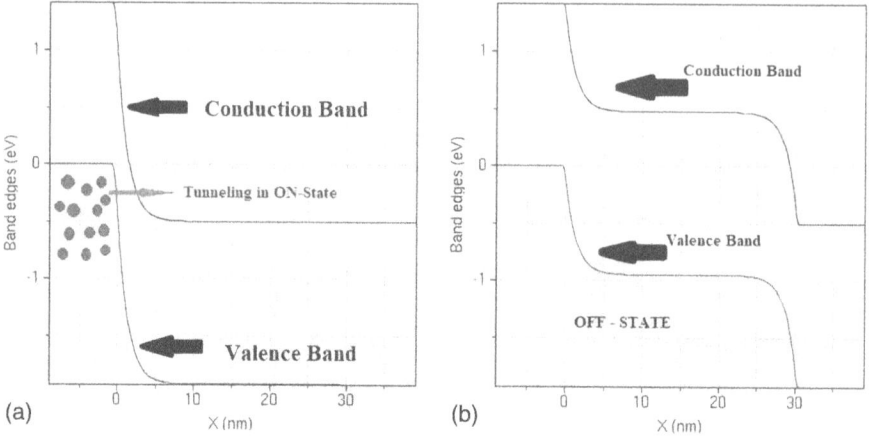

(a)   (b)

FIGURE 2.3 Energy band illustration: (a) ON-state and (b) OFF-state [1].

barrier with superior potential than it, it rebounds or tunnels via the barrier. The elevation, thickness, and geometry of the potential barrier determine the tunneling likelihood of charged particles (i.e., how much they are able to outsmart the potential barrier). The wave vector of any charged particle has imaginary value in the tunneling zone, whereas the real part falls in the incidence and the transmission domain (as shown in Figure 2.5). The chance of transmission is lowered as the exponential drop in the intensity of the wave function is generated by the wave vector's imaginary section around the tunneling zone.

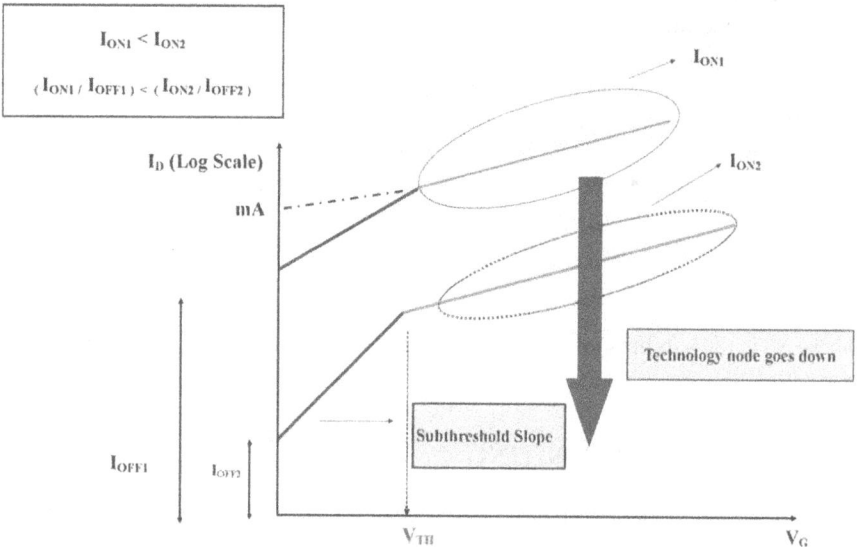

FIGURE 2.4 Subthreshold slope comparison curve for higher and lower technology nodes.

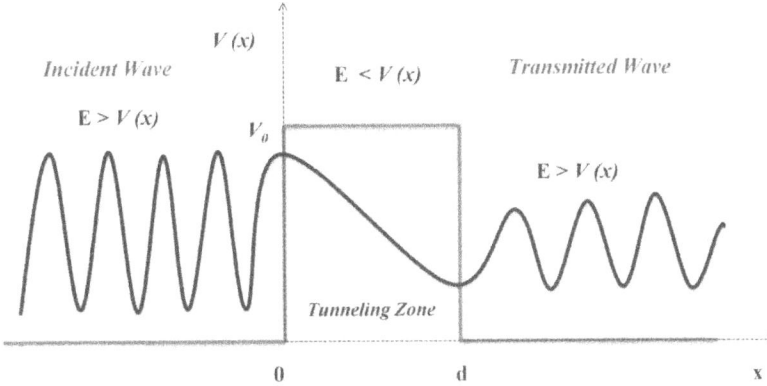

FIGURE 2.5 Tunneling phenomenon through a rectangular potential barrier [11].

The band-to-band tunneling rate can be calculated by Kane's tunneling theory as well as the Wentzel–Kramers–Brillouin (WKB) formulation; they are different in their own ways but, on simplification, they provide similar outcomes.

2.4   KANE'S TUNNELING THEORY

Zener tunneling among the conduction and valence bands (E_C) and (E_V), formulated by Evan O. Kane, keeps the base as the Fermi golden rule, which is to find the probability of transition per unit time between the two energy states, E_K and $E + dE_K$, with $\rho(k)$ as the energy density [12]. Taking ω as the transition probability and $a_k(t)$ as the perturbation constant, Equation (2.2) is the generalized expression:

$$\omega = \frac{1}{t}\int |\, a_k(t)|^2\, \rho(k)dE_k \qquad (2.2)$$

The finalized form, derived from time-dependent perturbation theory, where K is the constant and H' is the small time-dependent perturbation constant, is given on the basis of a matrix form in Equation (2.3):

$$G_{BTBT} = \frac{2\pi}{\hbar}\rho(k)|\langle k|H'|m\rangle|^2 \qquad (2.3)$$

The band-to-band tunneling formulated from Equation (2.3) stands as follows, having q, m^* as the charge and effective mass for the electron, whereas E_g is the band gap effective energy and E is the electric field:

$$G_{BTBT} = \frac{q^2 E^2 \sqrt{m^*}}{18\pi h^{-2}\sqrt{E_g}}e^{\left(-\frac{\pi\sqrt{m^*}E_g^{3/2}}{qhE\,2\sqrt{2}}\right)} \qquad (2.4)$$

Equation (2.4) depicts the direct exponential proportionality of the energy band tunneling rate with an electric field [11].

2.4.1 WENTZEL–KRAMERS–BRILLOUIN APPROXIMATION

The WKB approximation is used for solving Schrödinger's wave equation for one dimension (1D); it also calculates the potential of the particle in a bounded region and tunneling rate through the potential barrier [10, 13–15]. Consider the energy of a particle as E, encountering the potential V, and having a wave function ψ_w, as given by Equation (2.5):

$$\psi_w = Ae^{\pm ikx} \tag{2.5}$$

where k is the wave vector. If the sign is "+", this means the particle is traveling to the right; if it is "−", it is moving in the left direction. It is known from Equations (2.6) and (2.12) that the expression for total energy E [10]:

$$E = \text{Kinetic energy} + V \tag{2.6}$$

$$\Rightarrow E = \frac{1}{2}mv^2 + V \tag{2.7}$$

$$\Rightarrow E = \frac{P^2}{2m} + V \tag{2.8}$$

$$\Rightarrow E - V = \frac{P^2}{2m} \tag{2.9}$$

$$\Rightarrow P^2 = 2m(E - V) \tag{2.10}$$

$$\Rightarrow \hbar k = \sqrt{2m(E - V)} \tag{2.11}$$

$$\Rightarrow k = \frac{\sqrt{2m(E - V)}}{\hbar} \tag{2.12}$$

The WKB approximation (shown in Figure 2.6) fails at the vicinity of the classical point, when $E \approx V$, $k = 0$, and therefore $\frac{1}{k} = \infty$.

2.5 TFET STRUCTURE

The semiconductor industry has seen numerous advancements, from the early days of vacuum tubes to the MOSFETs and, in the modern days, the introduction of lower technology nodes. The scaling of MOSFETs has shown refinements in the switching to diminished gate capacitance (C_g) and lower power loss.

2.5.1 COMPARISON WITH MOSFETS

Every good thing comes at a cost, so by reducing the channel lengths and scaling the MOSFET, a lot of issues are faced, as the device is based on the thermionic emission and the drain current flow is supervised by the gate voltage whose variation toward

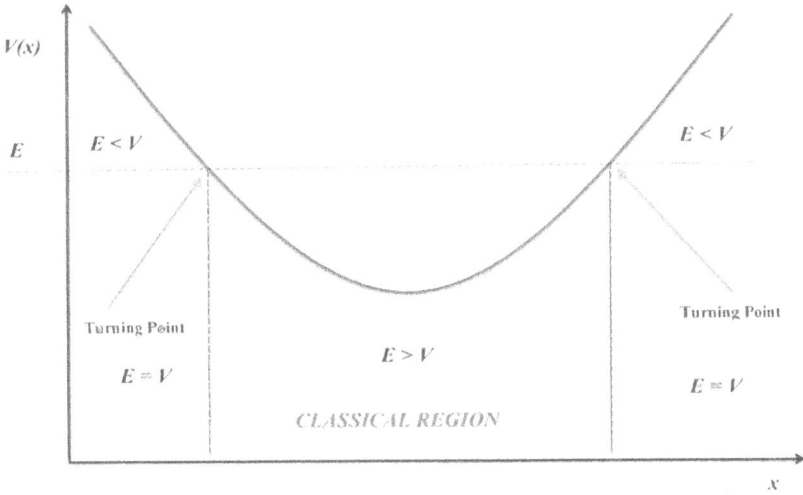

FIGURE 2.6 Wentzel–Kramers–Brillouin (WKB) approximation phenomenon.

increment reduces the potential barrier and, as a result, increases the OFF-current as the subthreshold slope. This expression is given in Equation (2.13), for the MOSFET cannot go lower than 60 mV [10, 16, 17].

$$SS = \left(\frac{d\left(log_{10}I_{DS}\right)}{dV_{Gs}} \right)^{-1} = 2.3 \frac{kT}{q} \left(1 + \frac{C_d}{C_{ox}} \right) \qquad (2.13)$$

Therefore, with a constant gate source voltage (V_{GS}), the current conduction remains fixed in the subthreshold region, whereas there is a constant conduction of drain current due to band-to-band tunneling in the TFETs. The subthreshold swing changes with the change in V_{GS} as the inverse proportionality of tunneling to an exponential function of an electric field (E), as per Equation (2.14):

$$I_{DS} \alpha e^{-\frac{1}{E}} \qquad (2.14)$$

As shown in Figure 2.4, lowering down the technology node, the subthreshold slope will lead to a lower value and an I_{ON}/I_{OFF} ratio of good shape. The comparison is shown in Figure 2.7, which also depicts the lower threshold voltage for the TFETs over their MOSFET counterparts, making the former a better switching device.

The major advantage of TFETs over MOSFETs is that their major working principle is band-to-band tunneling, where the charges seeps through a potential hindrance upon increasing the gate voltage. In the OFF-state shown in Figure 2.3(b), due to the high gap between the stacked energy levels, the I_{OFF} is shrunken, thus providing better device performance. The structural difference between a MOSFET and a TFET

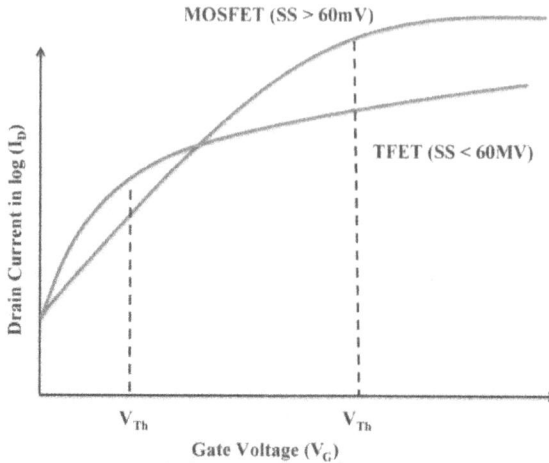

FIGURE 2.7   Subthreshold swing comparison for MOSFETs and TFETs.

is that the doping concentrations for the source and drain are different for the latter but similar in the former (Figure 2.8(a)–(c)). The body of the TFET has intrinsic type doping. In the case of n-type TFETs, the source is highly doped p-type, whereas in the other case it is n-type doping.

2.5.2   I-V Characteristics Analysis of TFETs

If we study the I-V characteristics of a double-gate TFET, as displayed in Figure 2.9, we see that the transfer characteristics in Figure 2.10(a), depicting a lower value of subthreshold slope (well below 60 mV/decade), result in superiority over regular CMOS circuits. Initially, when there is I_{DS} conduction, band-to-band tunneling is negligible. With the application of a significant amount of V_{GS}, the slow alignment of the Fermi level begins with the elevation of I_{DS} and goes on until the Fermi levels are fully aligned (Figure 2.9) [11]. Upon providing excessive amounts of V_{GS}, the I_{DS} saturates as the Fermi levels are further misaligned, moving down from the peak value. With each increase of V_{DS}, there is a depletion of free mobile charge carriers, thereby increasing the channel potential and the source channel electric field; finally, the I_{DS} increases to a saturated value with an unchanged V_{GS}. The I_{DS} is saturated due to the total depletion of channel potential.

2.6   VARIOUS TYPES OF TFETs

TFETs are categorized based on their framework as planar and three-dimensional (3D) architecture. Planar architecture has a current transfer surface that is planar in nature. The design can be built either on a silicon wafer with bulk nature or on silicon as the insulator surface. It outsmarts the bulk TFET substantial control around the channel region, and only the former has received substantial research.

(a)

(b)

(c)

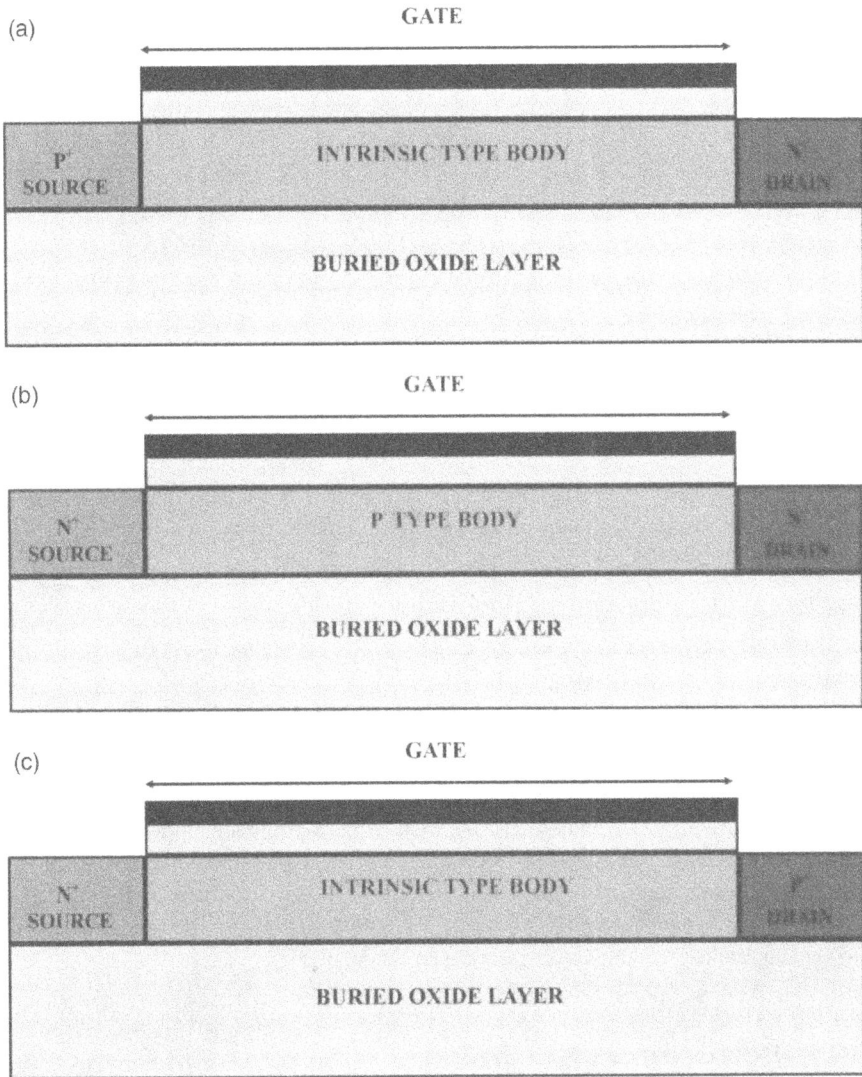

FIGURE 2.8   (a) Structure of an n-type TFET. (b) Structure of an n-type MOSFET. (c) Structure of a p-type TFET [10].

2.6.1   PLANAR TFETs

It consists of a slender silicon strip grown on a ~100 nm dense coating of embedded oxide on a silicon base. In a silicon-on-insulator TFET, the overall shallow silicon layer is exhausted, while the buried oxide layer blocks every source-to-drain oozing path across the bulk region. Furthermore, better channel stability in the gate is made possible by the small dimensions of the drain-body and source-body degradation sectors, as depicted in Figure 2.11.

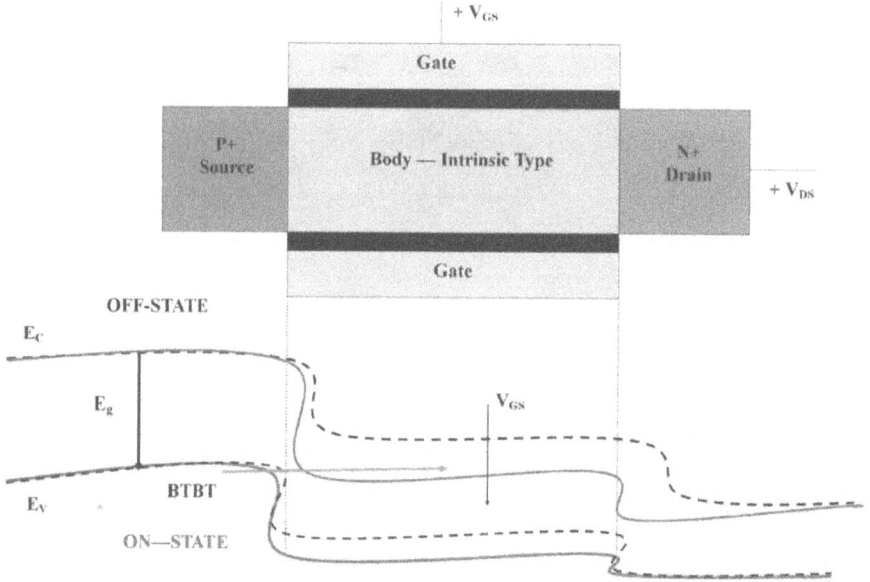

FIGURE 2.9 Energy band diagram showing band-to-band tunneling in ON and OFF circumstances for P-I-N double-gate TFETs [11].

2.6.2 DOUBLE-GATE TFETS

This structure comprises dual gates, one located at the pinnacle and one at the bottom. It ameliorates the electrostatic dominance of the gate on the channel, since the field lines from the gate generally finish there instead of in the channel (Figure 2.12). The I_{ON} is more than it usually is for a TFET with a single gate, because the device has two channels through which current can move.

2.6.3 DOUBLE MATERIAL GATE TFETS

Dual gates with differing work functions, evenly positioned throughout the length of the channel, make up the dual material gate (DMG) TFET (Figure 2.13). The channel segment closest to the source is covered by one gate, while the channel segment closest to the drain is covered by the other gate. In a p-type DMG TFET, the tunneling gate exhibits a healthier work function than the auxiliary gate. The surface potential in the I_{OFF} is wholly reliant on the auxiliary gate's work function if the tunneling gate's dimension is shorter than that of the auxiliary gate [19, 20].

The comparison of two single material gate (SMG) TFETs with various gate work functions to a DMG TFET shows that the I_{OFF} is reduced since the potential gap amid the source and channel is reduced in the OFF-condition, due to the lesser work function of the auxiliary gate. The I_{ON} of the SMG is greater with a higher work function (~5.0 eV) than with a curtailed work function (~4.5 eV). This increased I_{ON},

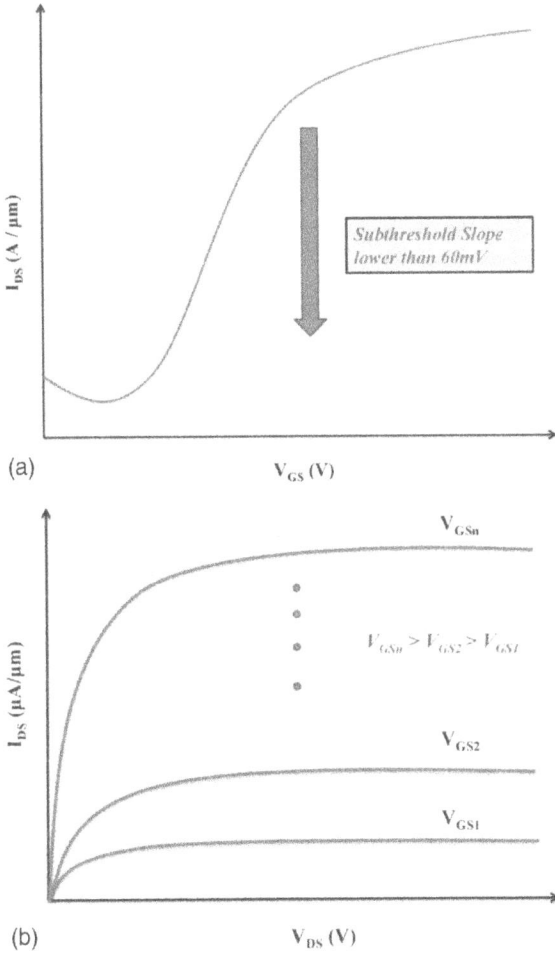

FIGURE 2.10   (a) $ID - VGS$, the transfer characteristics curve for TFETs and (b) the output characteristics curve for TFETs.

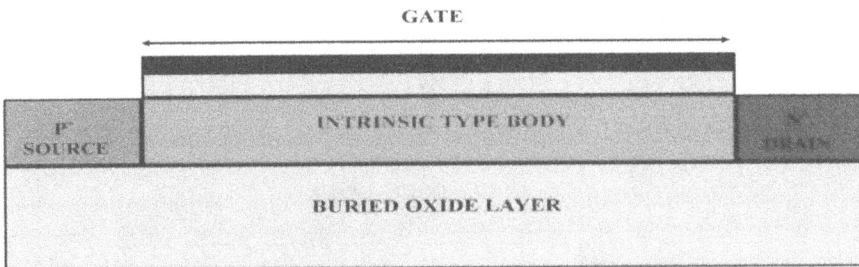

FIGURE 2.11   Silicon-on-insulator TFET (p-channel) planar structure [10].

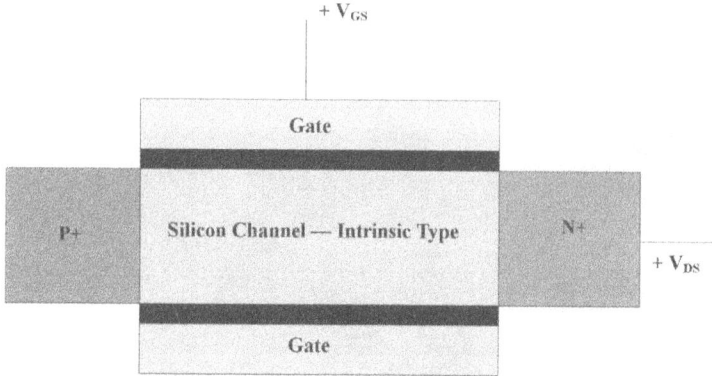

FIGURE 2.12   Double-gate n-channel TFET [18].

nevertheless, is accompanied by a corresponding rise in the I_{OFF}, ultimately providing a steeper subthreshold slope.

2.6.4 HETEROJUNCTION TFETs

A heterojunction is basically the conjunction of two dissimilar materials, which actually increases the flexibility of the energy band gap and in return enhances the performance over that of the homogeneous devices. Figure 2.14 shows the structure of a TFET with a hetero junction of materials *A* and *B*. The junction can be formed by calculation with Vegard's law, shown in Equation (2.5), in order for proper lattice matching:

$$\text{Vegard's law} = a_{A_{(1-x)}B_x} = (1-x)\,a_A + x_{aB} \tag{2.15}$$

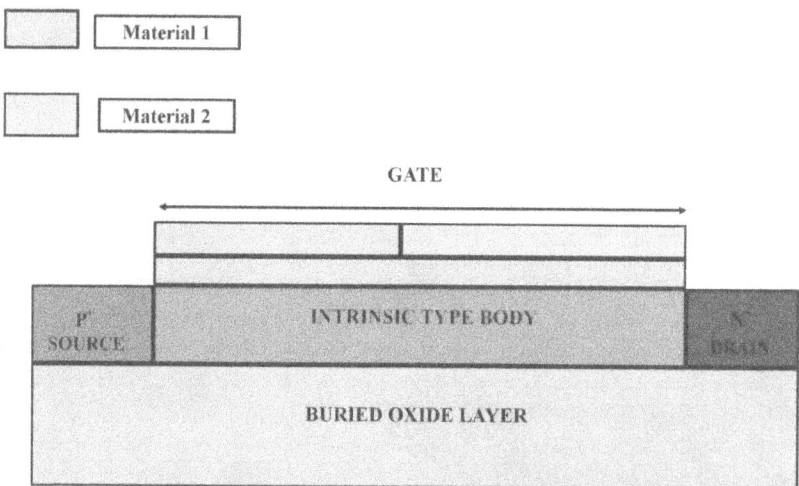

FIGURE 2.13   Double material gate n-channel TFET [10].

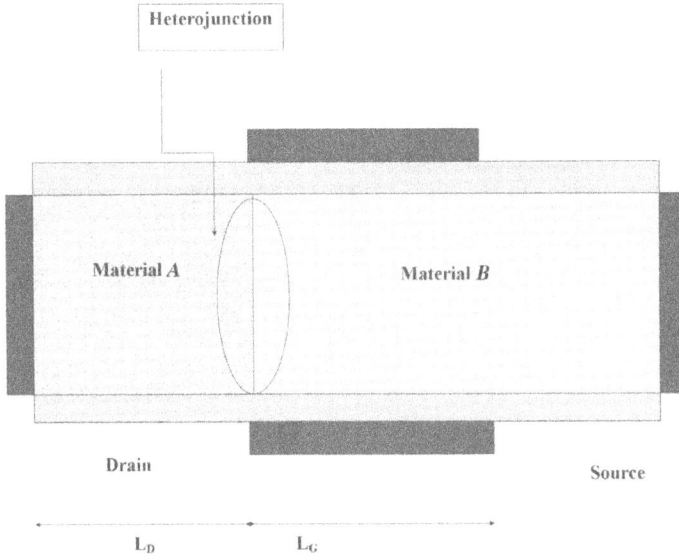

FIGURE 2.14   2D heterojunction TFETs with material A and material B.

Vegard's law presupposes that elements A and B have the same crystalline structure in their pure form (i.e., prior to mixing). In this case, $a_{A(1-x)}B_x$ is the solid solution's lattice value, a_A and a_B the pure elements' lattice parameters, and x is the solid solution's molar percentage of B.

Heterogeneous devices are not here to eradicate the tralatitious silicon, but to perform functionalities that are beyond the scope of silicon. A regular conventional silicon or germanium will be outsmarted by a SiGe device with a tuned lattice and an energy gap; this device can be used for different applications like sensors and light-emitting diodes (LEDs) [21–23].

2.6.5   FERROELECTRIC TFETs

The ferroelectric TFET is a structure that has been suggested for enhancing the subthreshold slope and the ON-current (I_{ON}). The ferroelectric material in the gate stack is polarized, which raises the voltage applied at the gate sensed by the channel (Figure 2.15). As a result, the I_{OFF}-to-I_{ON} transition becomes steeper, which enhances the subthreshold slope and the I_{ON}.

2.6.6   THREE-DIMENSIONAL TFETs

A 3D TFET is an architecture with a current-transferring surface around the three directions. The TFET with all-around gate control over the channel is one of the most crucial 3D architectures in respect to recent studies from the modern-day semiconductor industries. Figure 2.16 illustrates a 3D InGaAs–GaAs nanowire TFET.

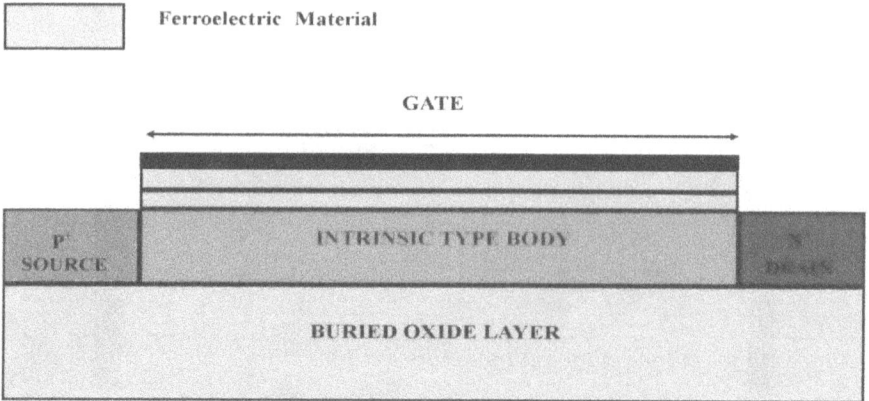

FIGURE 2.15   Ferroelectric TFET with p-channel [10].

2.7   TFET CHALLENGES AND IMPROVEMENTS

The major challenge for a TFET is to uplift the I_{ON} for advancement in the design and also adjust the subthreshold slope to less than 60 mV/decade, simultaneously lowering the I_{OFF}. Numerous remedies have been suggested to make a steeper sub-threshold slope and a healthier ON-current. Other notable challenges for TFETs are their ambipolar nature and upraised value of Miller capacitance.

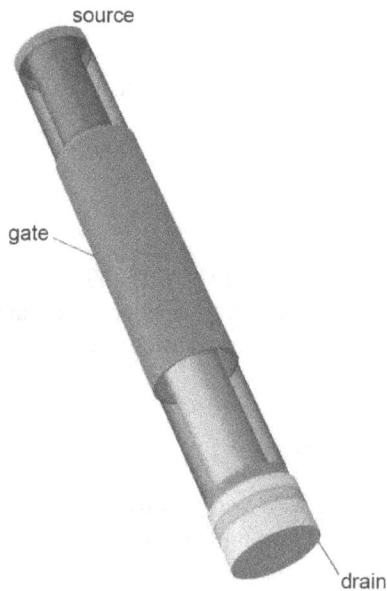

FIGURE 2.16   A 3D In GaAs–GaAs heterojunction nanowire [24] TFET.

2.7.1 High-K Gate Dielectric and Scaling Effect in TFETs

Direct tunneling is a very serious issue. Where the oxide capacitance (T_{OX}) is less than 5 nm and the operating voltage is 1 v or 0.5 v, there may be a tunneling current flowing through the gate oxide, which is defined as "direct tunneling," and it stands as a deviation from the Fowler–Nordheim theory. In other words, an oxide that should have an ideal property, that it should not conduct in case if a sufficiently large voltage has been applied, there is a very huge amount of electric field across the oxide causing it to be tilted by a large amount and the girth of the interface at the oxide and semiconductor is very small, that carriers can directly tunnel through the distance in the oxide and there would be a conduction in it.

If, in a MOS transistor with a SiO_2 gate dielectric, voltage is applied well below the breakdown voltage, Fowler–Nordheim tunneling can be observed [25, 26]. The band theory explanation of Fowler–Nordheim tunneling is depicted in Figure 2.17.

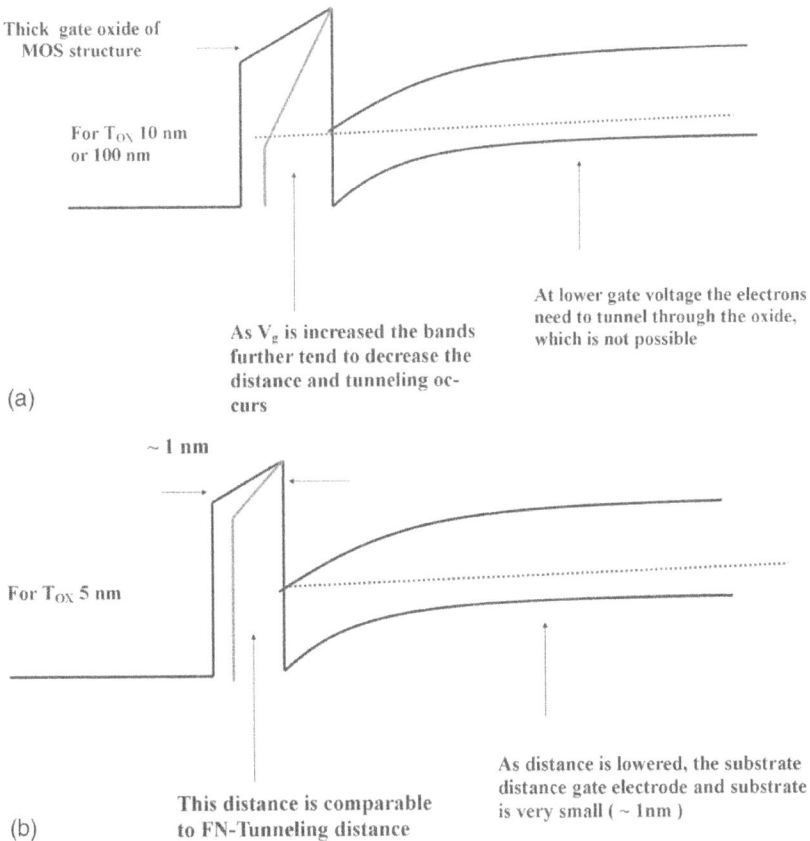

FIGURE 2.17 Fowler–Nordheim tunneling band-bending explanation: (a) Initial condition (b) Final result.

It the electric field is high, band bending occurs slowly, and a scenario may arise where the tunneling will occur by triangular barrier and the current will be predicted by Fowler–Nordheim tunneling. Therefore, it can be noted that direct tunneling occurs through a trapezoidal barrier, whereas Fowler–Nordheim tunneling occurs through a triangular barrier.

Another issue of direct leakage current arises when the $V_G < 1$ V, which in turn increases the leakage power and causes a fluctuation in the noise margin. During the OFF-state of the transistor, there is subthreshold leakage current and about 10% of I_{OFF} can be considered, but if the gate leakage current becomes comparable to subthreshold current, then the leakage component arises [27, 28].

$$\text{Preferable condition}: I_{Gate} \ll I_{OFF}$$

$$\text{Effective oxide thickness} = \left(\frac{\in_{SiO_2} * \textbf{\textit{Area}}}{C_{ox}} \right)$$

As the gate length is getting scaled and the drain is coming closer to the source, the gate electrode should be brought closer to the channel; else, the gate control will be lost. The field effect is due to the coupling capacitance (C) (Equation (2.16)), which increases as we scale down the gate oxide thickness (T_{OX}).

$$C = \left(\frac{\textbf{\textit{Area}} * \in_0 * \in_r}{T_{ox}} \right), \in_r = \textbf{\textit{Relative permitivity}} \tag{2.16}$$

Mathematically, if it is seen from the equation of effective oxide thickness (EOT) (given in Figure 2.18), and a high-K material with a thickness of 9.5 nm is used with

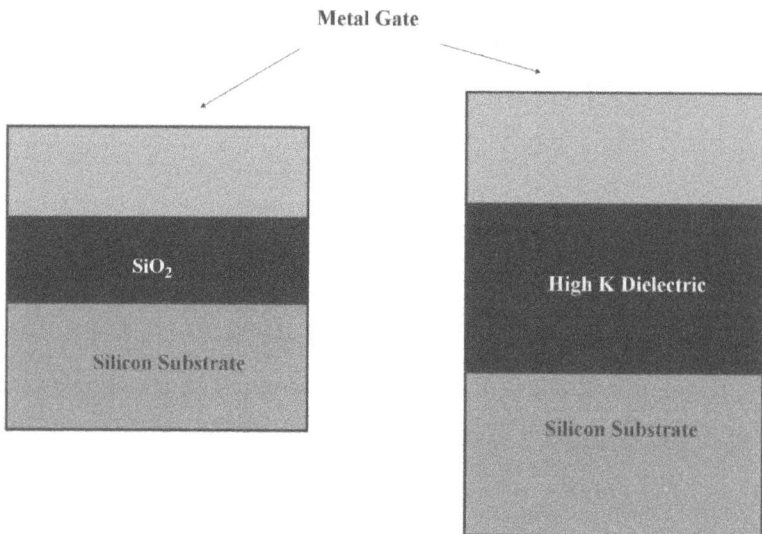

FIGURE 2.18 Comparison diagram between a SiO$_2$ and high-K dielectric in a transistor device.

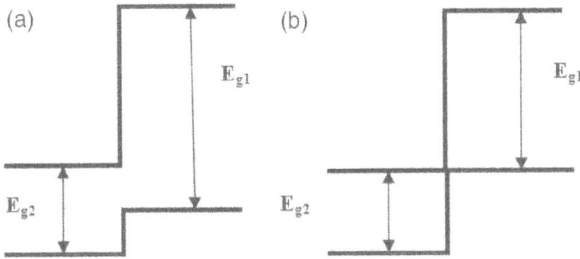

FIGURE 2.19   (a) A standard TFET and (b) an area-scaled TFET [11].

an arbitrary material with K = 40, the result will be ~1 nm, which is equivalent to the use of silicon dioxide of 1 nm.

The high-K dielectric materials provide K > 3.9, a wider bandgap, better band offset, and an excellent interface of silicon high k; adding to all these, there is an ease of process and reliability. If all of the above-mentioned features get summed up, HfO_2 can be a replacement for SiO_2.

At around 1 to 2 nm of the inversion layer, tunneling takes place in ordinary TFETs. The fundamental issue is that altering the method or procedure factor would not enhance the tunneling surface. Due to the need for a larger tunneling region, a new category of TFETs known as area-scaled TFETs has been suggested to address the issue of insufficient I_{ON}. This structure is shown in Figure 2.19. An expansion in the device's tunneling cross-sectional region accounts for the rise in I_{ON} in area-scaled devices. These devices employ a verticalline-tunneling technique to transmit current. Line tunneling takes place in a path parallel to an electric force's lines, whereas point tunneling is perpendicular to them. This principle is responsible for current passage in typical TFETs. In contrast to improving I_{ON} due to a significant spike, it simultaneously strengthens the subthreshold slope.

2.7.2 III–V COMPOUND SEMICONDUCTOR–BASED HETEROSTRUCTURE TFETs

The III–V semiconductors are a part of the compound semiconductor family (i.e., a combination of dissimilar elements from groups 3 and 5 of the Periodic Table to make a semiconductor), which are specifically non-silicon semiconductors. The compound semiconductors are responsible for a few applications that are not done by any elemental semiconductors (i.e., a single-element semiconductor, such as silicon or germanium). A smartphone is an appropriate example of an amalgam of the different compound semiconductors that bring it to life. With the purpose of escalating the functionality of a TFET, a heterostructure (Figure 2.14) is used for better band-to-band tunneling to a proper I_{ON}/I_{OFF} ratio. Here, we examine the combination of elements of group 3 (i.e., Al, Ga, and In) and group 5 (i.e., N, As, and P) to get the traditional III-V compound semiconductors. Various combinations can be had, such as GaAs GaP, AlAs, AlP, InAs, and InP, which exhibit properties resembling those of semiconductors. They have their own band gap and effective masses; for example, gallium arsenide will have a band gap of 1.4 eV, and aluminum arsenide has a band

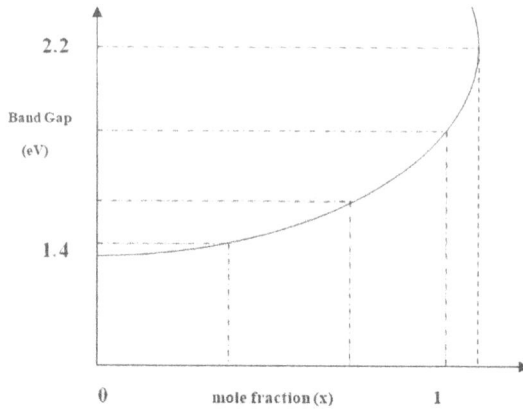

FIGURE 2.20   Band gap versus a mole fraction curve for a III–V compound semiconductor.

gap of 2.2 eV. The flexibility to tune the band gap is lacking in the elemental semi-conductors, and this makes it possible to harness compound semiconductors in many devices, improving their features. The III-V semiconductors will have their own lattice constant, and if it is needed to combine GaAs and AlAs for use in heterojunction TFETs, then the energy band gap can be tuned in between 1.4 and 2.2 eV.

If a heterostructure TFET is to be made with a combination of 30% GaAs and 70% AlAs, then the combination would be $Al_{0.7}Ga_{0.3}As$; the generic expression can be considered $Al_xGa_{1-x}As$, which depicts that X portion of gallium is replaced by aluminum [29–31]. The mole fraction versus band gap is shown in Figure 2.20, where it can be seen that by tuning the mole fraction x, the energy band gap (E_G) can be obtained. Here, the plot for AlGaAs is being shown, where, by tuning the mole fraction from 0 to 1, the E_G can be made flexible, hence making switching better [32, 33].

Alloys can be made by adapting the mole fraction. In order to construct a III–V compound TFET, depicted in Figure 2.21, proper selection of materials is needed, which is done by calculation with Vegard's law (Equation (2.15)) for minimal lattice mismatch and letter strain. Adding to all this, the heterojunction TFET exhibits the advantage of electron confinement, which in turn increases the gain. The major

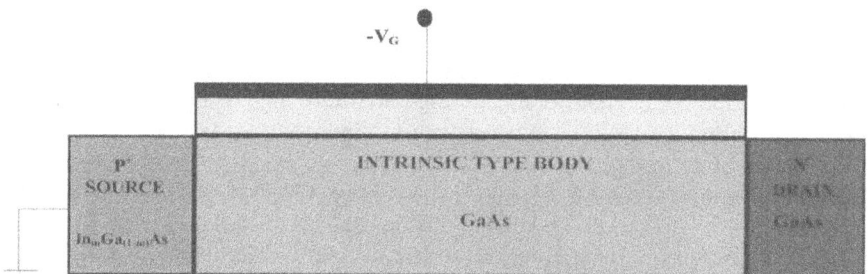

FIGURE 2.21   Schematic diagram of a III-V InGaAs–GaAs heterostructure TFET [10].

drawback is the density of states for III–V semiconductors, which lowers the tunneling probability and as a result diminishes I_{ON}.

2.7.3 AMBIPOLAR BEHAVIOR

Ambipolarity is a marking issue for the TFETs, with $VGS > 0$ for an n-type TFET and otherwise for its p-type counterpart. When $V_{GS} < 0$, the energy band is ascended, and as the channel is aligned the tunneling starts. For a p-type TFET, the source is n and the drain is p-doped, which is comparable for an n-type TFET in an OFF-state as very minute current is shown. Therefore, it can be said that the symmetric nature for $V_{GS} > 0$ and $VGS < 0$ is undesirable and is known as ambipolarity. A band diagram with tunneling is rendered in Figure 2.22(a).

This issue can be eradicated by making the drain and source asymmetric and making the source more highly doped over the drain side, which increases at the junction of the drain and channel (Figure 2.22(b)), thus reducing the band-to-band tunneling phenomenon or negative V_{GS}, but the I_{ON} remains undisturbed due to electron tunneling at the source–channel junction [34, 35].

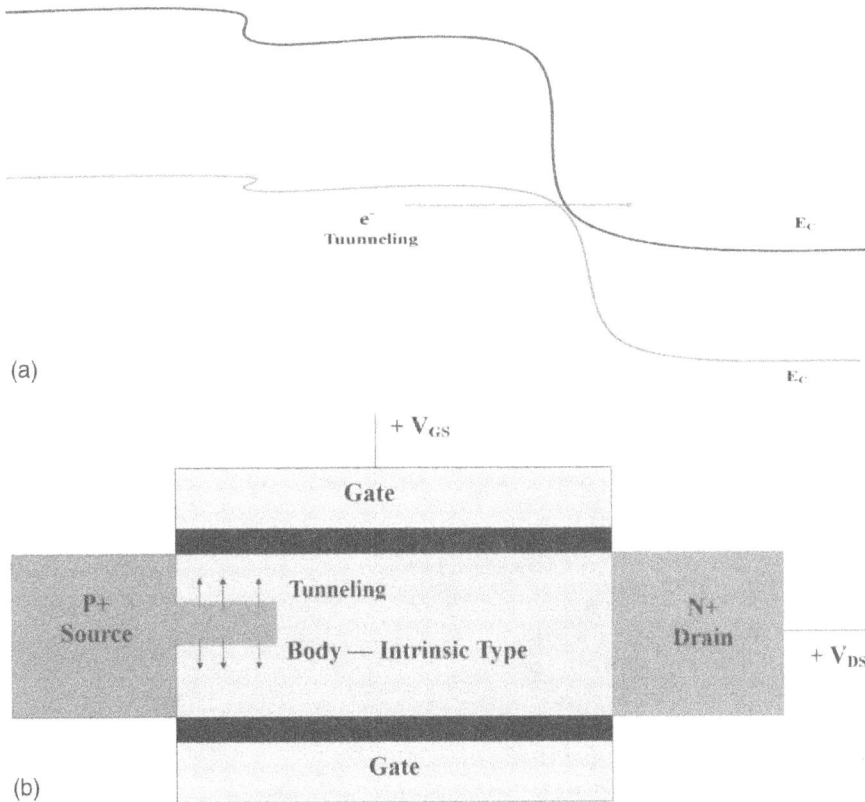

FIGURE 2.22 (a) Energy band diagram for an ambipolar condition in a double-gate TFET and (b) remedial measures for the ambipolar condition [11].

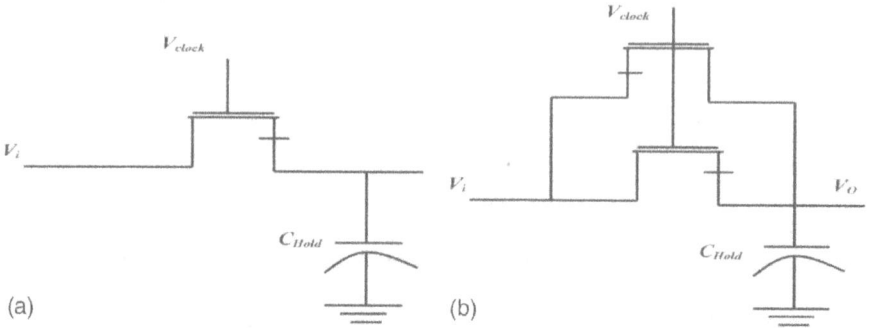

FIGURE 2.23 Track-and-hold nature design: (a) A single n-type TFET structure. (b) A double n-type TFET structure [11].

2.7.4 MILLER CAPACITANCE CONTROL

The TFET, having dissimilar doping in the source and drain, is a p-i-n nature diode with reverse biasing. There is unidirectional current conduction in TFET. A pass transistor logic designed by a logic transistor causes hindrance, so a bidirectional logic has been introduced for smooth conduction. Figure 2.23(a) shows a basic circuit with an n-TFET and track-and-hold nature. In this circuit, when the clock and the input voltage are high, the data is fed into the circuit and is tracked down, but as soon as the V_{DS} value is negative due to the lack of the bidirectional feature, there is no conduction, so Figure 2.23(b) is introduced [7, 8].

The output curve for the inverter circuit is given in Figure 2.24, which also shows the issue of Miller capacitance [36, 37] in the ON condition. In an n-TFET circuit, the overall gate capacitance is the C_{gd}, and whenever a certain voltage spike arises, there is a rise in circuit stabilizing time, power consumption, and delay time.

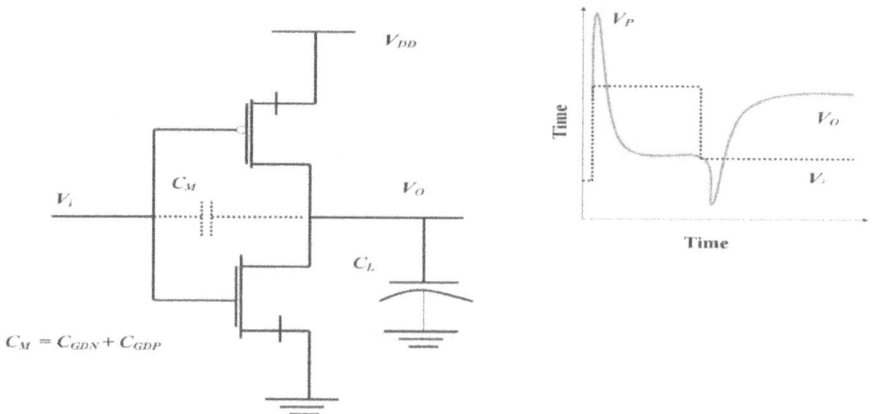

FIGURE 2.24 A TFET-based inverter circuit with an output curve depicting Miller capacitance in the ON-state [11].

This issue can be eradicated by the use of source drain engineering, diminishing the coupling capacitance by the application of diminished density-of-state materials in TFETs and the use of elevated value load capacitance.

2.8   CONCLUSION

This chapter describes the mechanisms of TFETs from scratch. It discusses topics from the basic tunneling phenomenon to the different classifications of the device, such as double material gate, heterojunction, ferroelectric, and so on. A deep-rooted survey is done from Kane's tunneling theory to the WKB approximation. Finally, a few challenges in TFET technology, such as issues with steeper slopes and a fluctuating ON-current resulting in ambipolarity, have been addressed along with their associated remedies, such as Fowler–Nordheim tunneling, the III-V compound semiconductor structure, and Miller capacitance control.

REFERENCES

1. S. Sen, A. Raman and M. Khosla, "A literature survey on tunnel field effect transistors," Proceedings of International Conference on Women Researchers in Electronics and Computing, WREC 2021, AIJR, September, 2021, pp. 506–512.
2. H. Lu and A. Seabaugh, Tunnel field-effect transistors: State-of-the-art. *IEEE Journal of the Electron Devices Society*, vol. 2, no. 4, pp. 44–49, 2014.
3. A. C. Seabaugh and Q. Zhang, Low-voltage tunnel transistors for beyond CMOS logic. *Proceedings of the IEEE*, vol. 98, no. 12, pp. 2095–2110, 2010.
4. F. Schwierz and J. J. Liou, "Status and future prospects of CMOS scaling and Moore's Law-a personal perspective," 2020 IEEE Latin America Electron Devices Conference (LAEDC), San Jose, Costa Rica, 2020, pp. 1–4
5. Y. Omura, A. Mallik and N. Matsuo, "Tunnel field-effect transistors (TFETs)," In MOS Devices for Low-Voltage and Low-Energy Applications, IEEE, 2016, pp. 53–80.
6. P. F. Wang, K. Hilsenbeck, T. Nirschl, M. Oswald, C. Stepper, M. Weis, D. Schmitt-Landsiedel and W. Hansch, Complementary tunneling transistor for low power-application. *Solid-State-Electronics*, vol. 48, no. 12, pp. 2281–2286, 2004.
7. S. Datta, H. Liu and V. Narayanan, Tunnel FET technology: A reliability perspective. *Microelectronics Reliability*, vol. 54, no. 5, pp. 861–874, 2014.
8. A. M. Ionescu and H. Riel, Tunnel field-effect transistors as energy-efficient electronic switches. *Nature*, vol. 479, no. 7373, pp. 329–337, 2011.
9. W. Cao, J. Jiang, J. Kang, D. Sarkar, W. Liu and K. Banerjee, "Designing band-to-band tunneling field-effect transistors with 2D semiconductors for next-generation low-power VLSI," 2015 IEEE International Electron Devices Meeting (IEDM), Washington, DC, USA, 2015, pp. 12.3.1–12.3.4
10. J. Saurabh and M. J. Kumar, Tunnel Field-Effect Transistors (1st ed.). CRC Press, 2016, pp. 39–77.
11. P. K. Dubey, Chapter1-tunnel FET: Devices and circuits, Editor(s): Brajesh Kumar Kaushik, In Advanced Nanomaterials, Nanoelectronics, Elsevier, 2019, pp.3–25, ISBN 9780128133538.
12. E. O. Kane, Theory of tunneling. *Journal of Applied Physics*, vol. 32, no. 1, pp. 88–91, 1961.
13. A. Mazurak and B. Majkusiak, "WKB approximation based formula for tunneling probability through a multi-layer potential barrier," 2012 15th International Workshop on Computational Electronics, Madison, WI, USA, 2012, pp. 1–3.

14. S. C. Miller Jr and R. H. Good Jr, AWKB-type approximation to the Schrödinger equation. *Physical Review*, vol. 91, no. 1, p. 174, 1953.
15. B. M. Karnakov and V. P. Krainov, WKB Approximation in Atomic Physics. Springer Science & Business Media, 2012.
16. I. A. Young, U. E. Avci and D. H. Morris, "Tunneling field effect transistors: Device and circuit considerations for energy efficient logic opportunities," 2015 IEEE International Electron Devices Meeting (IEDM), Washington, DC, USA, 2015, pp. 22.1.1–22.1.4.
17. U. E. Avci, R. Rios, K. J. Kuhn and I. A. Young, "Comparison of power and performance for the TFET and MOSFET and considerations for P-TFET," 2011 11th IEEE International Conference on Nanotechnology, Portland, OR, USA, 2011, pp. 869–872.
18. S. Sahoo, S. Panda, G. P. Mishra and S. Dash, "Tunneling path based analytical drain current model for double gate Tunnel FET (DG-TFET)," 2016 International Conference on Emerging Trends in Electrical Electronics & Sustainable Energy Systems (ICETEESES), Sultanpur, India, 2016, pp. 337–341
19. J. H. Kim, S. Kim and B. G. Park, Double-gate TFET with vertical channel sandwiched by lightly doped Si. *IEEE Transactions on Electron Devices*, vol. 66, no. 4, pp. 1656–61, Feb 27, 2019.
20. K. Boucart and A. M. Ionescu, Double-gate tunnel FET with high-κ gate dielectric. *IEEE Transactions on Electron Devices*, vol. 54, no. 7, pp. 1725–1733, 2007.
21. Y. Li, J. Zhang, Q. Chen, X. Xia and M. Chen, Emerging of heterostructure materials in energy storage: A review. *Advanced Materials*, vol. 33, no. 27, p. 2100855, 2021.
22. C. Convertino, C. B. Zota, H. Schmid, A. M. Ionescu and K. E. Moselund, III–V heterostructure tunnel field-effect transistor. *Journal of Physics: Condensed Matter*, vol. 30, no. 26, p. 264005, 2018.
23. Z. I. Alferov, Nobel lecture: The double heterostructure concept and its applications in physics, electronics, and technology. *Reviews of Modern Physics*, vol. 73, no. 3, p. 767, 2001.
24. Manual, ATLAS User'S. "Silvaco int." *Santa Clara, CA* 5 (2008).
25. J. C. Lee et al., "High-k dielectrics and MOSFET characteristics," IEEE International Electron Devices Meeting 2003, Washington, DC, USA, 2003, pp. 4.4.1–4.4.4.
26. A. N. Justeena, D. Nirmal and D. Gracia, "Design and analysis of tunnel FET using high K dielectric materials," 2017 International Conference on Innovations in Electrical, Electronics, Instrumentation and Media Technology (ICEEIMT), Coimbatore, India, 2017, pp. 177–180.
27. Z. Luo, H. Wang, N. An and Z. Zhu, A tunnel dielectric-based tunnel FET. *IEEE Electron Device Letters*, vol. 36, no. 9, pp. 966–968, 2015.
28. S. Thakre, A. Beohar, V. Vijayvargiya, N. Yadav and S. K. Vishvakarma, "Investigation of DC characteristic on DG-tunnel FET with high-K dielectric using distinct device parameter," 2016 IEEE International Symposium on Nanoelectronic and Information Systems (iNIS), Gwalior, India, 2016, pp. 124–128.
29. K. Boucart and A. M. Ionescu, "Double gate tunnel FET with ultrathin silicon body and high-k gate dielectric," 2006 European Solid-State Device Research Conference, Montreux, Switzerland, 2006, pp. 383–386.
30. N. Kumar and A. Raman, Low voltage charge-plasma based Dopingless tunnel field effect transistor: Analysis and optimization. *Microsystem Technologies* 26, 1343–1350, 2020.
31. A. Raman, K. J. Kumar and D. Kakkar et al. Performance investigation of source delta-doped vertical nanowire TFET. *Journal of Electronic Materials*, vol. 51, pp. 5655–5663, 2022.
32. T. Chawla, M. Khosla and B. Raj, Extended gate to source overlap heterojunction vertical TFET: Design, analysis, and optimization with process parameter variations. *Materials Science in Semiconductor Processing*, vol. 145, 2022, 106643

33. S. Bala and M. Khosla, Design and simulation of nanoscale double-gate TFET/tunnel CNTFET[J]. *Journal of Semiconductors*, vol. 39, no. 4, p. 044001, 2018.
34. S. Tiwari and R. Saha, Methods to reduce ambipolar current of various TFET structures: A review. *Silicon*, vol. 14, pp. 6507–6515, 2022.
35. S. Poria, S. Garg and S. Saurabh, "Suppression of ambipolar current in tunnel field-effect transistor using field-plate," 2020 24th International Symposium on VLSI Design and Test (VDAT), Bhubaneswar, India, 2020, pp. 1–6.
36. N. Bagga, N. Chauhan, D. Gupta and S. Dasgupta, A novel twofold tunnel FET with reduced miller capacitance: Proposal and investigation. *IEEE Transactions on Electron Devices*, vol. 66, no. 7, pp. 3202–3208, 2019.
37. T. Iizuka et al., "Miller-capacitance analysis of high-voltage MOSFETs and optimization strategies for lowpower dissipation," 2021 International Conference on Simulation of Semiconductor Processes and Devices (SISPAD), Dallas, TX, USA, 2021, pp. 44–47.

3 Modeling Approaches to Field-Effect Transistors

Abhay Pratap Singh, R.K. Baghel, Sukeshni Tirkey, and Bharat Singh Choudhary

3.1 INTRODUCTION

Modeling is a process of device development that allows engineers and designers to simulate and analyze the performance of a device before its fabrication. By creating a model, engineers can predict how a device will behave under different conditions, identify potential design flaws, and optimize its performance. There are several types of models used in device development, including physical models, mathematical models, and computer simulations. Physical models involve creating a physical prototype for a device, which is helpful for device simulation under a range of applications [1, 2]. A mathematical model's equations explore the behavior of a semiconductor device, and computer simulations use computer software to create a virtual model of the device and simulate its behavior. The use of modeling in device development has several benefits, including:

- *Reducing development costs*: Modeling of devices helps engineers and designers find potential design flaws at the beginning of the development process, which saves money and time by avoiding the need for expensive modifications later on.
- *Improving device performance*: Engineers may improve a device may improve and ensure that it will work as expected by simulating its response to a variety of scenarios.
- *Enhancing safety*: Modeling can help identify potential safety hazards and allow engineers to design devices that are safe for users and meet regulatory requirements.
- *Facilitating communication*: Modeling can help engineers and designers communicate their ideas and designs more effectively, allowing for more efficient collaboration and problem solving.

Overall, modeling is an essential tool in device development that helps engineers and designers create better-performing, safer, and more cost-effective devices. There are different modeling approaches to field-effect transistors (FETs), depending on the level of detail and complexity required.

DOI: 10.1201/9781003393542-3

3.2 ANALYTICAL MODELS

Mathematical equations are used in analytical models to characterize FET behavior. These models are simple and computationally efficient, and they provide insight into the fundamental physics of FETs. However, they do not capture all the complexities of FET behavior, such as device variability, and may require simplifying assumptions that limit their accuracy [3, 4]. The operation of a metal–oxide–semiconductor field-effect transistor (MOSFET) can be analyzed by using an analytical model that describes the device's behavior in terms of its physical parameters. The following subsections discuss some of the key parameters and equations that are used in MOSFET modeling.

3.2.1 THE THRESHOLD VOLTAGE (V_{TH})

A MOSFET's V_{th} is the gate-to-source voltage at which an FET begins to conduct; it allows current to flow in between the source and drain of the FET. The mathematical derivation of V_{th} can be obtained by considering the charge distribution in the oxide layer and semiconductor channel under the gate [5]. Let's assume that the MOSFET is in depletion mode, where the channel is initially formed without any potential applied to the gate terminal. Assuming a long-channel MOSFET, Poisson's equation is used to predict the potential distribution in the oxide and semiconductor areas.

3.2.2 POISSON'S EQUATION

Poisson's equation is a partial differential equation (PDE) used to define the electrostatic potential in the context of electromagnetism and physics. It describes how electric charges distribute themselves in space and how the electric potential (voltage) varies throughout a given region based on the charge distribution. Poisson's equation is used to define the static electrostatic of an object [6]:

$$\frac{d^2 2\phi}{dx^2} = \frac{-q}{\varepsilon} \tag{3.1}$$

where ε is the permittivity of the device material, ρ is the charge density of the device, ϕ is the electric potential, and x is the distance along the channel. Solving this equation by using appropriate boundary conditions for the electric potential in the semiconductor channel yields the following:

$$\phi(x) = \frac{qN_{DS}\left(2\phi_f - V_G - V(x)\right)}{2C_{ox}} \tag{3.2}$$

where N_{DS} is the doping concentration in the source–drain region, ϕ_f is the Fermi potential, VG is the gate voltage, $V(x)$ is the voltage drop, C_{OX} is gate oxide capacitance, and q is the electron magnitude eliminatory charge.

The voltage drop across the channel is calculated as:

$$V(x) = V_{DS} - x/L \tag{3.3}$$

where V_{DS} is the drain-to-source voltage, and L is the channel length. Substituting this expression for $V(x)$ in the equation for $\phi(x)$, we get:

$$\phi(x) = \frac{qN_{DS}\left(2\phi_f - V_G - V_{DS} - x/L\right)}{2C_{ox}} \tag{3.4}$$

The channel is formed when the electric potential present on the surface is nearly equal to the conduction band edge, which is given by:

$$\phi_s = E_C - E_F = \frac{kT}{q}\ln\left(\frac{N_a}{N_i}\right) \tag{3.5}$$

where E_C is the conduction band edge, E_F is the Fermi level, K is the Boltzmann constant, T is absolute temperature, N_a is the acceptor concentration present in the semiconductor, and N_i is the intrinsic carrier concentration.

Equating $\phi(x)$ to ϕ_s at the surface of the semiconductor, we get:

$$V_{th} = V_G - 2\phi_f - \left(\frac{qN_{DS} C_{ox}}{2\varepsilon}\right)\left(\frac{V_{DS}}{L} - \frac{kT}{q}\ln\left(\frac{N_a}{N_i}\right)\right) \tag{3.6}$$

This mathematical expression describes the threshold voltage of a MOSFET. It shows that V_{th} depends on several parameters, including the gate oxide capacitance, doping concentrations in the source–drain regions and semiconductor channel, temperature, and oxide thickness. This equation is valid for a long-channel (channel greater than 180 nm) MOSFET; for short-channel devices, additional effects like channel-length modulation (CLM) need to be considered.

3.2.3   DRAIN CURRENT (I_D)

The current flows in between the source and drain terminal of the MOSFET are known as the drain current (I_D). The expression for I_D is given by [7]:

$$I_D = \mu C_{ox}\left(\frac{W}{L}\right)(V_{GS} - V_t)V_{DS} - \frac{1}{2}\mu C_{ox}\left(\frac{W}{L}\right)V_{DS}^2 \tag{3.7}$$

where μ is electron mobility, W is the width of the channel, l is the length of the channel, V_{GS} is the gate–source voltage, and V_{DS} is the drain–source voltage. The I_D of a MOSFET is mathematically derived by using Equations (3.8), (3.9), and (3.10).

The charge present in the inversion layer is:

$$Q = Cox\left(V_{GS} - V_{th}\right) W/L \tag{3.8}$$

The electric field present in the channel is:

$$E = V_{ds}/L \tag{3.9}$$

The electron mobility in the channel is:

$$\mu_n = \mu_{n0}/(1+\alpha n\ E) \tag{3.10}$$

where μ_{n0} is the electron mobility present at zero electric field, αn is the constant related to the scattering mechanisms, and E is the electric field present in the channel.

The current density of an FET is defined as:

$$J = q_n \times \mu_n\ Q/W \tag{3.11}$$

where q is the magnitude charge of an electron, and n is the electron density present in the inversion layer. The drain-to-source (I_{DS}) current is:

$$I_{DS} = J \times W \times L \tag{3.12}$$

Combining these equations, the expression for the I_{DS} of a MOSFET is derived as:

$$I_{DS} = \mu C_{ox}\left(\frac{W}{L}\right)(V_{GS} - V_t)(V_{DS} - V_{DSat}) \tag{3.13}$$

where V_{Dsat} is the saturation voltage, which occurs when the electric field is strong enough in the channel to cause electrons to move at their maximum velocity. This equation is known as the "linear region equation," and it is valid as long as the V_{DS} is less than V_{Dsat}. Once the V_{DS} is equal to V_{Dsat}, the MOSFET enters the "saturation region" and the I_{DS} does not increase with V_{DS}. In the saturation region, the I_{DS} is expressed as:

$$I_{DSat} = q_n\mu_n\frac{W}{2L}(V_{GS} - V_{th})^2 \tag{3.14}$$

It is important to note that these equations assume that the MOSFET is operating in different regions like the linear, saturation, and cutoff regions. CLM, velocity saturation, and hot carrier injection also affect the I_{DS}. Figure 3.1 shows the I_{DS} curve for a conventional MOSFET.

A small-signal model can be obtained for a MOSFET by linearized device equations for a given operating point [8]. Mathematical expressions of small-signal models contain the following steps:

- Derive the direct current (DC) operating point of the MOSFET by solving the device equations at zero input signal.
- Perturb the DC operating point with a small input signal.
- Linearize the device equations around the perturbed operating point.
- Express the linearized equations by using small-signal parameters.
- Simplify the equations for obtaining the small-signal model.

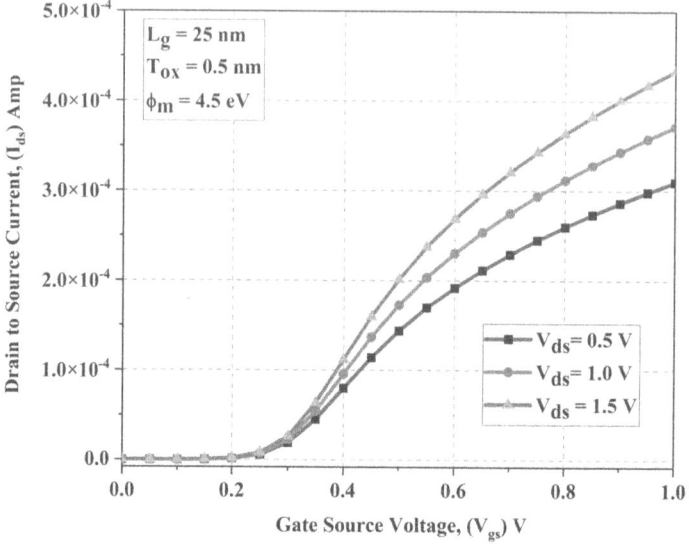

FIGURE 3.1 I–V (current and voltage) characteristics of a conventional metal–oxide–semiconductor field-effect transistor (MOSFET).

3.2.3.1 Derive the DC Operating Point

The DC operating point is obtained by setting the input signal to zero. This creates a fixed bias voltage at the gate, which affects the bias current flowing through the device. The DC operating point can be calculated by using Equations (3.15) and (3.16):

$$I_D = q_n \mu_n \frac{W}{2L} \left(V_{GS} - V_{th} \right)^2 \tag{3.15}$$

$$V_{DS} = V_{DD} - I_D \times R_D \tag{3.16}$$

where V_{DD} is the supply voltage, and R_D is the drain resistance.

3.2.3.2 Perturb the DC Operating Point

We introduce a small input signal $v_{gs}(t)$ to the gate of the MOSFET, which causes a change in the V_{GS} ($\Delta V_{GS}(t) = v_{gs}(t)$). This perturbation changes the drain current by a small amount, which can be expressed as:

$$\Delta I_D = g_m \Delta V_{GS} + g_{\{mb\}} \Delta V_{BS} \tag{3.17}$$

where g_m is the transconductance, $g_{(mb)}$ is the bulk conductance, and ΔV_{BS} is the changes in bulk–source voltage.

3.2.3.3 Linearize the Device Equations

Calculate the first derivative of the I_{DS} equation with respect to the perturbation ΔV_{GS}; the device equations may be linearized.

$$\Delta I_D = g_m \ \Delta V_{GS} \tag{3.18}$$

where g_m is the transconductance, defined as:

$$g_m = \partial I_D / \partial V_{GS} \qquad (3.19)$$

Similarly, the bulk conductance parameter can be defined as:

$$g_{\{mb\}} = \partial I_D / \partial V_{BS} \qquad (3.20)$$

3.2.3.4 Express the Linearized Equations in Terms of Small-Signal Parameters

The following small-signal parameters can be used to represent the MOSFET's small-signal model.

3.2.3.4.1 Transconductance (g_m) Parameter

This parameter represents the sensitivity of the I_D to changes in the V_{GS}, and it can be represented by [9]:

$$g_m = 2 \times I_D Q / \left(V_{GS} Q - V_{TH} \right) \qquad (3.21)$$

where $I_D Q$ and $V_{GS} Q$ are the DC values of the I_D and V_{GS}, respectively. Figure 3.2 shows the transconductance curve of the conventional MOSFET with variations in V_{DS} from 0.5 V to 1.5 V. When the V_{ds} is increased, the length of the channel is decreased; hence, increased electric field is obtained in the channel. This incremental increase in electric field leads to a higher velocity of the electrons, resulting in an increased g_m of the MOSFET.

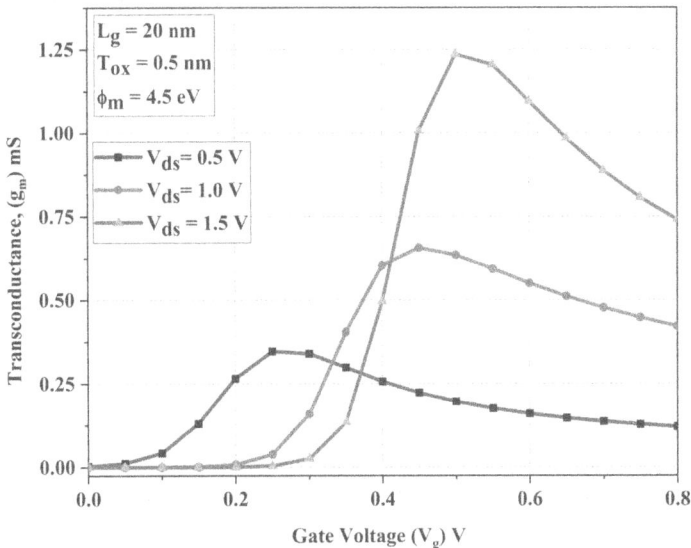

FIGURE 3.2 Transconductance curve of a conventional MOSFET.

3.2.3.4.2 Output Conductance (g$_{Ds}$) Parameter

This parameter represents the sensitivity of the I_D to changes in the V_{DS}, and it can be represented by:

$$g_{ds} = 1/(RD + \lambda) \qquad (3.22)$$

where λ is the CLM parameter.

3.2.3.4.3 Input Capacitance Parameter (C$_{gs}$)

This parameter represents the gate capacitance of the MOSFET, which is present in between the source and drain terminal of the device and is represented by:

$$C_{gs} = C_{ox} \times W \times L/(2 \times (V_{GS}Q - V_{TH})) \qquad (3.23)$$

3.2.3.4.4 Reverse Transfer Capacitance Parameter (C$_{gd}$)

This parameter represents the gate-to-drain capacitance of the MOSFET. It is represented by:

$$C_{gd} = C_{ox} \times W \times L/(2 \times V_{DS}Q) \qquad (3.24)$$

3.2.3.5 Simplify the Equations

This parameter represents the gate-to-drain capacitance (C_{gd}) based on the following parameters: C_{gd} is directly proportional to the oxide capacitance (C_{ox}), the width (W), and the length (L) of the channel. It is inversely proportional to the drain-to-source voltage (V_{DS}) and the charge (Q) stored in the channel. It affects various aspects of transistor operation, including switching speed, power consumption, and signal integrity, making it an important parameter in semiconductor device design.

3.3 EMPIRICAL MODELS

Empirical models are based on experimental data and are used to fit parameters that describe FET behavior [10]. Such models are more accurate than analytical models, but they require extensive experimental data and can be computationally expensive. Empirical models of MOSFETs are mathematical models that are used to describe the behavior of a device's different operating conditions. Empirical models of MOSFETs are essential for designing and simulating electronic circuits that use MOSFETs. These models provide accurate predictions of MOSFET behavior under different operating conditions, allowing circuit designers to optimize circuit performance and reliability.

3.3.1 BERKELEY SHORT-CHANNEL INSULATED-GATE FIELD-EFFECT TRANSISTOR (IGFET) MODEL (BSIM)

To accurately model the behavior of MOSFETs in circuit simulations, engineers have developed empirical models such as the BSIM [11, 12]. The BSIM has a desired set of equations that are used to describe the I–V characteristics of MOSFETs under various bias conditions and physical parameters.

The mathematical derivation of the BSIM is quite complex and involves several steps. Here, we provide a high-level overview of the main equations and parameters involved in the BSIM. The basic equation for the I_{DS} of a MOSFET is represented as:

$$I_{DS} = \mu C_{ox}\left(\frac{W}{L}\right)\left[(V_{GS} - V_t)V_{DS} - \left(\frac{V_{DS}^2}{2}\right)\right](1 + \lambda V_{DS}) \qquad (3.25)$$

To model short-channel effects, additional terms are added to the drain current equation:

$$I_D = I_{Dsat}\left[\frac{1 - (V_{GS} - V_{th})}{V_{DS} - \lambda V_{DS}}\right]\Psi(\phi, V_{DS}) \qquad (3.26)$$

where I_{Dsat} is the saturation current, ηF represents the field-dependent mobility degradation factor, $\Psi(\phi, V_{DS})$ denotes the surface-potential-dependent factor, and ϕ is the surface potential.

The threshold voltage (V_{th}) is given by:

$$V_{th} = V_{fb} + 2\phi F + \gamma\sqrt{(2\phi_F - V_{SB})} - 2\phi_F \qquad (3.27)$$

where V_{fb} is the flat band voltage, ϕ_F is the bulk surface potential, γ is the body-effect parameter, and V_{SB} is the source–body voltage.

The gate oxide capacitance (C_{ox}) is expressed as:

$$C_{ox} = \varepsilon_{ox}/t_{ox} \qquad (3.28)$$

where ε_{ox} is the dielectric constant of the oxide, and t_{ox} denotes the oxide thickness.

The CLM parameter (λ) is expressed by:

$$\lambda = \lambda_0 + \lambda_1(V_{DS} - V_{DSsat}) \qquad (3.29)$$

where λ_0 and λ_1 are the model parameters, and the saturation voltage is V_{DSsat}. These are just a few of the equations and parameters involved in the BSIM model. The full derivation is quite complex and involves a number of additional terms and assumptions. However, with these equations and parameters, engineers can accurately simulate the behavior of MOSFETs in a variety of circuit designs.

3.4 COMPACT MODELS

Compact models are a compromise between analytical and empirical models. These models use analytical equations (mathematical equations) to describe the fundamental physics of FETs and empirical data to fit parameters that capture device-specific behavior [13]. Compact models are widely used in circuit simulation tools and can accurately capture the behavior of FETs in complex circuits. There are several compact models of MOSFETs, but some of the most commonly used ones are given in this section.

3.4.1 SPICE Model

This is a popular MOSFET model used in circuit simulation software such as LTSpice, PSpice, and HSpice [14]. It is a semi-empirical model that uses a set of equations to model MOSFET behavior under different conditions, such as biasing, temperature, and frequency. The SPICE model can be understood by the following procedure.

3.4.1.1 Model Selection

In this context, selecting a SPICE model means selecting a suitable model that accurately represents the behavior of a specific electronic component in a circuit simulation. The selection of a suitable SPICE model is dependent on factors like the type of component being simulated, the accuracy required for the simulation, and the computational resources available. When selecting a model, it is important to consider its complexity and accuracy. More complex models may provide greater accuracy, but they also require more computational resources and can result in longer simulation times. Therefore, a trade-off between accuracy and simulation speed must be made. In general, the most commonly used SPICE models are the default models provided by the simulation software, which are often based on industry-standard models, such as the BSIM model for MOSFETs or the Ebers–Moll model for bipolar junction transistors (BJTs). However, in some cases, the default models may not be suitable, and it may be necessary to use more specialized or customized models. In these cases, it is important to carefully evaluate the accuracy of the model and validate it against experimental data before using it in a circuit simulation. Ultimately, the selection of a SPICE model depends on the specific requirements of the simulation and the available resources. Finding the optimal model for a given task may take some time and experimentation. Choose the appropriate MOSFET model based on the device characteristics and the simulation requirements.

3.4.1.2 Parameter Extraction

Parameter extraction in SPICE models involves the process of determining the values for model parameters that accurately represent the behavior of the actual device or circuit being modeled. This process is important because accurate parameter values are essential for reliable circuit simulation and analysis. There are several methods for parameter extraction in SPICE models, including the following.

3.4.1.2.1 Curve Fitting

This involves comparison of the SPICE model simulation to the data of the actual device or circuit. The parameters of the SPICE model are adjusted until the simulation results match the measured or simulated data.

3.4.1.2.2 Optimization

This involves minimizing the difference between the simulation results of the SPICE model and the simulated data by using an optimization algorithm. The algorithm adjusts the parameters of the SPICE model until the difference is minimized.

3.4.2   ENZ–KRUMMENACHER–VITTOZ (EKV) MODEL

This is a compact model of the MOSFET that was specifically developed for low-power and low-voltage circuit applications [15, 16]. It is a physics-based model that considers the non-uniform doping profiles and the effect of the V_G on the potential of the channel. This model presents a mathematical description for MOSFET behavior, including its DC characteristics and its alternating current (AC) behavior. The EKV model has the following assumptions:

- The channel of the MOSFET is assumed to be a uniformly doped semiconductor.
- Between the gate and the channel, an oxide layer is expected to exist as a uniform dielectric.
- It is assumed that the MOSFET is either in a weak inversion or strong inversion region.

The EKV model can be used to simulate MOSFET behavior in both DC and AC modes, making it a useful tool for circuit design and analysis. The EKV model includes the following parameters:

- *Threshold voltage (V_{th})*: The voltage that triggers the MOSFET to begin conducting current.
- *Body-effect coefficient (γ)*: The degree to which the body bias affects the sensitivity of the MOSFET threshold voltage.
- *Saturation current (I_{dsat})*: The current that flows through the MOSFET when it is fully turned on.
- *Subthreshold slope coefficient (K)*: The slope of the subthreshold region of the MOSFET characteristic.
- *Mobility (μ)*: The ability of the carriers to move through the MOSFET channel.
- *Channel-length modulation (λ)*: CLM is defined as the relationship between I_{ds} and the channel length of the MOSFET.
- *Early voltage (V_{EA})*: The voltage at which the MOSFET moves into the saturation zone of its operating range.

3.5   ANALYTICAL MODELING

This involves using mathematical equations to model the behavior of the device or circuit and deriving the parameters of the SPICE model from these equations. Method selection is dependent on parameters like the complexity of the device or circuit being modeled and the availability of measured or simulated data. Curve-fitting and optimization methods are generally used for more complex devices or circuits for which analytical modeling is not feasible or not accurate enough. Analytical modeling is usually reserved for simpler devices or circuits where mathematical equations can accurately capture the behavior of the device or circuit. Analytical modeling is usually reserved for simpler devices or circuits where mathematical equations can accurately capture the behavior of the device or circuit with the help of model parameters from the MOSFET datasheet, measurements, or simulation results.

3.5.1 DC ANALYSIS

DC analysis in SPICE simulation refers to the process of simulating the behavior of a circuit at a steady-state or static condition where all inputs are constant. This analysis is important because it helps in determining the operating point of a circuit, which is essential for circuit design and optimization. To perform DC analysis in SPICE, you need to define the circuit topology and provide values for all of the component parameters, such as resistors, capacitors, and voltage sources. You also need to specify the DC voltage or current values for any input sources in the circuit. Once the circuit is defined, you can run a DC simulation by using a SPICE simulator. During the simulation, the simulator solves a system of equations to determine the current and voltage at each node present in the circuit. The simulation results are typically presented in the form of a table or graph, showing the values of each node voltage and branch current at the steady-state condition. DC analysis in SPICE is useful for analyzing and optimizing circuits, such as amplifiers, filters, and power supplies. It helps in determining the biasing conditions for transistors and amplifiers, selecting component values for optimal performance, and optimizing the power consumption of the given circuit. The drain current and gate-source voltage are all part of the DC analysis used to find the MOSFET's operating point.

3.5.2 SMALL-SIGNAL ANALYSIS

In SPICE modeling, small-signal analysis refers to the linearization of the circuit around its operating point (also called the "DC bias point") in order to analyze its behavior in the small-signal regime [17, 18]. This is done by assuming that the circuit can be approximated by a linear model around the operating point, and that the signals of interest are small enough to be considered as perturbations around this point. To perform small-signal analysis in SPICE, the circuit must first be simulated in DC mode to obtain the operating point. Then, a small-signal AC analysis can be performed by applying a small AC signal to the circuit and measuring the resulting AC response. This involves specifying the AC amplitude, frequency, and phase of the input signal, and then running an AC simulation to obtain the AC response at various points present in the circuit. During the small-signal analysis, SPICE simulation computes the small-signal model for the circuit, which includes the small-signal parameters such as small-signal resistance, capacitance, and conductance. These parameters are used to compute the input impedance, output impedance, small signal, and other important parameters of the circuit. Small-signal analysis is useful for analyzing the stability, frequency response, and noise performance of analog circuits, such as amplifiers, filters, and oscillators. It allows designers to evaluate the performance of their circuits under small-signal conditions, which are typically the most relevant for analog signal-processing applications. Perform small-signal analysis to determine the MOSFET's small-signal behavior, including the transconductance, output resistance, and capacitances.

3.6 MODEL IMPLEMENTATION

Implement the MOSFET model in SPICE by using the extracted parameters and the results from the DC and small-signal analyses. To implement a model in SPICE, you typically need to follow the following steps:

- *Choose the appropriate model*: The first step is to choose the appropriate model for the component you want to simulate. SPICE supports a wide range of models for different types of components, such as diodes, transistors, and operational amplifiers. You can either use the built-in models in SPICE or create your own custom models.
- *Create a SPICE netlist*: A SPICE netlist is a text-based description of the circuit you want to simulate. You can either create the netlist manually or use a graphical user interface to generate it.
- *Define the component parameters*: Once the appropriate model is selected, you need to define the component parameters for the model. For example, if you are simulating a transistor, you need to specify the values of parameters like the threshold voltage, the drain–source resistance, and the gate–source capacitance.
- *Run the simulation*: Once you have defined the component parameters, you can run the simulation. SPICE will simulate the behavior of the circuit under the specified conditions and generate output waveforms for analysis.
- *Simulation and verification*: Simulate the MOSFET model, and verify the results against the expected behavior and the datasheet specifications. SPICE simulation is a computer program used for simulating and verifying the behavior of electronic circuits. The SPICE model is a mathematical representation of the electronic circuit, and it is used to simulate the behavior of the circuit under different conditions. SPICE models are used for analyzing and optimizing the performance of the circuit, and to verify the correctness of the circuit design [19–21]. Simulation in SPICE involves running a series of tests on the circuit model to analyze its behavior. These tests can be performed by using different inputs, such as voltage or current signals, to simulate different operating conditions. Simulation can help identify potential problems in the circuit design and optimize its performance. Verification in SPICE involves checking the correctness of the circuit model by comparing its simulation results with the expected behavior of the circuit. Verification can help ensure that the circuit model accurately represents the behavior of the physical circuit and that the design meets the required specifications. In summary, simulation and verification are critical steps in the design and analysis of electronic circuits using SPICE models. Simulation helps optimize circuit performance and identify potential problems, while verification ensures that the circuit model accurately represents the physical circuit and meets the required specifications.
- *Optimization*: Modify the MOSFET model settings to fine-tune simulation results. "Optimization" in SPICE modeling refers to a process that finds the suitable values for the model's parameters to match the actual behavior of

the component being modeled. The simulation accuracy depends on how well the SPICE model matches the real component's behavior, and optimization is necessary to ensure that the model is as accurate as possible. The optimization process typically involves adjusting the values of the model's parameters until the simulation results match the measured or expected behavior of the real component. The values of the parameters are adjusted based on a fitness function, which quantifies how well the simulation matches the real-world behavior. There are various optimization algorithms that can be used for SPICE modeling, such as gradient descent, genetic algorithms, and simulated annealing. These algorithms are used to find the best set of parameter values that minimize the difference between the simulated behavior and the real behavior. Overall, optimization in SPICE modeling is a crucial step in creating accurate circuit simulations and can help to reduce design time and costs.

- *Model validation*: Validate the MOSFET model against measurements or other simulations to ensure its accuracy and reliability. Model validation is a technique used to validate SPICE models by comparing their simulation results to measurements from actual hardware. It involves testing the model's ability to accurately predict the behavior of a device or circuit under different operating conditions. Model validation typically involves varying one or more operating parameters, such as temperature or voltage, and measuring the resulting behavior of the device or circuit. The simulation results are then compared to the measured data to determine the accuracy of the SPICE model. In order to perform model validation, the SPICE model must be carefully calibrated to the specific device being tested. This involves adjusting the model parameters, such as the transistor width and channel length, to match the device's physical characteristics. Once the model has been calibrated, it may be used to predict how the device will perform under a variety of scenarios and then checked against hard data. Model validation is an important step in the development of SPICE models, as it ensures that the models accurately represent the behavior of the devices they are intended to simulate. This is essential for designing and optimizing electronic circuits, as accurate models are necessary for predicting the behavior of a circuit before it is actually built.

- *Model documentation*: Document the MOSFET model, including the model equations, parameters, limitations, and assumptions, for future reference and use. To document a SPICE model, the following information should be provided:
  - *Model name*: The name of the model should be clear and descriptive.
  - *Model type*: The type of the component should be identified, such as MOSFET, diode, or resistor.
  - *Model parameters*: The parameters used in the model should be listed with their definitions, units, and default values. These parameters can include physical parameters such as length, width, and doping density for MOSFETs, or parameters such as forward voltage drop and reverse saturation current for diodes.

- *Equations*: The equations used to describe the behavior of the component should be provided. These equations can be derived from physical principles or empirical data.
- *Model limitations*: The limitations of the model should be described. For example, a MOSFET model may not accurately simulate the behavior of the component at high temperatures or high voltages.
- *Model version*: The version number of the model should be provided, along with the date of creation or modification.

  Documentation of SPICE models is important to ensure that they can be easily understood and used by other designers. A clear and comprehensive documentation can help in the debugging process and facilitate the design of complex circuits.

3.7   DEVICE SIMULATION

Device simulation uses numerical methods to solve the fundamental equations that describe the behavior of FETs. These methods can capture all the complexities of FET behavior, including device variability, and can be used to optimize device design [22, 23]. However, device simulation is computationally intensive and requires advanced computational resources. Device simulation can be understood by using the steps given in the following subsections.

3.7.1   DEFINE THE PHYSICAL STRUCTURE OF THE FET

This includes providing details on the FET's shape and its material qualities. The FET has the following regions: the source, drain, and gate terminal of the MOSFET. The source and drain are highly doped with the material and serve as the device's endpoints for the passage of current as it travels through the device. The gate area is an electrical barrier that consists of a very thin layer of an insulating substance, commonly silicon dioxide (SiO_2), which is used to divide two regions (source/drain) of the MOSFET the source region apart from the drain region. The material properties of the FET are critical to its performance. The semiconductor material should have high mobility, which means that electrons can move through the channel with minimal resistance. It is important for the insulating layer to have a high dielectric constant so that the voltage that is necessary to generate an electric field in the channel area is kept to a minimum. The gate metal should have a high work function to provide a good barrier for the electrons or holes in the channel.

3.7.2   DISCRETIZE THE DEVICE INTO SMALL ELEMENTS

Divide the FET into a mesh of small elements to create a numerical grid that covers the device. The process of discretization involves dividing the FET into a series of small, interconnected regions or cells [24]. Each cell is assigned a set of numerical values, which represent the relevant physical properties of the device at that location. These values are typically obtained from experimental data or theoretical models, and are used to determine the behavior of the device over time. The size and shape of each cell in the mesh depend on the specific requirements of the simulation. In general, smaller cells provide a more accurate representation of the device but require more computational resources to simulate. Conversely, larger cells are faster to simulate but may not accurately capture the behavior of the device in all situations.

Once the device has been discretized, the equations that govern its behavior can be solved numerically by using techniques such as finite element analysis or finite difference methods. These techniques involve breaking down the equations into smaller, more manageable pieces that can be solved iteratively. The resulting simulation provides a detailed picture of the behavior of the device, which can be used to optimize its design and performance.

3.7.3 SET THE BOUNDARY CONDITIONS

Boundary conditions are necessary to define the input and output of the device. The term "boundary conditions" refers to the inputs and outputs that define a device's operating environment. In order to set the boundary conditions for a device, it is important to consider its intended use and the properties of the inputs and outputs.

- *Inputs*: The inputs to a device can include physical quantities such as temperature, pressure, and voltage, as well as signals such as audio, video, or data. The boundary conditions for the inputs depend on the specific application of the device. For example, in a thermometer, one input boundary condition is the temperature range that it is designed to measure.
- *Outputs*: The outputs of a device can be physical quantities such as force, displacement, or heat, or they can be signals such as audio, video, or data. The boundary conditions for the outputs depend on the specific application of the device. For example, in a speaker, output boundary conditions include the frequency range and sound level that the speaker is designed to produce.

3.7.4 SOLVE THE GOVERNING EQUATIONS

Use numerical methods to solve PDEs that describe the behavior of the FET. This includes the continuity equation, the Poisson equation, and transport equations [25, 26]. An approach is to use finite element methods, which involve discretizing the domain into a mesh of elements and approximating the solution by using piecewise polynomials. This results in a set of linear equations, each of which may be solved by using iterative techniques as an alternative. In addition to the drift-diffusion equations, other equations may also need to be solved to fully describe the behavior of an FET, such as the Poisson equation for the electrostatic potential and the continuity equation for the current density. The specific equations and numerical methods used will depend on the specific device's geometry and operating conditions.

3.7.5 CALCULATE THE CARRIER DENSITIES AND ELECTRIC FIELDS

To calculate the electric field and carrier density in a MOSFET device simulation, we typically use a device simulation software such as technology computer-aided design (TCAD) [27]. The software uses numerical methods to solve semiconductor device equations such as the Poisson equation and carrier transport equations. The carrier densities and electric fields can be obtained from the solution of these equations. Carrier densities are the concentrations of holes and electrons in the semiconductor material, while the electric fields are the spatial variations in the electric potential across the device. In a MOSFET device simulation, the carrier densities and electric fields are typically calculated at different points in the device such as the source, drain, channel, and gate regions. These values are important for understanding the device behavior and optimizing its performance. In summary, the carrier densities and electric fields in a MOSFET device simulation can be obtained by using numerical methods that solve the semiconductor device equations. The values obtained are important for understanding and optimizing the device performance.

3.7.6 Determine the Device Characteristics

Use the carrier densities and electric fields to calculate device characteristics like the I–V characteristic curve, the C_{gg}, and the g_m. Here are the basic steps to calculate these device characteristics:

- *Simulation Parameters setup*: First, simulation parameters need be to defined, such as the material properties, doping profile, device dimensions, and operating conditions.
- *Solve the Carrier transport equations*: In order to derive the carrier densities and electric fields, the carrier transport equations, such as the drift-diffusion equations or Schrödinger's equation, must first be solved.
- *Calculate the I–V curve*: This curve is calculated for applied bias of the MOSFET, which results in current flow in the device. The I–V curve provides important information about the device performance, such as the V_{th}, the subthreshold slope (SS), and the saturation current.
- *Calculate the Capacitance*: The gate oxide capacitance of the MOSFET can be calculated from the charge density versus electric field by using the relationship $C = q/V$.
- *Calculate the Transconductance*: The transconductance is a measurement that indicates how much the device current shifts in response to a modification in the gate voltage. It is calculated from the I–V curve by using the equation $g_m = dI_{ds}/dV_g$, where g_m is the transconductance, I is the current flow through the device, and V_g is the gate voltage.

By following these given steps, we can obtain a comprehensive understanding of the device performance.

3.7.7 Simulated Results Validation

Simulation validation is a process in which the simulated results are compared with experimental data of the device. Validation of the simulation results in a TCAD simulation involves repeating the same. Here are some steps that can be taken to validate the simulation results in a TCAD simulation:

- *Establish the simulation conditions*: It's important to ensure that the simulation conditions are set up correctly, such as the device structure, material properties, doping profiles, and other relevant parameters. These should match the actual device being studied as closely as possible.
- *Compare simulation results to experimental data*: If experimental data is available, it can be used to compare against the simulation results. This can help to determine whether the simulation is accurately capturing the behavior of the device being studied. For example, if the simulation results show a higher current than the experimental data, there may be an issue with the simulation setup or model parameters.
- *Use established simulation results*: If there are established simulation results for similar devices or materials, these can be used to compare against the simulation results. This is helpful in the validation of simulated results.

3.7.8   OPTIMIZE THE DEVICE DESIGN

To optimize the device design of an FET for improved performance, the following steps can be taken:

- *Define the design parameters*: The first thing that has to be done is to determine the FETes design parameters, which include the gate length, gate oxide thickness, doping concentration, and channel material.
- *Simulate device performance*: Use a device simulation software to simulate the performance of the FET with the defined design parameters. This can help identify the areas where the device may not be performing optimally.
- *Analyze the simulation results*: Analyze the simulation results to identify any factors that limit the performance of the designed device. For example, if the simulation shows that the device is experiencing high leakage current, it may be necessary to adjust the doping concentration or gate oxide thickness.
- *Modify The design parameters*: Adjust the design parameters to improve the device performance based on the simulation results. For example, increasing the doping concentration profile in the channel region can reduce the resistance and improve the device speed.
- *Repeat the simulation and analysis*: Repeat the simulation and analysis process with the modified design parameters to evaluate their impact on device performance.
- *Verify and Fabricate*: Once the design parameters are optimized, verify the device performance through additional simulations and then fabricate the device using the optimized design.
- *Characterize the Device*: Finally, characterize the device to verify its performance and compare it with the simulation results. Any discrepancies can be used to further refine the design parameters.

By following the above steps, it is possible to optimize the design of the FET for improved performance.

3.7.9 FINALIZE THE DESIGN

When satisfied with the performance of the FET, finalize the design and prepare for fabrication.

3.8 EVALUATION OF A SINGLE-GATE MOSFET's PERFORMANCE

A fully depleted strained silicon-on-insulator (FD-S-SOI) single-gate MOSFET is a device that is made up of a thin strained silicon layer that is deposited on top of an insulator substrate [28, 29]. This device has a single-gate structure, which allows for higher carrier mobility and thus faster device operation. This device has lower V_{th}; therefore, it is suitable for low-power applications. The drain and source regions are connected to the gate, and the entire MOSFET is isolated from the substrate. The strained silicon layer provides enhanced electron mobility and lower leakage current, making this device more suitable for low-power applications. According to Moore's law, the number of transistors in a single chip should double every two years [30–32]. In the field of computing, this law is a rule of thumb that is widely recognized and has been used to define goals for research. Figure 3.3 shows the cross-sectional view of a FD-S-SOI single-gate MOSFET.

3.8.1 ANALYTICAL MODEL

The analytical model of a MOSFET consists of several mathematics equations to understand device behavior; these equations are based on the electrical and physical properties of the MOSFET. Analytical modeling is the process of creating a mathematical model based on axioms and principles. It involves deriving a mathematical

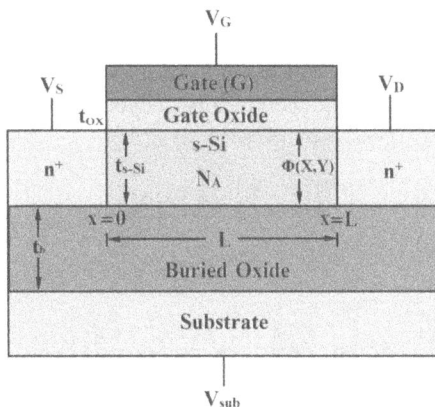

FIGURE 3.3 Cross-sectional view of a fully depleted strained silicon-on-insulator (FD-S-SOI) single-gate MOSFET.

equation that describes the system of interest and then solving the equation to generate a model [33]. It is a theory-driven approach where the model parameters can be determined by the principles and assumptions underlying the equation.

3.8.2 SURFACE POTENTIAL MODELING OF A SINGLE-GATE MOSFET

Poisson's equation is indeed a PDE that plays a fundamental role in modeling the behavior of a single-gate Metal-Oxide-Semiconductor Field-Effect Transistor (MOSFET). It describes the distribution of electric potential within the device and is crucial for understanding how the MOSFET operates. Poisson's equation, A PDE equations that model the surface potential of a single-gate MOSFET. This equation represents the variation of electrostatic potential in an area of space owing to the presence of charge, which can model the surface potential of a single-gate MOSFET [34]. This equation has been solved numerically with a finite difference method. The solution of the PDE gives the surface potential of the MOSFET.

3.8.3 EFFECT OF STRAIN ON A DEVICE'S BAND STRUCTURE

The type of strain that is imposed has a significant bearing on the kind of influence the strain has on the band structure of a material. It's possible for certain kinds of strain, like compressive strain, to cause a material's band gap to narrow, while other kinds of strain, like tensile strain, can cause the band gap to widen [35, 36]. In addition, strain can cause a material's valence and conduction bands to shift, resulting in a change in the material's electrical properties. For example, compressive strain can cause the valence and conduction bands to shift closer together, resulting in a decrease in the material's resistance. Conversely, tensile strain can cause the valence and conduction bands to shift further apart, resulting in an increase in the material's resistance.

3.8.4 FLAT BAND VOLTAGE (V_{FB}) OF THE FRONT CHANNEL

The voltage applied at the gate–source terminal of the MOSFET is referred to as the "flat band voltage" (V_{FB}). This voltage is measured when the I_{ds} is at zero. It is the voltage at which the MOSFET will begin to turn on, and it is usually measured at a V_{ds} of 0 V. The flat band voltage is typically between -0.5 and -2.5 V for a given MOSFET. V_{FB} can be expressed by [37]:

$$\left(V_{FB,f} \right)_{Si} = \phi_M - \phi_{si} \tag{3.30}$$

where:

$$\phi_{f-Si} = V_T \ln\left(\frac{N_a}{n_I} \right), \qquad \phi_{Si} = \frac{\chi_{Si}}{q} + \frac{E_{g,Si}}{2q} + \phi_{f-Si} \tag{3.31}$$

where ϕ_{Si} is the unstrained silicon work function, ϕ_M is the gate metal work function, ϕ_{f-si} is the Fermi potential present in unstrained silicon, χ_{si} is the electron affinity,

q is the magnitude of the electron charge, N_a is the acceptor-type doping profile, I is the intrinsic carrier concentration, and $E_{g,Si}$ is the band gap present in unstrained silicon.

3.8.5 FLAT BAND VOLTAGE OF THE BACKCHANNEL

The backchannel V_{FB} is affected by various parameters like doping concentration, bias applied, and temperature. V_{FB} is a useful parameter for consideration in device design and characterization of MOSFETs; this can be expressed as:

$$\left(V_{FB,b}\right)_{Si} = \phi_{sub} - \phi_{Si} \tag{3.32}$$

where:

$$\phi_{sub} = \frac{\chi_{si}}{q} + \frac{E_{g,Si}}{2q} + \phi_{f-sub}, \quad \phi_{f-Sub} = V_T \ln\left(N_{Sub}/n_i\right) \tag{3.33}$$

where ϕ_{Sub} is the substrate work function, and ϕ_{f-sub} is the Fermi work function with respect to substrate.

A built-in potential $(V_{bi,Si})$ is an electrical potential that exists across a material due to a difference in the concentrations of charge carriers on either side of the material. This potential arises due to the formation of an electric field across the material, which is usually caused by a difference in the number of electrons on either side of the material. $V_{bi,Si}$ is given by this expression [38]:

$$V_{bi,Si} = \frac{E_{g,Si}}{2q} + \phi_{f-Si} \tag{3.34}$$

3.8.6 MODEL FORMULATION

Electric potential contained within a MOSFET is modeled with the assistance of Poisson's equation on a two-dimensional (2D) scale. The electric potential is calculated using the distribution of charge carrier present in channel region of MOSFET. The Poisson equation calculates the surface potential of the MOSFET, which is the potential difference of the metal gate and semiconductor channel. This potential difference determines the flow of current through the MOSFET when potential is applied at the gate terminal. The equation can be expressed by [39]:

$$\frac{\partial^2 \phi_1(m,n)}{\partial m^2} + \frac{\partial^2 \phi_1(m,n)}{\partial n^2} = \frac{qN_A}{\varepsilon_{si}}, \text{ for } \quad 0 \le m \le L_1, 0 \le y \le t_{S-Si} \tag{3.35}$$

$$\frac{\partial^2 \phi_2(m,n)}{\partial m^2} + \frac{\partial^2 \phi_2(m,n)}{\partial n^2} = \frac{qN_A}{\varepsilon_{si}}, \text{ for } \quad L_1 \le m \le L, 0 \le n \le t_{S-Si} \tag{3.36}$$

A parabolic function provides a good approximation of the vertical potential profile.

$$\varphi_1(m,n) = \phi_{s1}(m) + a_{11}(m)y + a_{12}(m)n^2, \text{ for } \quad 0 \leq m \leq L_1, 0 \leq n \leq t_{S-Si} \quad (3.37)$$

$$\varphi_2(m,n) = \phi_{s2}(m) + a_{12}(m)n + a_{22}(m)n^2, \text{ for } \quad L_1 \leq m \leq L, 0 \leq n \leq t_{S-Si} \quad (3.38)$$

In order to solve Poisson's equation, first specify the appropriate boundary conditions.

The presence of the oxide layer does not cause a break in potential (displacement) across the strained silicon film and gate oxide contact.

$$\left. \frac{d\varphi_1(m,n)}{dn} \right|_{y=0} = \frac{\varepsilon_{ox}}{\varepsilon_{si}} \frac{\phi_{s1}(x) - V'_{Gs1}}{t_f} \quad (3.39)$$

$$\left. \frac{d\varphi_2(x,y)}{dn} \right|_{y=0} = \frac{\varepsilon_{ox}}{\varepsilon_{si}} \frac{\phi_{s2}(x) - V'_{Gs2}}{t_f} \quad (3.40)$$

where $V'_{GS_1} = V_{Gs} - \left(V_{FB_1,f}\right)_{Si}$, and $V'_{GS_2} = V_{Gs} - \left(V_{FB_1,f}\right)_{Si}$.

Here, the effect of the trapped charge is considered as:

$$\left(V_{FB_1,f}\right)_{Si} = \phi_M - \phi_{Si}, \left(V_{FB_2,f}\right)_{Si} = \phi_M - \phi_{Si} - \frac{qN_f}{C_{ox}} \quad (3.41)$$

Both the electric field and electric flux (displacement) at the trapped charge contact are continuous.

$$\left. \frac{d\varphi_1(m,n)}{dn} \right|_{n=t_{Si}} = \frac{\varepsilon_{ox}}{\varepsilon_{si}} \frac{-\phi_B(m) + V'_{Sub}}{t_b} \quad (3.42)$$

$$\left. \frac{d\varphi_2(m,n)}{dn} \right|_{n=t_{Si}} = \frac{\varepsilon_{ox}}{\varepsilon_{si}} \frac{-\phi_B(m) + V'_{Sub}}{t_b} \quad (3.43)$$

where $V'_{Sub} = V_{Sub} - \left(V_{FB,b}\right)_{Si}$.

At the location of the trapped charge, both the electric flux and the electric potential are continuous. This is a crucial point to keep in mind.

$$\left. \frac{d\varphi_1(m,n)}{dm} \right|_{m=L_1} = \left. \frac{d\varphi_2(m,n)}{dm} \right|_{m=L_1} \quad (3.44)$$

$$\phi_1(L_1, 0) = \phi_2(L_1, 0) \quad (3.45)$$

The surface potential measured from the source end of the MOSFET:

$$\phi_1(0, 0) = \phi_{s1}(0) = V_{bi,Si} \quad (3.46a)$$

The surface potential measured from the drain end of the MOSFET:

$$\phi_2\left(L_1+L_2,\,0\right)=\phi_{s2}\left(L_1+L_2\right)=V_{(bi,\,Si)}+V_{Ds} \tag{3.46b}$$

With the help of boundary condition (3.37–3.40), coefficient can be obtained and expression for $\phi_1(m,n)$ and $\phi_2(m,n)$. Put the value of $\phi_1(m,n)$ and $\phi_2(m,n)$ in Equations (3.33) and (3.34), respectively, and substitute the $y=0$ expression, resulting in:

$$\frac{d^2\phi_{s1}(m)}{dm^2}-\alpha\phi_{s1}(m)=\beta_1 \tag{3.47}$$

$$\frac{d^2\phi_{s2}(m)}{dm^2}-\alpha\phi_{s2}(m)=\beta_2 \tag{3.48}$$

where $\alpha=\frac{2\left(C_fC_{si}+C_fC_b+C_bC_{si}\right)}{t_{si}^2C_{si}\left(2C_{si}+C_b\right)}$.

$$\beta_1=\frac{qN_A}{\varepsilon_{si}}-2V'_{GS1}\frac{C_f\left(C_{si}+C_b\right)}{t_{si}^2C_{si}\left(2C_{si}+C_b\right)}-2V'_{SUB}\frac{C_b}{t_{si}^2C_{si}\left(2C_{si}+C_b\right)} \tag{3.49}$$

$$\beta_2=\frac{qN_A}{\varepsilon_{si}}-2V'_{GS2}\frac{C_f\left(C_{si}+C_b\right)}{t_{si}^2C_{si}\left(2C_{si}+C_b\right)}-2V'_{SUB}\frac{C_b}{t_{si}^2C_{si}\left(2C_{si}+C_b\right)} \tag{3.50}$$

It is possible to find a solution to Equations (3.45) and (3.46) by employing a straightforward nonhomogeneous differential equation of the second order, which, with the assistance of constant coefficients, may be stated as:

$$\phi_{s1}(m)=A\ e^{(nm)}+Be^{(-N*m)}-\frac{\beta_1}{\alpha} \tag{3.51}$$

$$\phi_{s2}(x)=Ce^{(N(m-L_1))}+De^{(-N(m-L_1))}-\frac{\beta_2}{\alpha} \tag{3.52}$$

where $N=\sqrt{\alpha}$, $p_1=\frac{\beta_1}{\alpha}$, and $p_2=\frac{\beta_2}{\alpha}$.

With the help of Equations (3.44) and (3.48), the boundary condition values for A', B', C', and D' can be obtained.

$$A'=\left\{V_{bi,Si}\left(1-e^{(-NL)}\right)+V_{DS}+\left(P_1-P_2\right)\cosh\left(NL_2\right)+P_2-P_1e^{(-NL)}\right\}/\left\{2\sinh(NL)\right\} \tag{3.53}$$

$$B'=\left[\left\{V_{bi,Si}\left(e^{(NL)}-1\right)+P_1e^{(NL)}-P_2-V_{DS}-\left(P_1-P_2\right)\cosh\left(NL_2\right)\right\}/\left\{(2\sinh(NL)\right\}\right] \tag{3.54}$$

$$C'=Ae^{(NL_1)+\frac{p_2-p_1}{2}} \tag{3.55}$$

$$D'=Be^{(-NL_1)+\left(\frac{p_2-p_1}{2}\right)} \tag{3.56}$$

It is possible to give an expression for the electric field present under the metal gate M_1, M_2 in terms of the horizontal component:

$$E_1(m) = A'Ne^{(Nm)} - B'Ne^{(-Nm)} \tag{3.57}$$

$$E_2(m) = C'Ne^{(N(m-L_1))} - D'Ne^{(-N(m-L_1))} \tag{3.58}$$

The minimal front-channel potential can be stated as:

$$m_{min} = \frac{1}{2n}\ln\left(\frac{B'}{A'}\right) \tag{3.59}$$

$$\phi_{s,min} = 2\sqrt{A'B'} - p_1 \tag{3.60}$$

Figure 3.4 shows variations of surface potential with changes in gate length. The parameters used are as follows: The work function of metal $\phi_M = 4.6$ eV; doping concentration $N_A = 1\times10^{16}$ cm^{-3}; length variation $L = 100, 50, 30$ nm; thickness of the gate oxide $t_{ox} = 2$ nm; $V_{ds} = 0.0$ V; and $V_{GS} = 0.1$ V.

3.8.7 Threshold Voltage (V_{TH}) Modeling

The V_{th} is the V_{GS} at which the MOSFET starts to conduct. The value of the V_{th} depends on various factors such as the T_{ox}, and the doping concentrations of the acceptors (N_a) and donors (N_d) in the substrate. Empirical models are mathematical equations that are derived from experimental data. Threshold voltage modeling is essential for designing and optimizing MOSFET-based circuits. It is helpful when forecasting the behavior of the device under a variety of various operating situations and for assuring the performance and dependability of the circuit. Modeling of V_{th} in

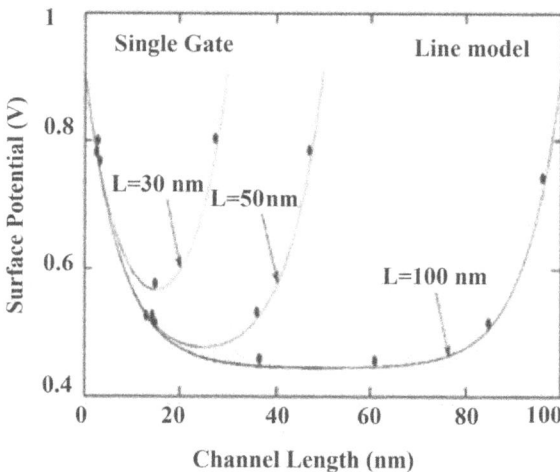

FIGURE 3.4 Variations of surface potential with changes in gate length.

a MOSFET is a process of obtaining the approximate values V_{th} by using empirical models. These models are based on physical parameters like oxide thickness, gate oxide capacitance, the substrate doping profile, and temperature [40]. It is possible to utilize the V_{th} model to optimize the design of the MOSFET by predicting the threshold voltage for a given set of parameters and by putting this information to use. Additionally, other models like the EKV model and BSIM4 model are used to analyze the threshold voltage of MOSFETs. These models are used for circuit simulations to accurately predict the performance of MOSFETs.

The V_{th} of a strained Si SOI MOSFET is modified by changing the condition at the front gate. Specifically, the threshold condition can be expressed as [41]:

$$\phi_{S,M} = \phi_{th} = 2\phi_{f,Si} \tag{3.61}$$

This expression shows that the V_{th} can be modified by controlling the surface potential at the interface between the silicon and the insulator, which can be achieved by adjusting the front-gate voltage. In a strained Si MOSFET, the strain in the silicon can affect the density of states (DOS) and band structure, which can in turn affect the V_{th}. By using the above expression (3.61), the V_{th} is optimized for performance in strained Si MOSFETs [42]:

$$V_{TH} = \frac{-\eta + \sqrt{\eta^2 - 4\sigma\xi}}{2\sigma} \tag{3.62}$$

Equation (3.62) provides a formula for calculating the V_{TH} based on parameters such as η, σ, and ξ. These parameters' material-related properties are used in MOSFETs, such as the doping concentration and the degree of strain in the channel.

Equations (3.63) and (3.64) relate to the voltage across the MOSFET channel:

$$V_{bi\_1} = V_{(bi,Si)}\left(1 - \gamma\right) + V_{Ds} - (u - v)\cosh\left(NL_2\right) - v + u\gamma \tag{3.63}$$

$$u = \frac{C_b}{C_f}V'_{SUB} - \frac{qN_At_{si}}{C_f} - V_{FB1,si}, v = \frac{C_b}{C_f}V'_{SUB} - \frac{qN_At_{si}}{C_f} - V_{FB2,si} \tag{3.64}$$

$$\text{where } \gamma = e^{(-NL)}, \sigma = \frac{1}{\gamma} + \gamma - 2 - \sinh^2(NL),$$

where V_{bi_1} is the built-in voltage of the source end of the channel, V_{Ds} is the drain–source voltage, V_{SUB} is the substrate voltage, t_{si} is the thickness of the Si channel, and N_A is the acceptor doping profile. Equation (3.64) considers various capacitances (C_b and C_f) and V_{FB} of the MOSFET.

$$\xi = V_{bi\_1}V_{bi\_2} - \sinh^2(NL)\left(\phi_{th} - u\right)^2, \eta$$

$$= V_{bi_1}\left(-\frac{1}{\gamma} + 1\right) + 2\sinh^2(NL)(\phi_{th} - u) - V_{bi\_2}\left(1 - \gamma\right) \tag{3.65}$$

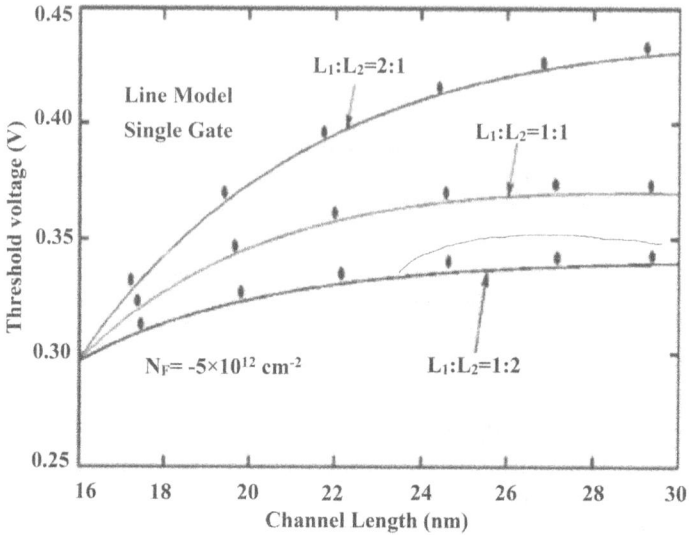

FIGURE 3.5 Threshold voltage V_{th} variations with changes in channel length L.

Equation (3.65) provides a formula for calculating the parameter η, which is used in Equation (3.62) to determine the threshold voltage. The parameter η depends on the built-in voltages (V_{bi_1} and V_{bi_2}) and the surface potential (ϕ_{th}) of the MOSFET.

Figure 3.5 shows variations V_{th} along with channel length L having different channel-length ratios (L_1/L_2 = 1:2, 1:1, 2:1; N_a = 1×10^{16} cm^{-3}, t_{ox} = 2 nm, V_{DS} = 0.1 V, and L = 100 nm).

3.9 CONCLUSION

Different types of modeling approaches, like analytical, empirical, and compact modeling processes, have been discussed. An analytical model for surface potential is useful for understanding the dynamics of surface charge and potential in various physical systems. It can be used to predict the behavior of surface potentials in different situations, such as electrolyte solutions in the presence of an electric field. The model can also be used to identify the effects of different parameters, such as the surface charge, surface dipole moment, surface potential, and dielectric constant, on the overall potentials of surfaces. Ultimately, this analytical model helps us to better understand how electrical charges interact with surfaces, which is important for many applications in science and engineering. The analytical model for threshold voltage can provide a reliable and efficient way to accurately predict and control the threshold voltage of MOSFET devices, and it provides a better understanding of the behavior of MOSFET devices and their performance under different conditions. Moreover, it helps to improve the design of MOSFETs and enhance device performance. Therefore, the analytical model for threshold voltage can be used as a powerful tool for the study and design of MOSFET devices.

REFERENCES

1. Fritzson, P., 2014. Principles of object-oriented modeling and simulation with Modelica 3.3: a cyber-physical approach. John Wiley & Sons.
2. Senturia, S.D., 1998. CAD challenges for microsensors, microactuators, and microsystems. Proceedings of the IEEE, 86(8), pp. 1611–1626.
3. Laursen, T.A., 2003. Computational contact and impact mechanics: fundamentals of modeling interfacial phenomena in nonlinear finite element analysis. Springer Science & Business Media.
4. Vaidya, V., Kim, J., Haddock, J.N., Kippelen, B. and Wilson, D., 2008. SPICE optimization of organic FET models using charge transport elements. IEEE Transactions on Electron Devices, 56(1), pp. 38–42.
5. Chhowalla, M., Jena, D. and Zhang, H., 2016. Two-dimensional semiconductors for transistors. Nature Reviews Materials, 1(11), pp. 1–15.
6. Pardoux, É. and Veretennikov, Y., 2001. On the Poisson equation and diffusion approximation. I. The Annals of Probability, 29(3), pp. 1061–1085.
7. Shim, C.H., Maruoka, F. and Hattori, R., 2009. Structural analysis on organic thin-film transistor with device simulation. IEEE Transactions on Electron Devices, 57(1), pp. 195–200.
8. Wenger, Y. and Meinerzhagen, B., 2019. Low-voltage current and voltage reference design based on the MOSFET ZTC effect. IEEE Transactions on Circuits and Systems I: Regular Papers, 66(9), pp. 3445–3456.
9. Singh, A.P., Baghel, R.K. and Tirkey, S., 2023, February. Enhanced low dimensional MOSFETs with variation of high K dielectric materials. In 2023 IEEE International Students' Conference on Electrical, Electronics and Computer Science (SCEECS) (pp. 1–5). IEEE.
10. Angelov, I., Rorsman, N., Stenarson, J., Garcia, M. and Zirath, H., 1999. An empirical table-based FET model. IEEE Transactions on Microwave Theory and Techniques, 47(12), pp. 2350–2357.
11. Zhao, W. and Cao, Y., 2006. New generation of predictive technology model for sub-45 nm early design exploration. IEEE Transactions on Electron Devices, 53(11), pp. 2816–2823.
12. Gildenblat, G., Li, X., Wu, W., Wang, H., Jha, A., Van Langevelde, R., Smit, G.D., Scholten, A.J. and Klaassen, D.B., 2006. PSP: An advanced surface-potential-based MOSFET model for circuit simulation. IEEE Transactions on Electron Devices, 53(9), pp. 1979–1993.
13. Root, D.E., Xu, J., Horn, J., Iwamoto, M., Rudolph, M. and Fager, C., 2011. The large-signal model: Theoretical foundations, practical considerations, and recent trends. Nonlinear Transistor Model Parameter Extraction Techniques, pp. 123–170.
14. Han, D., Noppakunkajorn, J. and Sarlioglu, B., 2013, June. Efficiency comparison of SiC and Si-based bidirectional DC-DC converters. In 2013 IEEE Transportation Electrification Conference and Expo (ITEC) (pp. 1–7). IEEE.
15. Enz, C.C. and Vittoz, E.A., 2006. Charge-based MOS transistor modeling: the EKV model for low-power and RF IC design. John Wiley & Sons.
16. Sinha, A.K., 2017. Bias-point calculation and parameter extraction using EKV compact MOS model equation. IETE Journal of Education, 58(1), pp. 42–48.
17. Bradde, T., Grivet-Talocia, S., Toledo, P., Proskurnikov, A.V., Zanco, A., Calafiore, G.C. and Crovetti, P., 2021. Fast simulation of analog circuit blocks under nonstationary operating conditions. IEEE Transactions on Components, Packaging and Manufacturing Technology, 11(9), pp. 1355–1368.
18. Bonani, F., Guerrieri, S.D. and Ghione, G., 2003. Physics-based simulation techniques for small-and large-signal device noise analysis in RF applications. IEEE Transactions on Electron Devices, 50(3), pp. 633–644.

19. Chen, M., Rosendahl, L.A., Condra, T.J. and Pedersen, J.K., 2009. Numerical modeling of thermoelectric generators with varing material properties in a circuit simulator. IEEE Transactions on Energy Conversion, 24(1), pp. 112–124.
20. Chen, M. and Rincon-Mora, G.A., 2006. Accurate electrical battery model capable of predicting runtime and IV performance. IEEE Transactions on Energy Conversion, 21(2), pp. 504–511.
21. Salazar, L. and Joos, G., 1994. PSPICE simulation of three-phase inverters by means of switching functions. IEEE Transactions on Power Electronics, 9(1), pp. 35–42.
22. Bank, R.E., Rose, D.J. and Fichtner, W., 1983. Numerical methods for semiconductor device simulation. SIAM Journal on Scientific and Statistical Computing, 4(3), pp. 416–435.
23. Singh, A.P., Shankar, P.N., Baghel, R.K. and Tirkey, S., 2023, February. A Review on Graphene Transistors. In 2023 IEEE International Students' Conference on Electrical, Electronics and Computer Science (SCEECS) (pp. 1–6). IEEE.
24. Ketterhagen, W.R., am Ende, M.T. and Hancock, B.C., 2009. Process modeling in the pharmaceutical industry using the discrete element method. Journal of Pharmaceutical Sciences, 98(2), pp. 442–470.
25. Buturla, E.M., Cottrell, P.E., Grossman, B.M. and Salsburg, K.A., 1981. Finite-element analysis of semiconductor devices: The FIELDAY program. IBM Journal of Research and Development, 25(4), pp. 218–231.
26. Bank, R.E., Coughran, W.M., Fichtner, W., Grosse, E.H., Rose, D.J. and Smith, R.K., 1985. Transient simulation of silicon devices and circuits. IEEE Transactions on Computer-Aided Design of Integrated Circuits and Systems, 4(4), pp. 436–451.
27. Hellings, G., Eneman, G., Krom, R., De Jaeger, B., Mitard, J., De Keersgieter, A., Hoffmann, T., Meuris, M. and De Meyer, K., 2010. Electrical TCAD simulations of a germanium pMOSFET technology. IEEE Transactions on Electron Devices, 57(10), pp. 2539–2546.
28. Sahu, P.K., Pradhan, K.P., Mohapatra, S.K., Sahu, P.K., Pradhan, K.P. and Mohapatra, S.K., 2014. A study on SCEs of FD-s-SOI MOSFET in nanoscale. Universal Journal of Electrical and Electronic Engineering, 2(1), pp. 37–43.
29. Özben, E.D., Lopes, J.M.J., Nichau, A., Lupták, R., Lenk, S., Besmehn, A., Bourdelle, K.K., Zhao, Q.T., Schubert, J. and Mantl, S., 2010. Rare-earth scandate/TiN gate stacks in soi mosfets fabricated with a full replacement gate process. IEEE Transactions on Electron Devices, 58(3), pp. 617–622.
30. Schaller, R.R., 1997. Moore's law: Past, present and future. IEEE spectrum, 34(6), pp. 52–59.
31. Theis, T.N. and Wong, H.S.P., 2017. The end of Moore's law: A new beginning for information technology. Computing in Science & Engineering, 19(2), pp. 41–50.
32. Cavin, R.K., Lugli, P. and Zhirnov, V.V., 2012. Science and engineering beyond Moore's law. Proceedings of the IEEE, 100(Special Centennial Issue), pp. 1720–1749.
33. Xie, Q., Xu, J. and Taur, Y., 2012. Review and critique of analytic models of MOSFET short-channel effects in subthreshold. IEEE Transactions on Electron Devices, 59(6), pp. 1569–1579.
34. Shen, C., Ong, S.L., Heng, C.H., Samudra, G. and Yeo, Y.C., 2008. A variational approach to the two-dimensional nonlinear Poisson's equation for the modeling of tunneling transistors. IEEE Electron Device Letters, 29(11), pp. 1252–1255.
35. Peelaers, H. and Van de Walle, C.G., 2012. Effects of strain on band structure and effective masses in MoS 2. Physical Review B, 86(24), p. 241401.
36. Li, J., Shan, Z. and Ma, E., 2014. Elastic strain engineering for unprecedented materials properties. Mrs Bulletin, 39(2), pp. 108–114.
37. Yamamoto, Y., Kita, K., Kyuno, K. and Toriumi, A., 2007. Study of la-induced flat band voltage shift in metal/HfLaOx/SiO2/Si capacitors. Japanese Journal of Applied Physics, 46(11R), p. 7251.

38. Lyu, J.S. and Lee, K.S.N., 1992. Determination of flat-band voltages for fully depleted silicon-on-insulator (SOI) metal-oxide-semiconductor field-effect transistors (MOSFET's). Japanese Journal of Applied Physics, 31(9R), p. 2678.
39. Schwartz, P., Barad, M., Colella, P. and Ligocki, T., 2006. A cartesian grid embedded boundary method for the heat equation and Poisson's equation in three dimensions. Journal of Computational Physics, 211(2), pp. 531–550.
40. Jean, Y.S. and Wu, C.Y., 1997. The threshold-voltage model of MOSFET devices with localized interface charge. IEEE Transactions on Electron Devices, 44(3), pp. 441–447.
41. Dasgupta, A. and Lahiri, S.K., 1986. A novel analytical threshold voltage model of MOSFETs with implanted channels. International Journal of Electronics Theoretical and Experimental, 61(5), pp. 655–669.
42. Tsormpatzoglou, A., Dimitriadis, C.A., Clerc, R., Pananakakis, G. and Ghibaudo, G., 2008. Threshold voltage model for short-channel undoped symmetrical double-gate MOSFETs. IEEE Transactions on Electron Devices, 55(9), pp. 2512–2516.

4 Dynamics of Trap States in Organic Thin-Film Transistors (OTFTs)

Farkhanda Ana, Haider Mehraj, and Najeeb-ud-Din

4.1 INTRODUCTION

4.1.1 BACKGROUND

The success story of the semiconductor industry is undoubtedly silicon. The enormous growth of the electronics market can be attributed to the device scaling proposed by Gordon Moore in 1965, resulting in high-speed integrated circuits. The design of circuits in the nanometer regime coupled with the high cost and complexity of lithographic equipment has led to research for alternative materials to silicon. The pathbreaking discovery to alternate semiconducting materials was made in 1978 when polyacetylene, an intrinsically insulating organic conjugated polymer, demonstrated electrical conductivity. This discovery of electrically conducting polymers led to Alan Heeger, Alan MacDiarmid, and Hideki Shirakawa winning the Nobel Prize in Chemistry in 2000 [1,2]. Since then, research has focused on designing electronic devices with the advantages of carbon-based materials, which is not otherwise possible using the inorganic silicon.

The term "organics" evokes the thought of plastics that are regarded as insulators. It would have been ludicrous to think that polymers could be made to conduct electricity. In fact, the electronics sector has used plastics a lot as inactive packaging and insulating material. The unprecedented performance improvement in semiconductors, storage, and displays at steadily declining costs that we witness today has been made possible by new organic materials. However, most of these organic materials are used either as passive insulators or as sacrificial stencils (photoresists), and thus play no active part in how an electronic device operates [1]. The current cost/performance ratio of logic devices is made possible by photoresists and insulators, two important groups of passive organic materials. The essential components that characterize chip circuitry and permit continuous device size reduction are photoresists. New resists must be developed to keep lithographic scaling as optical tools improve due to special lens design and light sources. In order to meet the resolution, sensitivity, and processing requirements of each new generation of chips, chemists developed special photosensitive polymers. The combination of enhanced photoresist resolution capabilities and better optical tools has enabled the production of transistors with feature sizes less than 100 nm, which is substantially smaller than the 193 nm exposure wavelength of the existing optical exposure tool [3, 4].

DOI: 10.1201/9781003393542-4

4.1.2 WHY ORGANICS?

The enormous growth of the electronics industry is based on silicon-based transistors with increased miniaturization per device per unit area, as suggested by Moore's law. But the thought-provoking question is "How many Moore generations can we have?" Intel's 10 nm Super Fin transistor is in high-volume production and Intel 7 previously known as Enhanced Super Fin shows 10–15% performance per Watt gain over previous generation due to transistor optimization. [5]. Transistor sizes so small that they resemble an atom will have different device behavior, and we may end up having higher costs because of the unique photolithographic equipment required. This is currently a serious limitation in the scaling of silicon. By 2025, it's feasible that we'll hit the upper bounds of performance advancements in reasonably priced silicon devices, magnetic storage, and screens [1, 4]. Therefore, fundamental studies of materials may open the door to new product form factors. Organic materials have emerged as promising for electronic applications because of their semiconductive properties. Organic materials offer mechanical flexibility and toughness compared to inorganic silicon. The use of low-quality amorphous silicon for storage applications still requires a glass substrate to grow, whereas if the low-end applications of amorphous silicon are replaced with organic materials, we would be able to achieve new device functionalities that would be completely impossible with silicon [4]. The purpose of organic-based electronic devices is not to match or outperform silicon-based technologies in performance, but rather to enable wholly new device functions that are difficult to realize with silicon [5, 6]. These factors make organics the materials of the twenty-first century.

4.1.3 ADVANTAGES AND DISADVANTAGES

The use of organic materials in the field of electronic applications offers many benefits, especially the properties of mechanical flexibility, optical transparency, and toughness, making these materials viable candidates in the field of emerging electronics. Organic transistors can be fabricated using inkjet printing, making roll-to-roll manufacturing possible, in contrast to the high-cost multistep photolithography required for silicon transistors. Organic materials are processed at low temperatures, resulting in low processing costs and reduced capital investments. The fabrication processes are simple and affordable based on inkjet printing, stamping, and so on [7]. In one day, typically a sheet-feed printing machine can process the same area of material as a Si wafer production plant can process in one year. Conjugated polymers have enabled the development of organic circuits that are amenable to plastic substrates, thereby facilitating the creation of electronic devices that are both compact and lightweight, as well as structurally robust and flexible. Organic materials are biodegradable, as they can be manufactured on substrates using paper, plastic, cloth, and so on, resulting in the development of more sustainable and environmentally conscious electronic technologies with an enhanced level of capabilities [4]. The potential of these materials to broaden our electronic domain in manners that will significantly transform the mode in which society engages with technology is immense.

Despite many advantages, organic materials suffer from low charge mobility compared to their inorganic counterparts. They have shorter lifetimes and are more susceptible to environmental conditions. The air stability of n-type organic materials is an issue, especially when they are exposed to oxygen and humid conditions; hence, the applications of organic electronics are dominated by p-type materials [8].

4.1.4 Organic Materials

Organic materials employed in electronic applications are mostly p-type because of their better air stability and higher conductivity. The organic materials that are of interest in electronics belong to the class of organic solids that contain π electrons. Conjugated polymers used for electronic applications are sp^2 hybridized [6, 7]. Sp^2 orbitals on each carbon atom form a σ bond. The unhybridized p_z on each carbon atom shares electrons off the internuclear axis and forms a π bond. Conjugated materials exhibit a distinctive structural pattern in terms of alternate single bonds followed by a double bond. The conduction electrons are composed of loosely bound electrons within the π bonds [7, 8]. The delocalization of the π electron cloud occurs throughout the conjugation length of the polymer. The molecular orbitals undergo splitting into bonding and antibonding states that are conventionally referred to as HOMO (highest occupied molecular orbital) and LUMO (lowest unoccupied molecular orbital), respectively, as depicted in Figure 4.1 [7, 8]. Many organic materials like polyacetylene, poly(3-octylthiophene) (P3OT), poly(3-hexylthiophene) (P3HT), rubrene, tetracene, and pentacene have been identified for electronic applications [6, 8]. A novel derivative of pentacene, triisopropylsilyl (TIPS) pentacene, which is

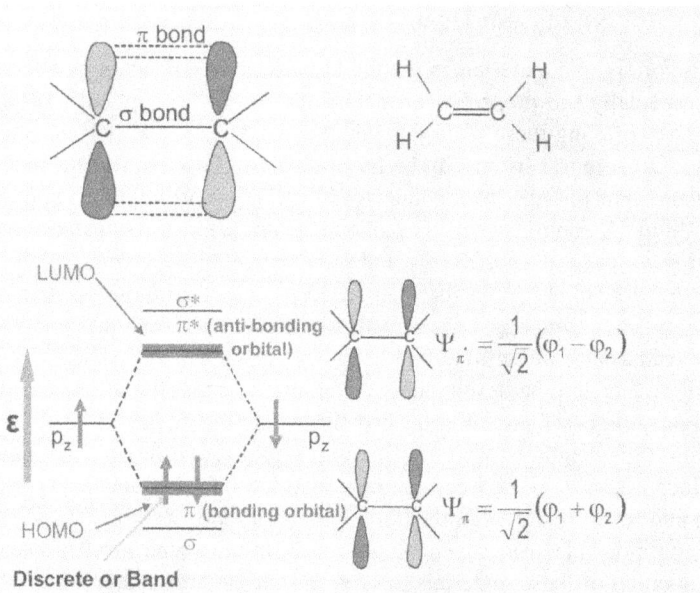

FIGURE 4.1 Ethylene molecule with σ and π bonds.

solution processible, has also been developed. For thin-film transistor (TFT) applications, pentacene has unquestionably been regarded as a high-mobility material. This chemical compound is classified as a polycyclic aromatic hydrocarbon and is composed of five benzene rings that are fused linearly [8, 9]. Research has led scientists to synthesize new organic compounds that are air stable and have an improved mobility of >1 cm^2/V-s, like dinaptho-thieno-thiophene (DNTT) [10]. Figure 4.2 shows the chemical structure of various organic semiconductors.

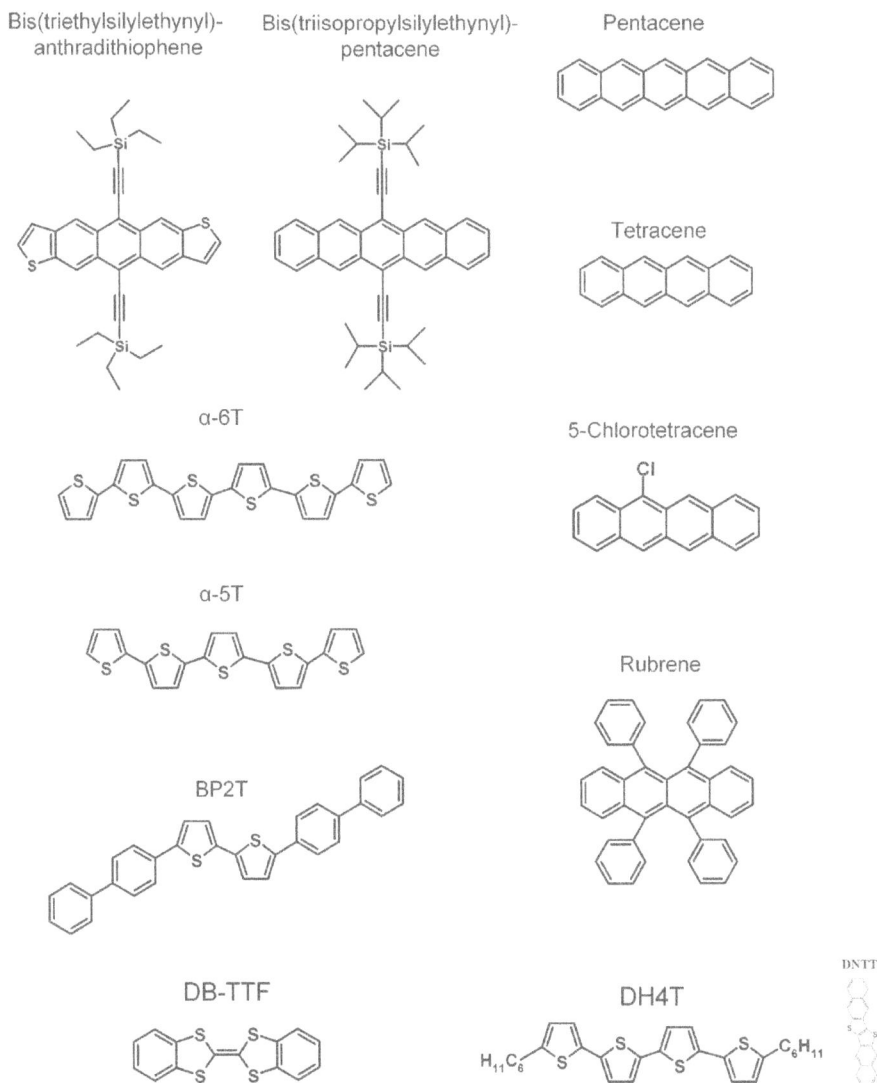

FIGURE 4.2 Typical chemical structure of different organic semiconductors [6, 8, 10].

4.2 APPLICATIONS

Organic technologies that are popular and commercialized are organic light-emitting diodes (OLEDs), organic solar cells, and organic thin-film transistors (OTFTs) [2, 4]. OLEDs are used as display elements in large-screen active-matrix displays by many leading companies in their touchscreen mobile handsets. Organic photovoltaics (OPVs) represent a significant domain of application for organic materials. They are not intended to supplant silicon-based photovoltaics but, rather, to leverage the distinctive properties of OPVs, such as their flexibility, extensive coverage, and affordability. OTFTs present a cost-effective option to silicon transistors in the context of large-area OTFT-based arrays. These arrays are commonly utilized as driver circuits in display applications, where the integration density and switching rate are not critical factors [9]. Circuits utilizing OTFTs that rely on conjugated polymers are amenable to flexible substrates, thereby enabling the production of circuits that are compact, lightweight, flexible, and structurally resilient. According to studies of recent research advancements and enhanced operational capabilities, it is now feasible to implement OTFTs in intricate areas such as radiofrequency (RF), telecommunications, and aerospace [1, 4]. The studies also emphasized the optimization of OTFT performance to facilitate the integration of circuits made entirely of plastic materials.

Organic materials have opened up a new future of flexible and wearable electronics based on the mechanical flexibility offered by the substrates on which they are fabricated [1]. Organic materials are used in display arrays and in many other applications, such as electronic newspapers, skin sensors such as artificial skin and muscles, radiofrequency identification (RFID) tags, smart textiles, electronic noses, batteries and supercapacitors, and memory devices. OTFTs are dominating the display market, where they are used in large matrix display backplanes. The research is currently focused on improving their performance for use as driver elements in large matrix displays [2, 3]. Organic electronics has made possible the realization of a wide array of innovative applications, such as wearable sensors and 4K displays using light-emitting diodes (LEDs). Furthermore, fabrication using flexible substrates makes organic applications cheaper in cost compared to their inorganic counterparts. Toward the end of 2025, it is estimated that the organic market will be a $250 billion business without replacing much of the inorganic semiconductors in existing electronic products [4]. OLEDs, organic solar cells, and organic transistors are examples of commercially available organic materials utilized in electronics. OTFTs are not often employed because of their poor carrier field-effect mobilities. OTFTs are the most frequently researched application of organic materials in the electronics industry. OTFTs offer a cost-effective alternative to applications using amorphous silicon-based TFTs, such as large-area display units.

Despite the poor physical and electrical properties of organic materials, research on these materials has led to them attaining performance at par with that of hydrogenated amorphous silicon. Because of the unique properties offered by organic materials, the *International Technology Roadmap for Semiconductors* (ITRS) finds organic-based devices to be one of the pioneering and promising technologies in the area of emerging electronics [11].

Pentacene has demonstrated a high mobility of around 1 cm^2/V-s (three orders of magnitude less than what monocrystalline silicon transistors can attain) and has been mostly used in TFT structures as an active semiconductor. TIPS pentacene is a solution-processible derivative of pentacene and is thus more popular than pentacene. DNTT has also been reported to be used as an active semiconductor in OTFTs [10]. One downside of OTFTs is that they cannot be used in high-speed applications. Nonetheless, OTFTs have prospective uses, such as in displays and smart devices, because of their simple manufacturing processes, structural flexibility, and cheap cost. Experimental synthesis and imaging techniques have demonstrated that organic materials have an irregular, granular structure. The granular arrangement leads to the formation of gap states in between, which are known as "trap states." These trap states trap the charge carriers, thereby decreasing the effective charge density and limiting mobility.

4.3 ORGANIC THIN-FILM TRANSISTORS

A potential application of organic materials in the electronics industry is transistor technology. Transistors using organic materials as the active semiconductor are finding applications in digital signal processors (DSPs), memory devices, mixer circuits, and the like. Organic transistors are designed on the schematic of TFTs, which has demonstrated compatibility with low-mobility materials [12]. They are especially appealing for uses in microelectronics applications (RFID tags, sensors, etc.), contrary to the exorbitant cost of packing traditional Si circuits, which makes them unaffordable. Moreover, OTFT circuits based on conjugated polymers can be fabricated on plastic substrates, allowing for the creation of small, light, structurally sturdy, and flexible electronic devices [4, 12]. An OTFT has three contact terminals: the gate, source, and drain contact. OTFTs do not have a body contact and, hence, no body effect. OTFTs can be designed in staggered and inverted staggered device configurations, depending upon the placement of the source and drain with respect to the semiconductor [13–15]. In a staggered structure, gate contact is always on top with respect to source–drain contacts. Organic semiconductors can be deposited on the top or bottom of the source–drain contacts in staggered geometry. In the inverted staggered structure, the gate contact is always below the source–drain contact regions. The structural schematics that can be realized based on the placement of contact areas in OTFTs are shown in Figures 4.3 and 4.4.

The deposition of metal for the source–drain on the organic semiconductor leads to a metal-semiconductor junction, resulting in poor grain structure. This leads to reduced-mobility regions near source–drain contacts [16]. These low-mobility regions around source–semiconductor and drain–semiconductor contacts lead to high-resistance areas. The top-contact devices have less area of contact between the source–drain region and organic semiconductor compared to the bottom-contact devices. The small area of the low-mobility region in top-contact devices leads to small contact resistance. Hence, top-contact devices show better performance than the bottom-contact organic transistors [17]. However, the bottom-contact transistors are preferred, as the top-deposited organic semiconductor is protected during etching, high-temperature processing, and metal penetration in the fabrication

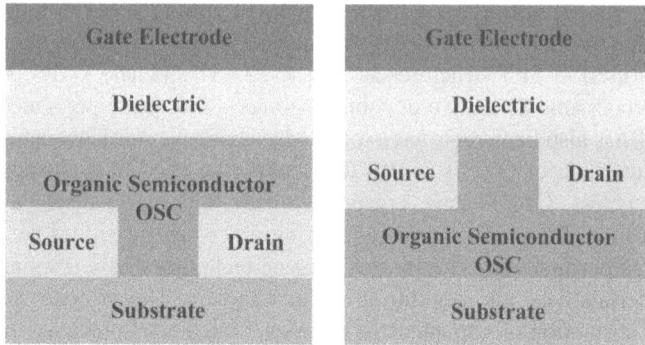

FIGURE 4.3   Staggered organic thin-film transistor (OTFT) structure [13].

process [6, 17]. The selection of the device configuration is thus predominantly determined by the application for which the device is designed. A TFT behaves similarly to a typical MOSFET, with the exception that TFTs only operate in the accumulation regime due to the absence of a space-charge region. Using a p-type semiconductor as a TFT's active layer, and if a negative gate voltage is supplied, holes are induced in the semiconductor to produce an accumulation layer. When an n-type semiconductor is used, however, the accumulation layer is created because electrons are induced by a positive gate voltage.

4.4   EXISTENCE OF TRAP STATES

Organic materials possess a granular structure wherein the grain boundaries act as trapping centers. The structure of pentacene (an organic semiconductor) under scanning tunneling microscopy (STM) is shown in Figure 4.5. The grain boundaries act as trapping centers, resulting in a density of defect states in addition to a density of states in the valence band and conduction band [18]. The defect states are distributed within the forbidden gap, as assumed by the multiple trap and release model [19, 20]. The multiple trap and release model explains that the charge carriers can move within the bands and between the HOMO to a trap level or from one trap level to

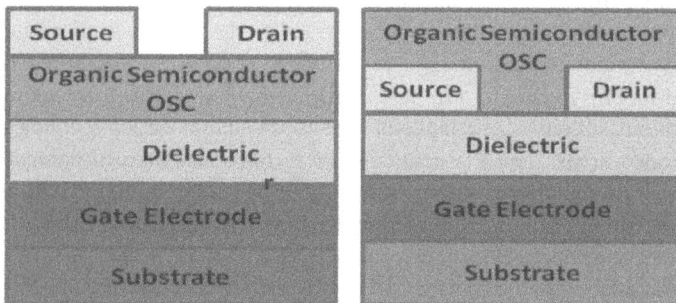

FIGURE 4.4   Inverted staggered OTFT structure [13].

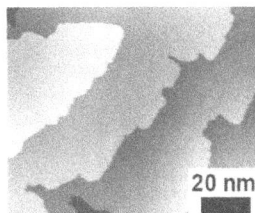

FIGURE 4.5 Pentacene on styrene/Si under scanning tunneling microscopy (STM) [21].

another. Thus, movement of charge carriers within the organic semiconductors is more complex compared to the inorganic semiconductors. The states of the trap exhibit an energetic division that spans both the HOMO and the LUMO [20]. The trap states are distributed throughout the band structure of the organic semiconductor and are classified as shallow traps or deep traps, depending upon their distance from band edges, as shown in Figure 4.6. The traps can be deep or shallow based on their distribution in the energy band.

The shallow traps play a significant role in charge transport, as they are close to band edges and are responsible for trapping and detrapping. The deep states, being in the middle of the band gap, do not influence the electrical characteristics significantly. Figure 4.7 depicts a Gaussian distribution explaining trap density in deep states and an exponential distribution explaining trap density in shallow states.

4.5 CHARGE TRANSPORT THEORY

Charge transport in an organic electronic material has been described in several theories, and many models have been proposed. The band transport in inorganic semiconductors describes that the atoms in a crystal are held together tightly by covalent bonds. As a result, electronic interactions between atomic orbitals in an inorganic crystal are strong, and broad bands with bandwidths of a few eV are formed. This facilitates high-mobility charge transfer. Organic crystals are characterized by the presence of van der Waals forces that hold the constituent atoms together. These forces are relatively weak in nature. The feeble electronic interactions that occur

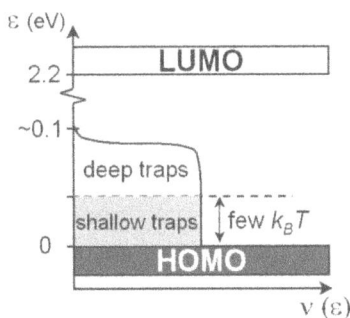

FIGURE 4.6 Deep and shallow trap states.

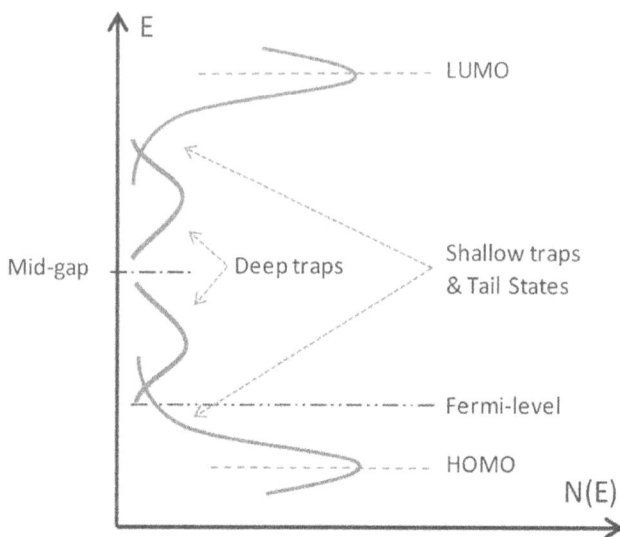

FIGURE 4.7   Model for trap states within bands [18].

between atomic orbitals of adjacent lattice sites lead to the formation of narrow bands, which possess bandwidths that are lower than 500 meV [22].

Charge transport models are shown in Figure 4.8. Charge carrier transport has been popularly described by "hopping," which defines carrier transport within local-ized levels. The hopping of carriers is influenced by temperature. The theory of variable-range hopping (VRH) has been described for charge transport. Based on the activation energy, a charged particle may jump over a short distance with high energy or over a long distance with low energy [11, 23, 24]. The Poole–Frenkel (PF) conduction model is based on the concept that a trapped carrier jumps to the band from a defect state and moves freely within the band. The transport in organic mate-rials can be more accurately described by considering the hybrid charge transport picture, wherein the movement of charge carriers is described by the combination of the PF and band transport models [25]. This describes the dependence of mobility on temperature, bias, and transitory effects. The ratio of trapped to free charges at thermal equilibrium results in a mobility that is dependent only on temperature. The variation in applied bias causes Fermi-level shifts, and the trapped charge can thus vary accord-ingly. This phenomenon results in the dependence of charge mobility on applied bias.

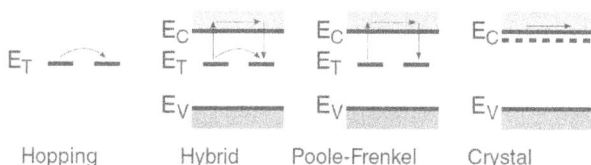

FIGURE 4.8   Charge conduction models for organic semiconductors [11].

4.6 ANALYTICAL MODELING OF TRAP STATES

Designing and simulation of OTFT-based circuits are necessary for the commercialization of organic electronics as a marketable technological resource. This calls for the creation of analytical models and procedures that are applicable to organic materials. It has been established that using MOSFET models to simulate and develop OTFTs is a practical way to forecast how effectively these devices perform. However, the device characteristics substantially deviate from the MOSFET theory's predicted behavior [26]. As the current conduction mechanism in OTFTs differs from that of inorganic transistors, the MOSFET models are unreliable for precisely predicting the performance of OTFTs. Metal–oxide–semiconductor (MOS) models do not account for the defect states that are unavoidable in disordered organic semiconductors. Consequently, it becomes crucial to consider the material and behavioral characteristics of organic materials while analyzing and designing the device models for OTFTs.

4.6.1 THRESHOLD VOLTAGE (V_{TH}) AND TRAP DENSITY OF STATES

Most of the simulation studies on OTFTs have been carried out using pentacene as an active semiconductor owing to its nearly crystalline structure. An acceptor trap density of states has been reported to exist in pentacene [16]. When free holes are occupied, these states can be neutral (N_T^0) or carry a positive charge (N_T^+) [27, 28]. The deep trap density of states approximated by the Gaussian function can be expressed as [28]:

$$g_{GA}(E) = N_{GA}exp\left[-\left(\frac{E_{GA} - E}{W_{GA}}\right)^2\right] \qquad (4.1)$$

where N_{GA} is the Gaussian density at equilibrium, E_{GA} is the peak energy, and W_{GA} is the characteristic decay energy.

The shallow trap density of states approximated by the exponential distribution can be expressed as [28]:

$$g_{TA}(E) = N_{TA}exp\left[\frac{E_V - E}{W_{TA}}\right] \qquad (4.2)$$

where N_{TA} is the valence band intercept edge density, and W_{TA} is the characteristic energy.

Since trap states are responsible for occupying holes, the threshold voltage equation for OTFTs gets modified as [29]:

$$V_{TH} = V_{FB} - \frac{Q_{it}}{C_i} - \frac{qp}{C_i} - \frac{qN_t^+}{C_i} \qquad (4.3)$$

where V_{FB} is the flat band voltage and has a negative value for pentacene; flat band voltage is the gate voltage that must be supplied to prevent band bending in a semiconductor. C_i is the gate oxide capacitance (F/cm²), qp corresponds to the channel hole carriers, and qN_t^+ is the aggregate amount of confined electric charge per unit surface area. For a trap-free semiconductor, the last term of Equation (4.3) becomes zero.

The inclusion of trap states in the simulation studies on OTFTs has shown an increase in the value of V_{TH} in accordance with Equation (4.3) [29, 30]. The holes induced in the channel may get trapped for an initial increase in values of gate–source voltage (V_{GS}). But for sufficiently high values of V_{GS}, trap states are almost filled, and the additional carriers induced contribute to the increase in device currents.

4.6.2 Mobility Modeling

The PF mobility model defines the observed crowding phenomenon in OTFTs at small values of drain–source voltage (V_{DS}). This behavior is because of the enhancement in mobility due to increased values of the electric field. The PF model defines the charge mobility, which is mathematically represented as [22, 31]:

$$\mu(E) = \mu_0 exp\left[-\frac{\Delta}{kT} + \left(\frac{\beta}{kT} - \delta \right)\sqrt{E} \right] \tag{4.4}$$

where $\mu(E)$ is field-dependent mobility, μ_0 is intrinsic mobility, Δ is zero field activation energy, β is the electron PF factor, δ is the fitting parameter, k is the Boltzmann constant, and T is temperature. The PF mobility model offers a theoretical framework to account for the variations in carrier mobility observed in OTFTs, which are influenced by temperature and electric field.

The trap states may contribute to surface roughness, and charge carriers may be subjected to the scattering phenomenon. The presence of structural imperfections, interruptions in the material's continuity, and interactions between charges and surface phonons at the interface between pentacene and oxide may have a negative impact on the mobility of the device [32]. Research works have reported that OTFTs show mobility reduction at an increased magnitude of vertical electric fields [32]. Therefore, it is necessary to simulate the characteristics of devices, as well as models of mobility degradation, in conjunction with the PF mobility model. The surface roughness causes phonon scattering, resulting in mobility degradation [31]. The mobility simulation of OTFTs using various mobility models has shown a significant impact on device parameters, such as channel mobility, threshold voltage, and transconductance [32]. Results have shown that the mobility degradation behavior in OTFTs has been accurately modeled by the surface mobility model (SURFMOB) given by Jeffrey T. Watt. To include trap states, mobility has been analytically modeled by modifying the SURFMOB proposed by Watt [28, 32, 33]:

$$\frac{1}{\mu_{eff,p}} = A\left(\frac{10^6}{E_{eff,p}} \right)^{-0.29} + B\left(\frac{10^6}{E_{eff,p}} \right)^{-1.62} + C\left(\frac{10^{18}}{N_B} \right)^{-1}\left(\frac{10^{12}}{N_i} \right)^{-1} \tag{4.5}$$

$$E_{eff,p} = E_\perp + ETAP.WATT(E_0 - E_\perp) \tag{4.6}$$

where $E_{eff,p}$ is the net field, N_B denotes the density of trapped charges located at the surface, and N_i is the density of charges under inversion. The hole mobility is defined in terms of three scattering mechanisms in Equation (4.5). The phonon scattering, surface roughness, and charged impurity scattering are represented by the first, second, and third terms on the right-hand side of Equation (4.5), respectively. Each of the first two variables is of relevance for simulating mobility degradation behavior in organic semiconductors because they account for the impacts of the electric field. The universal field-mobility relation may be expressed in these two terms. *ETAP. WATT* is a fitting parameter with a value of 0.33 in Equation (4.6). At the insulator–semiconductor interface, E_0 is the electric field orthogonal to the oxide–semiconductor interface, and $E_\perp$ is the electric field orthogonal to the current direction. The Watt model provides values for the pre-exponents in the three components of Equation (4.5) at $T = 300$ K.

The trap states concentration is directly proportional with the trapped charge density in OTFTs. Therefore, by replacing the N_B in Equation (4.5) with N_V and omitting the N_i term:

$$\frac{1}{\mu_{eff,p}} = A\left(\frac{10^6}{E_{eff,p}}\right)^{-0.29} + B\left(\frac{10^6}{E_{eff,p}}\right)^{-1.62} + C\left(\frac{10^{18}}{N_V}\right)^{-1} \tag{4.7}$$

The effective charge mobility of OTFTs can be combined using Matthiessen's rule [30]. The rule states that the various independent functions having an influence on the same parameter can be combined as follows:

$$\frac{1}{U} = \frac{a_x}{X} + \frac{b_x}{Y} - \frac{a_z}{Z}\left(\pm\frac{a_v}{V}\cdots\right) \tag{4.8}$$

where a_x, a_y, a_z, ... are weighting parameters, often considered to be unity. When limiting or enhancing the value of U, the (+) and (−) signs are respectively utilized. Different functions X, Y, Z, (V), and so on are defined for the same quantity (which can be something like mobility or current), and U is the output value. The effective carrier mobility using Matthiessen's rule for mobility is given as [34]:

$$\frac{1}{\mu} = \frac{1}{\mu_a} + \frac{1}{\mu_b} \tag{4.9}$$

where μ_a and μ_b are mobilities from different unrelated mechanisms.

The net mobility of OTFTs can be obtained as:

$$\frac{1}{\mu_{OTFT}} = \frac{1}{\mu_0} + \frac{1}{\mu_{PF}} + \frac{1}{\mu_{eff,P}} \tag{4.10}$$

where μ_0 is the zero field carrier mobility, μ_{PF} is the PF mobility, and $\mu_{eff,P}$ is given by Equation (4.7). PF mobility demonstrates the field-dependent mobility behavior of OTFTs [16, 25].

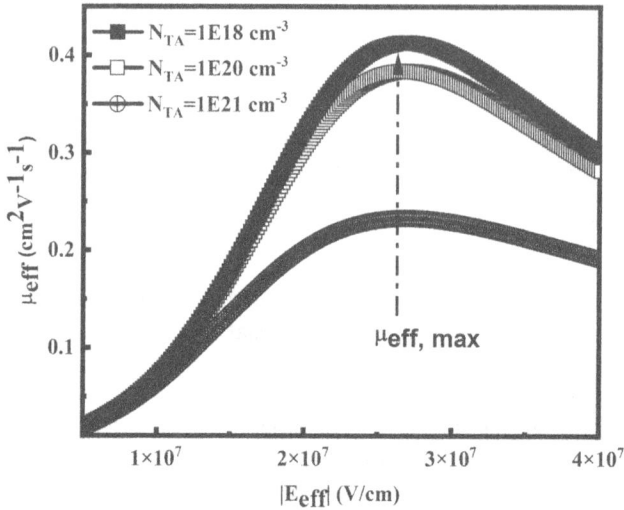

FIGURE 4.9 Effective mobility variation with the effective electric field [29].

$\mu_{eff,p}$ as a function of $E_{eff,p}$ has been plotted in Figure 4.9 using Equation (4.10). The results have been plotted for different values of acceptor trap density. The results indicate that the effective mobility of charge carriers rises linearly as the electric field increases. Nonetheless, the mobility reaches a maximum for a given electric field value before declining. The mobility reduces as the number of trap states rises. This result confirms that more charge carriers are trapped as the trap states per unit volume increases, whereas the free charge carriers per unit volume fall, resulting in a reduction in mobility. The observed current crowding behavior of OTFTs at low field values can be modeled by the PF mobility model alone. But the PF mobility model is not sufficient to predict the mobility behavior of OTFTs at large values of gate field. The mobility reduction observed through device simulations in OTFTs proves the fact that the traps cause surface roughness and phonon scattering, leading to mobility degradation of holes in organic semiconductors. It has also been found that mobility degradation results in reductions of channel mobility, threshold voltage, and transconductance.

4.6.3 DRAIN CURRENT MODELING

To develop a complete electrical model of OTFTs in the presence of the trap states, an energy band diagram is shown in Figure 4.10. E_{FM} is the metal Fermi level, E_{FS} is the pentacene Fermi level, E_i is the intrinsic level, E_{tr} is the trap distribution (maximum near HOMO and minimum near E_{FS}), and $q\Phi_m$ and $q\Phi_s$ are the metal and semiconductor work functions, respectively. For positive values of V_{GS}, the trap level (E_{tr}) is located below the Fermi energy; therefore, the hole carriers are not trapped, and trap states are neutral. At negative values of V_{GS}, the holes are captured by the trap states (being close to HOMO), thus reducing the drain current. As the magnitude of gate voltage is increased, the trap states are occupied with holes and filled.

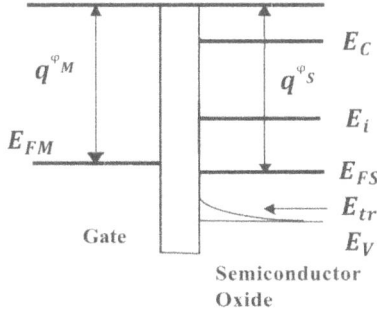

FIGURE 4.10 Energy-band diagram showing the position of Fermi level eFS with respect to trap distribution etr at equilibrium.

Furthermore, an increase in V_{GS} increases free charge density, as the trap volume is fixed and the current at the drain terminal increases linearly.

Thus, the threshold voltage (V_{TH}) increases in the presence of trap states [29, 30]. At the oxide–semiconductor interface, the Fermi level falls and becomes pinned close to HOMO. Hence, the trapped charge density and the free charge carriers created in the channel must be balanced by the gate charge. The trap density exhibits Fermi-level position dependency, as specified in [27]:

$$N_T^+(E_F) = N_0(T)\exp\left[\frac{(E_V - E_F)}{kT_2}\right] \qquad (4.11)$$

and

$$N_0(T) = \alpha(T)g_{T0}\frac{k^2 T_2^2}{kT_2 - kT} \qquad (4.12)$$

The temperature-dependent nature of $N_0(T)$ is independent of the Fermi level's spatial position. The function denoted by $\alpha(T)$ is a constant, slowly varying function that is exclusively dependent on temperature and serves to rectify the integration error. g_{T0} represents the density of defect states when $E = E_V$, and T_2 represents the rate of decay or characteristic energy. $\alpha(T)$ oscillates between 1 and 0.8, according to numerical simulations [27].

The general equation for calculating the potential in FETs can be solved to accommodate the influence of trap states as a parameter [29]:

$$F(\psi) = \sqrt{\left(e^{\frac{-q\psi(x)}{kT}} + \frac{q\psi}{kT} - 1\right) + \frac{n_i^2}{N_V^2}\left(e^{\frac{q\psi(x)}{kT}} - \frac{q\psi}{kT} - 1\right) + \frac{N_0(T).T_2}{N_V.T}\left(e^{\frac{-q\psi(x)}{kT_2}} - 1\right)} \qquad (4.13)$$

Equation (4.13) can be used to calculate the electric field and total charge in the semiconductor [29]. The drain current (I_D) has been solved with the inclusion of traps for OTFTs [29]:

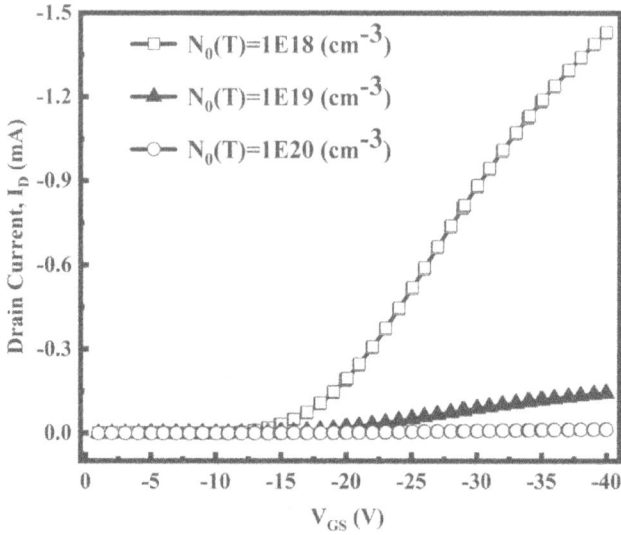

FIGURE 4.11 A plot of $id\text{-}V_{GS}$ characteristics with variation in acceptor trap density $n_0(t)$ and a characteristic decay rate of $\gamma = 1$.

$$I_D = -\frac{\mu_p W N_V . C_{OX}^{(\gamma+1)}}{L N_o(T)^\gamma}\left((V_{GS} - V_{TH})^{\gamma+1}.V_{DS} - \frac{V_{DS}^{\gamma+2}}{\gamma+2}\right) \qquad (4.14)$$

where $\gamma = \frac{T_2}{T}$ is a constant.

γ is the parameter that accounts for the effect of trap states on the current–voltage characteristics. If trap density is not present, $\gamma = 0$, and Equation (4.14) reduces to the basic MOS equation. A plot of OTFT transfer characteristics using Equation (4.14) with the variation in values of trap density of states is shown in Figure 4.11. The decline in I_D at high values of $N_0(T)$ is seen in Figure 4.11. The current reduction due to increasing trap density causes a decrease in free mobile charge carriers, such as holes.

4.6.4 CAPACITANCE MODELING

Charge accumulation or depletion results in channel capacitance. The changes in surface potential from positive to negative values cause the charge carriers to be trapped and detrapped, and result in trap capacitance per unit area C_T, defined by [29, 35, 36]:

$$C_T(\psi_s) = \frac{dQ_T(\psi_s)}{d\psi_s} \qquad (4.15)$$

The capacitance model for OTFTs has been shown in Figure 4.12, where C_{OX} appears in series with the parallel combination of C_C and C_T. The trapping and

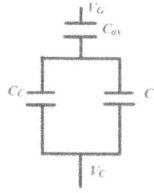

FIGURE 4.12   Capacitance model of an OTFT with trap capacitance [29].

detrapping of charge carriers lead to trap capacitance in parallel with channel capacitance.

The total capacitance between gate and body terminals (C_{GB}) can be represented in Equation (4.16) [29]:

$$\frac{1}{C_{GB}} = \frac{1}{C_{OX}} + \frac{1}{(C_C + C_T)} \tag{4.16}$$

where C_{ox} is the gate oxide capacitance, C_C is the channel capacitance, and C_T is the trap capacitance.

Channel capacitance $C_c = C_{ox}$ for transistors in accumulation mode, and thus Equation (4.16) can be rewritten as:

$$dQ_S = (C_{ox} + C_T)d\psi_S = \eta C_{ox}d\psi_S \tag{4.17}$$

where $\eta = 1 + \frac{C_T}{C_{ox}}$ is the slope factor.

Device capacitance has been studied for its effect of traps using Equation (4.17), and the results are shown in Figure 4.13. For the concentration of trap states less

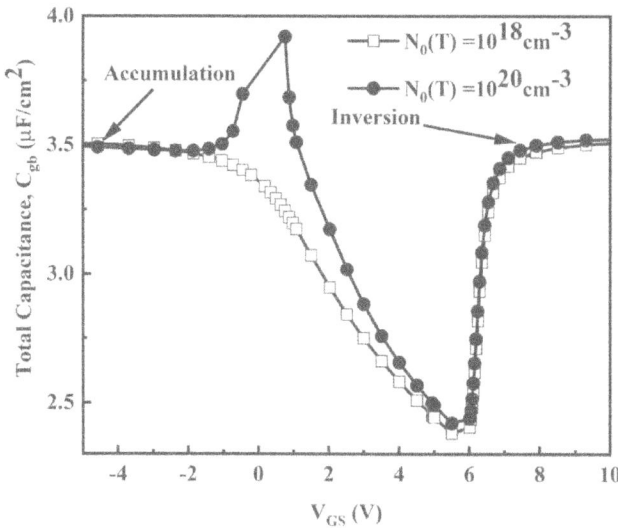

FIGURE 4.13   Plot of C_{gb} versus V_{GS} with varying trap densities [29].

than trap states in the valence band ($N_0(T) < N_V$), the capacitance is comparable to that of a device without traps. Yet the C-V curves show a strong jump for values of $N_0(T) > N_V$. The traps being charged and discharged via charge carriers are the root of this impulsive jump in the C-V characteristics. The distribution of trap states in shallow states is assumed to be exponential; thus, maximum trap states are close to the band edge. Traps at localized levels may generate or annihilate carriers. The trap states, even under flat band circumstances, may be filled and empty. When V_{GS} decreases and becomes negative, the Fermi level drops downward and becomes stuck at HOMO at the oxide–semiconductor interface [37, 38].

This results in trapping and detrapping, which cause a bump in the capacitance curves. In inversion and accumulation modes, the device capacitance is dominated by gate oxide capacitance. The C-V displays a bump at the beginning of accumulation, which is caused by additional capacitance introduced by traps while getting filled [39, 40]. A further detailed description has been provided in Figure 4.14, which plots C_{gb} normalized with C_{ox}. The normalized capacitance has a value of 1 in both modes. Only at the start of the device accumulation mode of operation is a small jump in capacitance observed. The trap capacitance has momentarily contributed 0.1 μF/cm^2 of capacitance to the existing C_{ox} for an increase in traps from 10^{18} per cm^3 to 10^{20} per cm^3 (i.e., for a 10^2 cm^{-3} increase in N_{TA}, capacitance per unit area increases by just 0.1, which is not a substantial change). Furthermore, the capacitance is not a fixed quantity and gets influenced by input signal, doping concentration, and trap concentration [37, 40]. Capacitance also changes under applied stress and is a function of time. Thus, trap states under the application of voltage behave dynamically and influence device capacitance also.

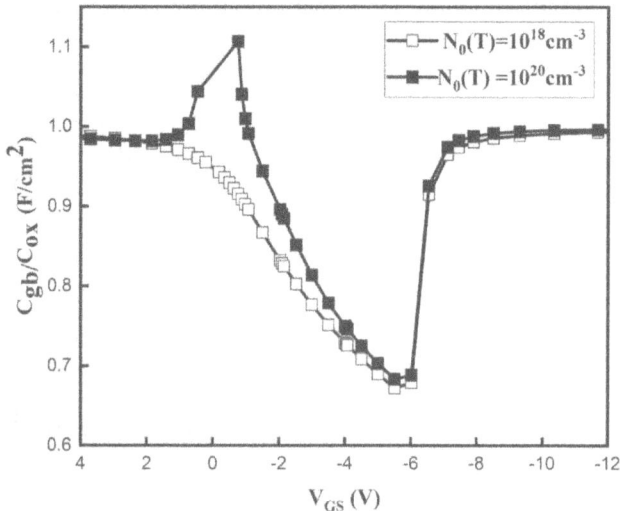

FIGURE 4.14   Plot of normalized capacitance with v_{GS} [29].

4.7 CONCLUSION

Although organic electronics is opening new dimensions of applications in the area of flexible electronics, the efficiency of TFTs is limited by their noncrystalline structure, resulting in trap states. In this chapter, the dynamic nature of trap states has been discussed in detail, and their effect on device characteristics has been modeled analytically. Under no trap conditions does the current–voltage model reduce to a simple MOS transistor equation. The trap states have an influence on mobility behavior, and as a result a combined mobility model using the PF mobility and surface mobility models has been derived. The outcomes derived from the modeling equations exhibit a high level of concurrence with the variations in trap parameters. The observed fluctuations in device performance parameters in response to alterations in trap parameters suggest that the modeling of organic semiconductor-based transistors would be incomplete without trap states. The modeling results have shown that the lower the density of trap states, the better is the performance. A capacitance model has been reviewed using a unified charge control model. Thus, in order to develop a compact model for organic transistors, the trap states must be included as a modeling parameter. The utilization of said models can offer a comprehensive solution to the compact modeling and characterization of OTFTs that does not require dependence on MOSFET models for the purpose of designing and advancing organic-based circuits toward a commercialized platform. Future advancements in organic electronics must be coupled with the development of correct modeling techniques predicting device behavior. Furthermore, the reduction of trap states is of prime importance if the mobility of organic materials is to be enhanced. This necessitates fabricating defect-free organic materials or exploring nature to discover hidden gold for the electronics industry.

REFERENCES

1. Gargi Konwar, Pulkit Saxena, Sachin Rahi and Shree Prakash Tiwari, "Edible Dielectric Composite for the Enhancement of Performance and Electromechanical Stability of Eco-Friendly Flexible Organic Transistors," *ACS Applied Electronic Materials*, vol 4, no. 10, pp. 5055–5064, 2022. http://doi.org/10.1021/acsaelm.2c01082.
2. Mohan V. Jacob, "Organic Semiconductors: Past, Present and Future," *Electronics*, vol. 3, no. 4, pp. 594–597, 2014.
3. S. Mijalkovic D. Green, A. Nejim, G. Whiting, A. Rankov, E. Smith, J.Halls and C. Murphy, "Modelling of Organic Field-Effect Transistors for Technology and Circuit Design," in *Proc. 26th International Conference on Microelectronics (Miel 2008)*, Nis, Serbia, 2008.
4. Leslie A. Pray, "Organic Electronics for a Better Tomorrow: Innovation, Accessibility, Sustainability," White paper in *Chemical Sciences and Society Summit (CS3)*, San Francisco, California, United States, 2012. Available: http://www.rsc.org/globalassets/04-campaigning-outreach/policy/research-policy/global-challenges/organic-electronics-for-a-better-tomorrow.pdf.
5. Ian Cutress and Anton Shilov, "*Intel Details Manufacturing through 2023: 7 nm, 7+, 7++, with Next Gen Packaging*," 2019, Available: https://www.anandtech.com/show/14312/intel-process-technology-roadmap-refined-nodes-specialized-technologies
6. Antonio Facchetti, "π-Conjugated Polymers for Organic Electronics and Photovoltaic Cell Applications," *Chemistry of Materials*, vol. 23, pp. 733–758, 2011.

7. Terje A. Skotheim and John R. Reynolds, *"Handbook of Conducting Polymers: Conjugated Polymers, Processing and Applications,"* Third edition, New York, NY, USA: Taylor and Francis, CRC Press, 2007.

8. Simone Locci, "Modeling of the Physical and Electrical Characteristics of Organic Thin Film Transistors," Ph. D. thesis, Dept. of Electrical and Electronic Engineering, University of Cagliari, Cagliari CA, Italy, 2009.

9. Dipti Gupta, Pradipta K. Nayak, Seunghyup Yoo, Changhee Lee and Yongtaek Hong, "Importance of Simulation Studies in Analysis of Thin Film Transistors Based on Organic and Metal Oxide Semiconductors," in *Numerical Simulations of Physical and Engineering Processes*, Croatia: Intech, pp. 79–100, 2011.

10. U. Zschieschang et al., "Dinaphtho [2,3-b:2′,3′-f]thieno[3,2-b]thiophene (DNTT) Thin-Film Transistors with Improved Performance and Stability," *Organic Electronics*, vol. 12, no. 8, pp. 1370–1375, 2011.

11. Heinz Bassler and Anna Kohler, "Charge Transport in Organic Semiconductors," *Topics in Current Chemistry*, vol. 312, pp. 1–66, 2012.

12. Poornima Mittal, Anuradha Yadav, Y. S. Negi, R. K. Singh and Nishant Tripathi, "Parameter Extraction and Analysis of Pentacene Thin Film Transistor with Different Insulators," in *International Conference on Advances in Electronics, Electrical and Computer Science Engineering-EEC*, 2012, pp. 458–462.

13. G. Konwar, P. Saxena, V. Raghuwanshi, S. Rahi and S. P. Tiwari, "Solution-Processed Biopolymer Dielectric Based Organic Field-Effect Transistors for Sustainable Electronics," *2022 6th IEEE Electron Devices Technology & Manufacturing Conference (EDTM)*, Oita, Japan, pp. 107–109, 2022. http://doi.org/10.1109/EDTM53872.2022.9798041.

14. Chang Hyun Kim, Yvan Bonnassieux and Gilles Horowitz, "Fundamental Benefits of the Staggered Geometry for Organic Field-Effect Transistors," *IEEE Electron Device Letters*, vol. 32, no. 9, p. 1302, 2011.

15. Poornima Mittal, B. Kumar, Y.S. Negi, B.K. Kaushik and R.K. Singh, "Channel Length Variation Effect on Performance Parameters of Organic Field Effect Transistors," *Microelectronics Journal*, vol. 43, pp. 985–994, 2012.

16. Dipti Gupta, Namho Jeon and Seunghyup Yoo, "Modeling the Electrical Characteristics of TIPS-Pentacene Thin-Film Transistors: Effect of Contact Barrier, Field-Dependent Mobility, and Traps," *Organic Electronics*, vol. 9, pp. 1026–1031, 2008.

17. Dipti Gupta, M. Katiyar and Deepak Gupta, "An Analysis of the Difference in Behavior of Top and Bottom Contact Organic Thin Film Transistors Using Device Simulation," *Organic Electronics*, vol. 10, pp. 775–784, 2009.

18. Soonjoo Seo, "Structural and Electronic Properties of Pentacene at Organic-Inorganic Interfaces Pentacene on Styrene," Ph.D. thesis, Materials Science, University of Wisconsin, Madison, 2010.

19. Akansha Sharma, Sarita Yadav, Pramod Kumar, Sumita Ray Chaudhuri and Subhasis Ghosh, "Defect States and Their Energetic Position and Distribution in Organic Molecule Semiconductors," *Applied Physics Letters*, vol. 102, p. 143301, 2013.

20. Stephen R. Forrest, "The Path to Ubiquitous and Low-Cost Organic Electronic Appliances on Plastic," *Nature*, vol. 428, pp. 911–918, 2004.

21. Poornima Mittal, B. Kumar, B.K. Kaushik, Y.S. Negi and R.K. Singh, "Analysis of Pentacene Based Organic Thin Film Transistors through Two Dimensional Finite Element Dependent Numerical Device Simulation," *Special Issue of International Journal of Computer Applications on Optimization and on-Chip Communication*, no. 3, Feb. 2012.

22. J. A. Carr and S. Chaudhary, "The Identification, Characterization and Mitigation of Defect States in Organic Photovoltaic Devices: A Review and Outlook," *Energy & Environmental Science*, vol. 6, pp. 3414–3438, 2013, http://doi.org/10.1039/C3EE41860J.

23. Benjie N. Limketkai, "Charge-Carrier Transport in Amorphous Organic Semiconductors," Ph.D. thesis, Massachusetts Institute of Technology, 2008.
24. M. C. J. M. Vissenberg and M. Matters, "Theory of the Field-Effect Mobility in Amorphous Organic Transistors," *Physical Review B*, vol. 57, no. 20, pp. 964–967, 1998.
25. P. Stallinga, "Electronic Transport in Organic Materials: Comparison of Band Theory With Percolation/(Variable Range) Hopping Theory," *Advanced Materials*, vol. 23, pp. 3356–3362, 2011.
26. Antonio Valletta et al., "A Compact SPICE Model for Organic TFTs and Applications to Logic Circuit Design," IEEE Transactions on Nanotechnology, 2016, http://doi.org/10.1109/TNANO.2016.2542851.
27. P. Stallinga and H.L. Gomes, "Modeling Electrical Characteristics of Thin-Film Field-Effect Transistors II. Effects of Traps and Impurities," *Synthetic Metals*, vol. 156, pp. 1316–1326, 2006.
28. ATLAS Manual. Device simulation Software (Atlas TCAD device simulator, Silvaco TCAD software), Santa Clara: Silvaco, 2012.
29. F. Ana and Najeeb-ud-Din, "An Analytical Modeling Approach to the Electrical Behavior of the Bottom-Contact Organic Thin-Film Transistors in Presence of the Trap States," *Journal of Computational Electronics*, vol. 18, pp. 543–552, 2019. https://doi.org/10.1007/s10825-019-01314-6.
30. F. Ana and Najeeb-ud-Din, "Simulation Study of the Electrical Behavior of Bottom contact Organic thin Film Transistors, " *2014 IEEE 2nd International Conference on Emerging Electronics (ICEE)*, Bengaluru, India, pp. 1–4, 2014. http://doi.org/10.1109/ICEmElec.2014.7151133.
31. Mohammad Mottaghi and Gilles Horowitz, "Field-Induced Mobility Degradation in Pentacene Thin-Film Transistors," *Organic Electronics*, vol. 7, pp. 445–606, 2006.
32. F. Ana and Najeeb-ud-Din, "Effect of Mobility Degradation on the Device Performance of Organic Thin-Film Transistors," *2016 IEEE Region 10 Conference (TENCON)*, Singapore, pp. 3261–3264, 2016. http://doi.org/10.1109/TENCON.2016.7848654.
33. J. Watt, Modeling the performance of liquid-nitrogen-cooled CMOS VLSI, Desertation Thesis, Standford University, 1989.
34. Ognian Marinov and M. Jamal Deen, "Organic Thin-Film Transistors: Part I—Compact DC Modeling," IEEE Transactions on Electron Devices, vol. 56, no. 12, pp. 2952–2961, 2009.
35. Yuan Taur and Tak H. Ning, *"Fundamentals of Modern VLSI Devices,"* Second edition, Cambridge University Press, 1998.
36. Yannis Tsividis, *"Operation and Modeling of the MOS Transistor,"* Second edition, Oxford University Press.
37. Wolfgang L. Kalb, Simon Haas, Cornelius Krellner, Thomas Mathis and Bertram Batlogg, "Trap Density of States in Small-Molecule Organic Semiconductors: A Quantitative Comparison of Thin-Film Transistors With Single Crystals," *Physical Review B*, vol. 81, p. 155315, 2010.
38. M. G. Helander, Z. B. Wang, J. Qiu and Z.H. Lu, "Band Alignment at Metal/Organic and Metal/Oxide/Organic Interfaces," *Applied Physics Letters*, vol. 93, p. 193310, 2008.
39. Z. Tang and C.R. Wie, "Capacitance-Voltage Characteristics and Device Simulation of Bias Temperature Stressed a-Si:H TFTs," *Solid-State Electronics*, vol. 54, pp. 259–267, 2009.
40. Slah Hlali, Neila Hizem, Liviu Militaru, Adel Kalboussi and Abdelkader Souifi, "Effect of Interface Traps for Ultra-Thin High-k Gate Dielectric Based MIS Devices on the Capacitance-Voltage Charcateristics," *Microelectronics Reliability*, vol. 75, pp. 154–161, 2017. http://doi.org/10.1016/j.microrel.2017.06.056.

5 An Insightful Study and Investigation of Tunnel FET and Its Application in the Biosensing Domain

Priyanka Goma and Ashwani K. Rana

5.1 INTRODUCTION

It is well-known that the phenomenon of particle tunneling is a quantum mechanical process. Many semiconductor devices have used this mechanism of carrier transport to obtain better figures of merit (parameters such as threshold voltage, ON and OFF-current, subthreshold swing, etc.) in several applications. Tunnel diodes and tunnel field-effect transistors (TFETs) work in reverse bias mode, as tunneling is a junction breakdown process. This chapter will help the reader to understand the basics of tunneling by using energy band diagrams (EBDs), which provide a qualitative analysis of the device while following the transport and continuity equations. One can calculate the tunneling probability through EBDs, which ultimately helps one to calculate the net current flowing from one terminal to the other in a TFET device. The OFF- and ON-state band diagrams provide knowledge of the transport of carriers through the tunneling junction. These diagrams also help to obtain a technique to calculate the threshold voltage of TFETs, which is unique as well as more accurate than the conventional constant-current method of threshold-voltage determination. The basic TFET architecture is explained in this chapter, followed by the challenges that it may face at the simulation and fabrication levels. TFETs are considered when an application demands low standby power due to less OFF-current or leakage. Biosensing, on the other hand, is one of the expanding domains incorporating the usage of TFETs. Therefore, this chapter illustrates a biosensor device based on the TFET architecture and some perspectives on the implementation of such devices in the future.

5.2 BASIC TFET ARCHITECTURE

Chip miniaturization designed at deep subnanometer technology nodes requires low standby power leakages. This is an important capability that a device must have while undergoing scaling. It is evident from the literature that when we scale down the metal–oxide–semiconductor field-effect transistor (MOSFET) to smaller than a 50 nm gate length, leakage issues arise that are responsible for high I_{off} currents. This can also be observed by studying the subthreshold region of the device. It can be noticed that MOSFETs with larger I_{off} tend to have greater subthreshold slopes (SS).

DOI: 10.1201/9781003393542-5

Mathematically, 2.3 kT/q gives the lowest possible value of SS, which is approximately equal to 60 mV/decade [1]. Therefore, there is a need to address such issues, and one possible efficient candidate is a tunnel mechanism–based field-effect transistor (FET) known as the TFET. A TFET has a wider barrier energy band in its OFF-state, which obstructs the flow of carriers. This helps to obtain a diminished leakage in the OFF-state, thereby inducing lower SS. Due to a different carrier transport in TFETs, the ON-current is affected. This drawback can be mitigated by using multi-gate structures [2, 3], pocket engineering [4, 5], hetero-materials in the body [6, 7], high-k oxides [8], multi-material gates [2, 3], and more [9, 10].

This section describes a basic double-gate TFET architecture. If the electrons are responsible for the conduction, then it is called an n-channel TFET (or n-TFET), and if the conduction is because of holes, then it is called a p-channel TFET (or p-TFET). Figure 5.1(a) and (b) present the simplest structure of n-TFETs and p-TFETs based on a double-gate geometry. Double-gate structures are explained in this chapter to help the reader understand the need to use multiple gates to significantly improve the ON-current of these devices. As with MOSFETs, there are three main regions in any basic TFET device: the source, drain, and channel. For an n-TFET, the source is a

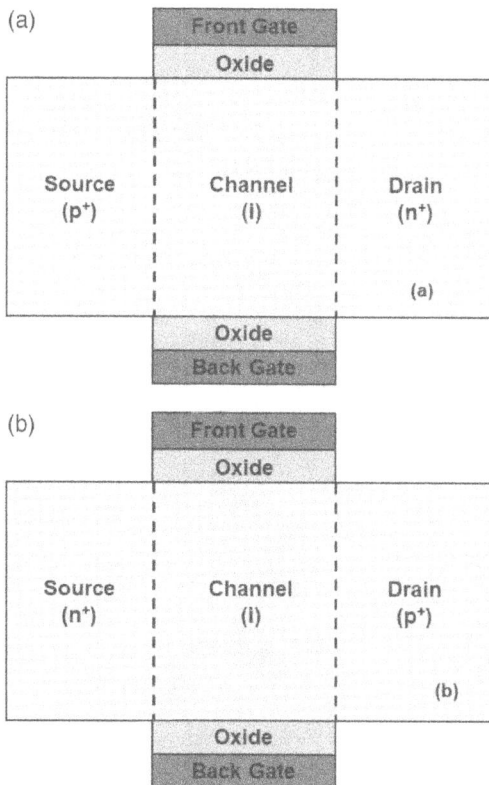

FIGURE 5.1 Basic structure of a tunnel field-effect transistor (TFET): (a) n-channel and (b) p-channel.

highly p-doped region, the drain is a moderately n-doped region, and the channel is an intrinsic region. Similarly, for a p-TFET, the source is a highly n-doped region, the drain is a moderately p-doped region, and the channel is an intrinsic region. Let us take, for instance, the case of an n-channel TFET. When there is no gate voltage ($V_{gs} > 0$) and the drain voltage rises above zero ($V_{ds} > 0$) (OFF-state), there is a negligible current flow. This current flows as a result of the presence of a few electrons in the conduction band, which moves to the drain side upon the applied drain bias. Also, it should be remembered that the p-type valence band has more electrons than the conduction band of the same. When the drain voltage is greater than zero ($V_{ds} > 0$), and the gate voltage is increased ($V_{gs} > 0$) (ON-state), the source–channel interface is impacted. As soon as the gate voltage increases, the junction width begins to decrease as the energy bands of the source and channel start to align themselves. Under the impact of an applied electric field, electrons migrate from the source to the channel when the valence band of the source aligns with the conduction band of the channel. This barrier width resembles the potential well, and the drifting of electrons is happening because of tunneling across the barrier. As V_{gs} increases, it impacts the probability of tunneling along with electron concentration in the channel. This phenomenon is explained in Section 5.3.

5.2.1 CHARACTERISTICS OF TFETS

The generalized transfer characteristics of a TFET are shown in Figure 5.2. Three major visible regions are the ambipolar region, OFF-state region, and ON-state region. Ambipolarity is the behavior of n-TFETs for negative gate voltages.

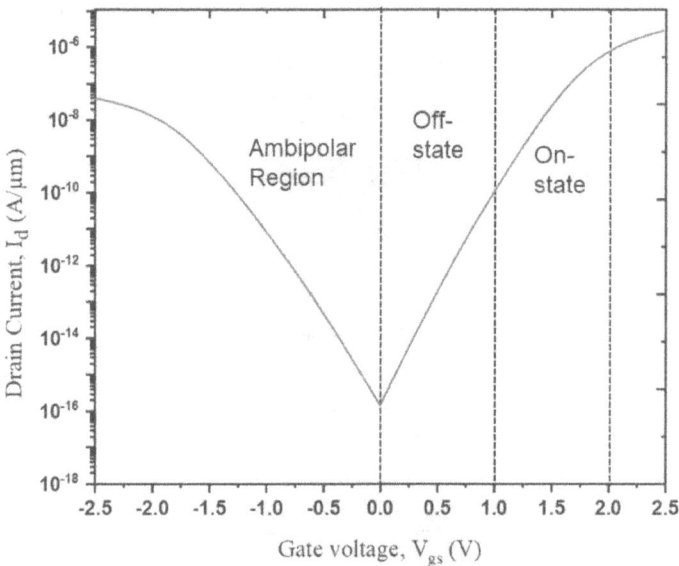

FIGURE 5.2 Transfer characteristics of an n-TFET.

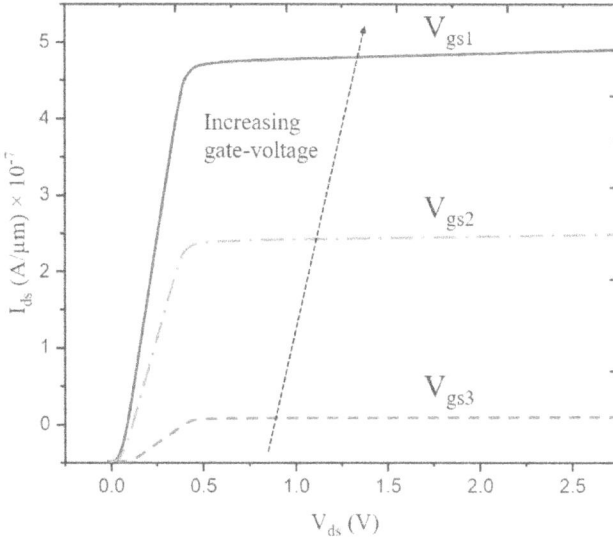

FIGURE 5.3   Output characteristics of an n-TFET.

The source–drain symmetrical doping makes the magnitude of the current in the ambipolar state rise. In this state, the majority carriers from the drain side drift toward the drain–channel area and generate a current. Because the doping of the drain side is more than that of the channel region, the presence of vacant energy states starts to fill, contributing to an ambipolar current. An OFF-state exists when the gate voltage is nonzero but smaller than the threshold voltage (V_{th}) of the device. At zero gate voltage, there are no energy states present in the channel region, and hence OFF-current is much less. Now that the conduction band of the channel has been lowered by lowering the gate voltage, current has begun to grow. If V_{gs} is increased more than V_{ds}, the channel potential is pinned at the ON-state; consequently, no appreciable increase in current is achieved.

Figure 5.3 depicts the device's output characteristics. The channel potential is stuck at the drain potential as the gate potential is larger, so when the drain potential increases, it increases the channel potential and as a result the current surges. As soon as the drain potential drops off, the channel potential becomes independent of the gate potential. This means that the drain current is practically unbothered by an increase in the biasing of the drain. The output properties in this instance are saturated.

5.3   UNDERSTANDING TUNNELING WITH ENERGY BAND DIAGRAMS

The prime focus of this section is to learn the tunneling mechanism, and for that band diagrams must be drawn to understand the carrier transport occurring within the bands. These diagrams give an idea of the qualitative behavior of the device. In order to comprehend how electrons are moved from the source side to the channel

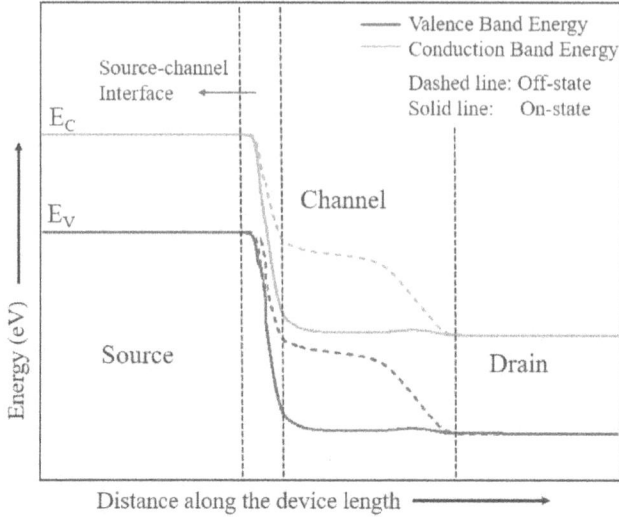

FIGURE 5.4 Energy band diagram for the OFF- and ON-states of an n-TFET.

side, n-TFETs are used as a point of reference. When the device is activated, electrons can tunnel between the source and channel bands, or between the drain and channel bands. The latter scenario is undesirable, as it contributes to the ambipolar behavior of TFETs. This means that there is a cross-conduction of electrons from the drain to the channel, which gives rise to the leakage in the device. The width of the barrier is also a critical factor.

To facilitate electron tunneling and improve conduction and ON-current, the barrier width must be reduced. As the V_{gs} and V_{ds} are increased, the barrier width is decreased. Because of these differences, TFETs have gate-voltage-dependent and drain-voltage-dependent threshold voltages. In order to prevent tunneling from taking place in the OFF-state, the TFET should be designed so that the barrier width decreases. Figure 5.4 depicts the OFF-state and ON-state barrier widths and the tunneling process. The only conduction that occurs in the OFF-state is due to the small number of electrons that are drifted from the channel to the drain, as there are no energy levels available in the conduction band of the channel to tunnel through. When a positive potential is applied at the gate terminal, more energy levels are available for conduction, and the barrier width decreases due to lowering of the channel conduction band.

It is evident from these results that rising V_{gs} reduces the barrier width, and therefore the electrons present at lower energy states can tunnel through it significantly. When we make the gate and drain voltage equivalent to each other, the channel potential tends to saturate. As soon as the V_{gs} is increased beyond the V_{ds}, the channel potential is pinned up, and there is no significant shortening of barrier width. Thus, the overall tunneling probability is improved at higher gate voltages but saturates beyond a certain point. A drop in the barrier width with an increase in V_{gs} is visible in Figure 5.5.

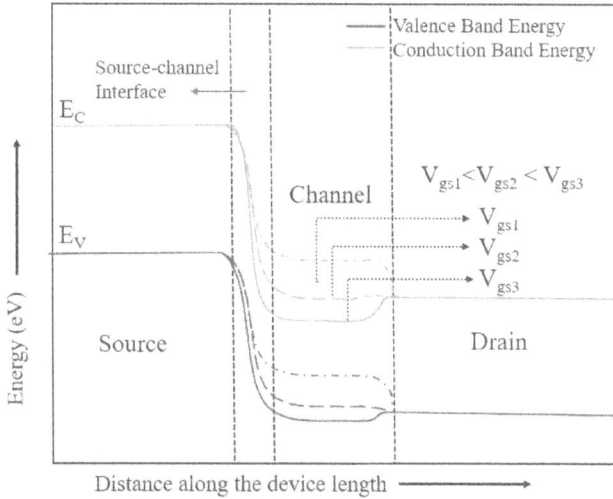

FIGURE 5.5   Energy band diagram for varying gate voltages in an n-TFET.

5.4   CHALLENGES FOR TFETs

5.4.1   SIMULATION METHOD

The TCAD tools help to study through simulation the response of a device during real-time operation. Many of these tools are evolving to achieve results that are closer to their experimental/real-time operation. They help researchers to study the 1D, 2D, and 3D variations in many domains, such as electric field, current, energy bandgap distribution, and potential. These tools are referred to as TCAD (technology computer-aided design) tools. The fabrication of a semiconductor device involves various steps (lithography, deposition of layers, annealing, etc.), and the fabricated device can have variations in the results in comparison to the simulation of the same device. This happens due to the inefficiency of many TCAD tools in replicating the fabricated structure before its actual fabrication. This becomes a challenge for the engineers.

Accuracy is one of the vital parameters that ensure the candidacy of any device. In this section, a widely used tool for the simulation and analysis of TFETs known as Silvaco ATLAS is discussed. There are some key points that one should consider while performing TFET simulations in general:

1. Devices such as TFETs require a careful simulation study, as they involve quantum equations. Therefore, their analysis must contain models that include the effects of such equations.
2. Simulation of TFETs must incorporate the bandgap narrowing effects. Bandgap narrowing can occur due to different materials used to design the TFET, temperature variations, and so on.
3. Electric field distribution in TFETs is a nonlocal tunneling phenomenon. Thus, an electric field calculation in tunnel devices should include

the variations involving nonlocal tunneling. Thus, such models must be considered.

4. The numerical simulation methods must be specified correctly to have a converged solution. It is observed that at short channel lengths, some TFET simulations do not converge, and hence, the behavior of such devices during scaling remains unknown. Thus, there is a need to perform such simulations on specific software that incorporates non-equilibrium green functions to study the device behavior. Such simulations are called "atomistic simulations."

5. Device fabrication involves wear and tear on the edges of the device. Pre-analysis of such effects can mitigate challenges related to the device's accuracy. Such analyses are known as "variability analyses." Many tools do not accurately provide variability analysis of TFETs.

6. Another challenge faced by the simulation tools is the analysis of 2D materials. If the TFET involves the use of 2D materials in its design, there are chances that the results are not accurate with respect to the experimental point of view.

7. Consideration of device parameters according to the *International Technology Roadmap for Semiconductors* (ITRS) guidelines is difficult for TFET devices due to their different carrier transport mechanism. Along with this constant, field scaling is also not applicable on tunnel devices. Therefore, simulating and designing an optimized TFET device needs more attention.

5.4.2 Fabrication Method

Silicon is an obvious choice for TFETs due to its compatibility with complementary metal–oxide–semiconductor (CMOS) fabrication procedures, and due to the fact that a less defective and high-quality interface between the Si and the gate dielectric can be formed. Some of the reported works give an insightful perspective on TFET fabrication [11–13]. A fabrication scheme for manufacturing of a basic TFET device is shown in Figure 5.6. The process can begin with a clean piece of Si wafer (i), over which the oxide can be grown with a local oxidation process known as LOCOS (local oxidation of silicon) (ii). Using chemical vapor deposition, a layer of poly-silicon can be deposited, which works as the gate material in the device. After its deposition, gate patterning can be achieved with the help of photolithography (iii). In this, an ultraviolet (UV) light is exposed over the area, which needs to be etched out later in step (iv). The diffusion of p^+ and n^+ source and drain can be done with an ion implantation technique. For the activation of dopants in these regions, a laser spike annealing (LSA) methodology is practiced that often utilizes chemical treatments to repair any damage from implant dosage (v). The next steps involve metallization for biasing of the device (vi) and wafer bonding to form a double-gate structure (vii). The final device is shown in step (viii), which is a conventional structure of a double-gate tunnel field-effect transistor (DGTFET). The overall ON-current of TFETs is much less than that of ordinary MOSFETs at higher supply voltages. It is evident that TFETs cannot replace MOSFETs unless they offer a subthreshold current range

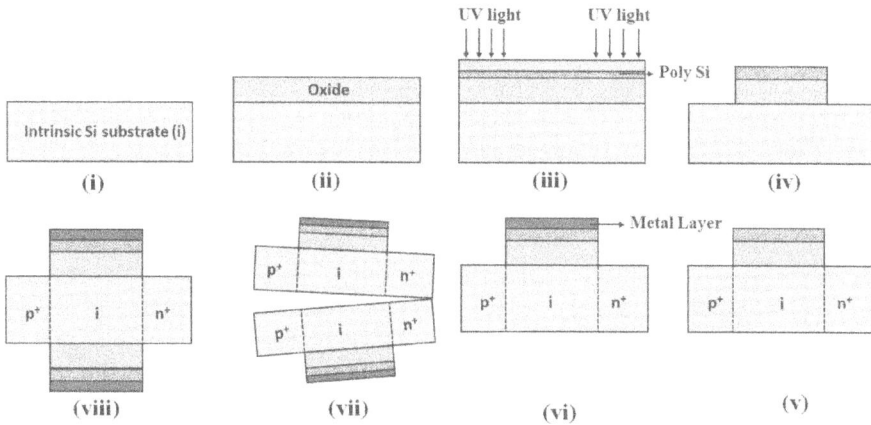

FIGURE 5.6 Basic fabrication scheme for a double-gate TFET device.

with a much smaller SS (60 mV/dec) than MOSFETs. This goal has not been reached in many experiments, which may be because of basic problems with materials and device design as well.

Nonetheless, the TFET fabrication process is getting more meticulous, and incentivizing variability analyses are being reported for statistically significant as well as experimental scenarios. There are still some fabrication challenges that must be addressed beforehand.

5.4.3 NOISE ISSUES

Random telegraph noise (RTN) exists in devices when a single trap center captures the electrons out of a few carriers. Additionally, it can be found in gadgets that are made using a flawed crystal lattice. Temperature, radiation, and mechanical stress influence the RTN to a significant extent. Chen et al. [14] published an experimentally determined RTN amplitude on TFET characteristics. According to the study, the high amplitude of RTN is caused by the non-uniform variation of the band-to-band tunneling (BTBT) generating rate along the device width direction. Reduced critical route density as a result of decreasing device width leads to increased RTN amplitude.

Low-frequency noise known as "flicker" is caused by charge carriers being trapped and released in trap states present at the gate oxide around the quasi-Fermi level. This process is also referred to as "pink noise" and has 1/f as a power spectral density. When semiconductor devices are biased, this noise occurs. With increasing frequency, the intensity of noise lessens. It is mostly produced at the junction between the gate oxide and the Si substrate. Dangling bonds can be seen at the Si substrate's interface. There are other energy states created by these dangling connections. Flicker noise is caused by charge carriers that are trapped at certain points after crossing between energy levels and then released in a chaotic fashion.

Flicker noise becomes highly unwanted at the nanoscale in the case of the produced device because this noise grows as the device dimensions decrease.

By measuring the amount of flicker noise, one may assess the quality and dependability of a gadget. The RTN is dominant, having a slope of $1/f^2$ in large-channel TFETs; the low-frequency noise in TFETs follows a $1/f$ frequency dependence [15]. A study by Huang et al. [16] presents the findings of an experimental investigation to understand the influence of low-frequency noise on TFETs utilizing various source junction topologies. Due to the nonlocal BTBT process, the active traps that are situated in the region where the BTBT is creating electron–hole pairs (EHPs) are seen to be the cause of the noise in TFETs. The traps situated where the junction electric field is greatest have only a minimal effect on TFET noise.

Half static random-access memory (HSRAM) cells were presented by Luong et al. [17] in order to investigate the potential of TFETs for 6T-SRA. With Si nanowire, this described structure has been stretched. Even without the ambipolar behavior, the proposed TFET structure falls short of expectations. The static figure of merit has also been constrained as a result of the analysis of the proposed structure. The impact of a trapped charge causing RTN to occur in a heterojunction TFET (HTFET)-based SRAM was examined by Pandey et al. [18]. The investigation of a 10T SRAM based on a Schmitt trigger mechanism was the main topic of this study. The investigation has made it evident that HTFET-based SRAM cells operate admirably, even when RTN is present. For 0.2 V of applied bias, the presented design improves performance by 15% compared to a silicon-based fin field-effect transistor (FinFET).

5.5   ROLE OF TFETs IN BIOMOLECULE SENSING

Many industries such as biomedical, defense, food environmental, etc. make extensive use of biosensors.. When designing a biosensor, detection speed and sensitivity are critical parameters. When conventional biosensors are used to detect biomolecules, they have poor sensitivity and are also time- and money-intensive. As a result, the concept of a dielectrically modulated field-effect transistor (DMFET) is proposed to address the primary problems regarding detection processes [19–21]. There are basically two types of biosensors: label-based and label-free biosensors. Magnetic, electrochemical, fluorescent, and other label detection methods are used, and they alter the inherent properties of biomolecules. Such label-based detection techniques produce inaccurate results and take a long time to produce results. Because of this, more accurate label-free electrical detecting techniques have become a fascinating study area. Biosensors composed of FETs can identify biomolecules without the use of labels or enzymes. FET-based biosensors profit from their compact design, low price, and capacity for mass production. The ion-sensitive field-effect transistor (ISFET), first presented by Bergveld in 1970 [21], is a FET-based biosensor that can detect biomolecules (charged such as DNA) but is incompatible with neutral biomolecules. ISFET works on the pH-based properties of the electrolyte/analyte solutions, and its sensitivity depends upon the pH of the same. The maximum sensitivity obtainable by the Nernst equation is 59 mV per unit change in the pH of the analyte. Also, the size of an ISFET is large enough for any in vivo application. Thus, there is a need to find a better, more sustainable, and more efficient biosensor device.

DMFETs based on FinFETs [22], nanotube FETs [23–24], negative-capacitance FETs (NCFETs) [25], TFETs [26], and impact ionization FETs [27] can sense bio-molecules. The biomolecules are addressed as either charged biomolecules (e.g., DNA) or neutral biomolecules (e.g., keratin). DMFETs observe the electrical charac-teristics changing as these biomolecule properties change.

To meet the demands of today's technology and fabrication criteria, device dimensions must be downscaled. A smaller transistor needs a small amount of ana-lyte to detect (the size of the analyte must be considered as well). Miniaturization of MOSFET dimensions beyond a certain limit results in drawbacks such as increased power consumption, short-channel effects (SCEs), more leakage, and a degraded I_{on}/I_{off} ratio. TFETs have been implemented efficiently for sensing biomolecules and solving the mentioned problems. TFETs have a steeper SS (<60 mV/decade), a low value of leakage current and low power requirement, a quick response, and, most importantly, a higher sensitivity toward biomolecule detection. Even though TFETs are the most efficient MOSFET substitute, they have unavoidable drawbacks such as ambipolar conduction and poor ON-current. But these issues can be addressed by utilizing different device geometries and different materials.

Today, many reported TFET-based biosensors have outperformed the conven-tional FET biosensor, considering their sensing metrics, low power consumption, and lesser reliance on the fill-in factor of biomolecules inside the nanocavity. Although the TFET biosensor has shown stable sensitivity values, it can be challenging to detect biomolecules when the fill-in factor, which is the proportion of a filled cavity to the overall cavity area, is low. TFET biosensors often show poor performance when biomolecules are placed distantly from the tunneling interface because the binding probability of the biomolecules is low within the cavity.

The ON-state current in TFETs can be amended by utilizing various mechanisms, such as the incorporation of lower-bandgap (low E_g) materials for line and vertical tunneling (e.g., $Si_{(1-x)}Ge_{(x)}$), metal work-function variations of the gate electrodes, a double-gate or FinFET gate structure, multi-dielectric materials (stacking up or horizontal placement of different dielectrics), and so on. The conduction of negative or ambipolar current can be mitigated by using a higher E_g material in the drain region, lower drain doping than the source, metal work-function adjustments of the drain electrode, and other techniques. The concentration of carriers such as electrons in the region below the gate can be sufficiently increased by increasing the carrier concentration in the source in comparison to that of the drain (asymmetrical doping). This increases the electron tunneling rate and ON-current. Along with this, a cross-conduction from the drain to the channel is also reduced.

Because of random dopant fluctuations (RDFs), achieving an abrupt doping profile of the carriers at the junctions of the source–channel or drain–channel is extremely difficult. Consequently, physical doping has fabrication costs and difficulties. The charge plasma dopingless strategy can overcome the shortcomings of conventional TFETs. The charge plasma technique, which uses appropriate metals to build source and drain regions, can make fabrication relatively easy.

A dopingless TFET based on charge plasma induction is explained in the remain-der of this chapter. The device resembles the studies shown by P. Goma et al. [26]. A reduction in RDFs can be obtained by using the charge plasma method; a surge in the I_{on} is expected to occur because of the pocket under the cavity near the junction of the source and channel, and also because of the tunnel gate engineering.

5.5.1 Device Architecture and Key Parameters

The device architecture shown in Figure 5.7 incorporates a Si-Ge pocket present at the source–channel junction. The device is designed considering all the important factors that may increase the ON-current. For example, the gate region of the device is modified by altering the ratio of the tunnel gate length and thickness with respect to the auxiliary gate length and thickness. The region below the nanocavity is modified using a high-k dielectric material that may increase the ON-current and modulate the threshold voltage. A 45 nm technology node is utilized in this device. The thickness of the nanocavity is adjusted in such a way that the biomolecules of 7 nm × 5 nm size can be efficiently incorporated within it. The nanocavity is designed near the source–channel junction because the quantum mechanical effects affect the device's performance.

The key parameters to design this biosensor are given in Table 5.1. The separation between the channel and drain metals is set to 22 nm in order to obtain a very less ambipolar current because this length, also known as "underlap length," affects the ambipolarity and leakage in the biosensor.

5.5.2 Functionality of the Presented Biosensor

The source–channel junction, called the "tunneling junction," is not far from the nanocavity. When a biomolecule is placed in a nanocavity, the current alters as the dielectric modulation affects the transport of electrons over the tunneling junction. For an empty nanocavity, the controlling voltages, namely V_{gs}, V_{ds}, and pocket, are

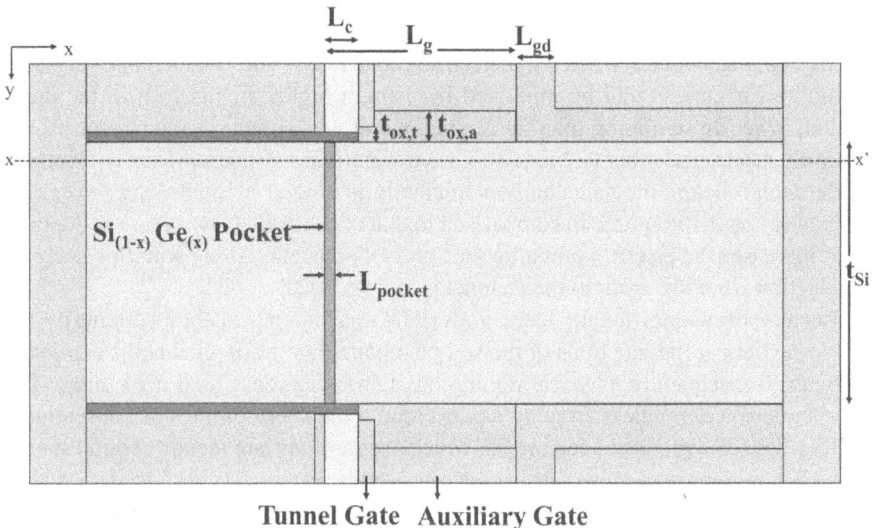

FIGURE 5.7   Schematic of the dopingless TFET biosensor.

TABLE 5.1

Some Key Parameters for Dopingless Biosensor Design

Physical Design Parameter	Value	Symbol
Thickness of body (nm)	10	t_{Si}
Thickness of cavity (nm)	5	t_{cav}
Cavity length (nm)	7	L_{cav}
Gate length (nm)	45	L_g
$Si_{(1-x)}$–$Ge_{(x)}$ pocket length (nm)	2	L_{pocket}
High-κ oxide thickness (nm)	0.8	t_1
Work function of drain metal gate (eV)	3.9	Φ_{DM}
Tunnel-gate oxide thickness (nm)	1.8	t_2
Tunneling Gate (TG) work function (eV)	3.9	Φ_{TG}
Auxiliary Gate (AG) oxide thickness (nm)	3	t_3
Length of source/gate underlap (nm)	3	L_{sg}
Length of gate/drain underlap (nm)	22	L_{gd}
AG work function (eV)	4.1	Φ_{AG}
Work function of source metal gate (eV)	5.93	Φ_{SM}

responsible to determine the width of the tunneling as well as the BTBT rate. This initiates the change in the capacitive relation of the gate and the region below it. As we increase the K value of the biomolecule, there is a resultant increase in the I_{on} of the biosensor. A change in the drain current with biomolecules is shown in Figure 5.8, which follows Landauer's formula [28] given by Equation (5.1):

$$I_{ds} = \frac{2q}{\hbar} \int_{E_{C,Ch}}^{E_{V,S}} \left(f_{V,S}(E_i) - f_{C,Ch}(E_i) \right) T_{BTBT} dE_i \qquad (5.1)$$

The Fermi energy levels in the source valence band, $f_{V,S}(E_i)$, and the channel conduction band, $f_{C,Ch}(E_i)$ are used to calculate the occupancy of tunneling electrons at energy E_i. The energies corresponding to the valence band of source and conduction band of channel regions are given by $E_{V,S}$ and $E_{C,Ch}$; q is the charge, $\hbar$ is reduced plank's constant, and T_{BTBT} is the tunneling probability.

Drain current is maximum for the biomolecules whose K value is 12 and is least for the empty nanocavity. The maximum attainable value of ON-state current is almost 1.5×10^{-5} A/μm for the keratin biomolecule. A further potential step to try to stop carrier flow from drain to gate is provided by adjusting Φ_{AG} to 4.1 eV. Another parameter that needs to be observed is the drain-induced barrier thinning (DIBT) in the TFET biosensor. DIBT occurs when the drain voltage varies in the device, and this drain voltage also impacts the barrier width. It is shown in Figure 5.9 that the barrier width is lowered at higher drain voltages; therefore, the tunneling rate is also impacted, as it depends majorly on the energy band barrier width.

The Henderson–Hasselbalch equation, depending on the pH and molar concentration of the biomolecules, calculates the charge on them. It can be observed that

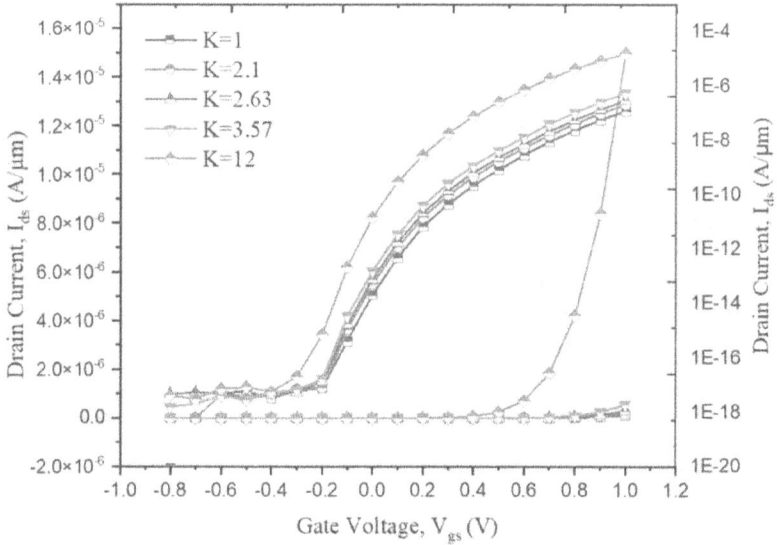

FIGURE 5.8   Transfer characteristics of the dopingless TFET biosensor.

the behavior of charged biomolecules at the nanocavity–oxide interface varies. The charge on a biomolecule is positive if the isometric point exceeds pH, and vice versa for a negative biomolecule, where the isometric point should be lower than pH. The electric field changes depending on the physical parameters, such as the dielectric constant (K) and charge density (ρ) of various biomolecules as they inhabit the nanoscaled cavity region.

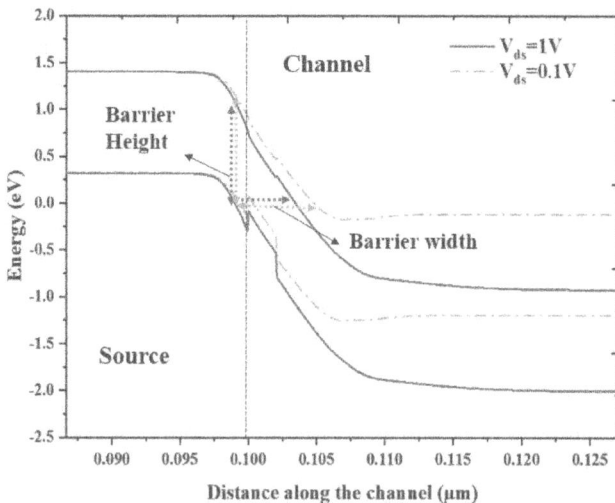

FIGURE 5.9   Impact of drain voltage on the energy bands of the presented biosensor.

The electric field is critical to understanding our device's functionality. The tunneling width influences how quickly electrons cross from the source energy band to the channel region. The tunneling mechanism operates at the source–channel (tunneling) interface, and this only happens when the electrons have the minimum required energy to pass the barrier. Extra carriers present in the pocket region aid in increasing the generation rate as the pocket is doped. Tunneling begins as soon as the required amount of drift is obtained. The electrons gain a high electric field value, which depends on the applied voltage, in addition to the source and pocket doping.

In the absence of biomolecules, the majority of electrons have a low concentration under the cavity region. When a biomolecule with a dielectric constant greater than one (K > 1) is immobilized in the cavity region, it changes the net concentration of electrons in this region, raising the potential with the rise in K values. Initially, the potential for the source region is constant, and it begins to vary as soon as the biomolecules are introduced.

The band alignment leads to a thin barrier (gap) that seems to be thin enough for the drifted electrons to cross. Band bending is only observable in the source–channel area, hence the pocket should be placed near this junction. Increased coupling between metal and semiconductor is caused by high-k oxides, which raise the V_{th} requirement. Even when the K value is changed little, the threshold voltage changes noticeably because of the high dielectric layer of oxide covering the source and nanocavity area. It displays a large shift in threshold voltage since these changes are taking place right above the tunneling junction.

V_{th} and I_{on} sensitivity are given in the equations for neutral and charged biomolecules [29–31]. The expressions for these quantities are given by Equations (5.2) and (5.3):

$$\Delta V_{th} = V_{th}\mid_{K=1} - V_{th}\mid_{K>1(charged)} \tag{5.2}$$

$$S_{I_{on}} = \frac{I_{on}\mid_{K>1(charged)} - I_{on}\mid_{K=1}}{I_{on}\mid_{K=1}} \tag{5.3}$$

Figure 5.10 shows how the V_{th} varies with different K values. A huge shift in the V_{th} of the device for an empty and a keratin-filled nanocavity is evitable from the results. The sensitivity associated with the V_{th} value is 550.4 mV, which is very good and a lot more than the ideal Nernst limit for label-based biosensor devices. In addition to this, the sensitivity associated with the drain current is 1.438×10^2, which is better than many related works and has a scope for improvement as well.

5.6 FUTURE OF TFETs IN BIOSENSING

TFETs have shown tremendous performance in biomolecule sensing despite their low ON-current and ambipolarity. Conventional TFET biosensors had many drawbacks; therefore, variations in their device architecture have proven to be quite efficient. Point- and line-tunneling TFETs [32–34], pocketed TFETs, gate-all-around TFETs [35], negative-capacitance TFETs, vertical TFETs [2], and so on have been researched and hold great promise for future biosensors.

Optimization of TFET biosensors must also be addressed in order to present their efficient practical usage. Another issue associated with TFET biosensors is their

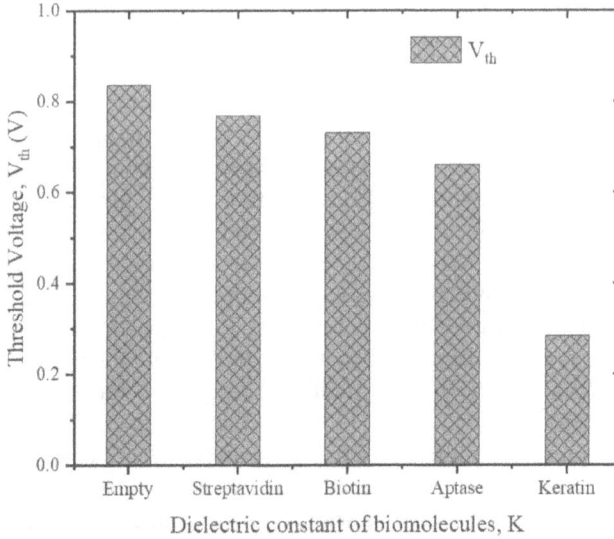

FIGURE 5.10   Threshold voltage variation with different biomolecules.

downscaling, as discussed in this chapter. Addressing these issues can resolve the present challenges. 2D material–based devices and biosensors have very few studies and investigations to date [36]. These devices require quantum software to analyze their function properly, and designing a biosensor using 2D materials and TFETs requires huge computational power. Such calculations involve the use of non-equilibrium green functions and can outperform the existing research on TFET biosensors.

REFERENCES

1. D. Esseni, M. Pala, P. Palestri, C. Alper and T. Rollo, "A Review of Selected Topics in Physics Based Modeling for Tunnel Field-Effect Transistors," in Semiconductor Sci Technology, vol. 32, article 083005, 2017.
2. M. Verma, S. Tirkey, S. Yadav, D. Sharma and D. S. Yadav, "Performance Assessment of A Novel Vertical Dielectrically Modulated TFET-Based Biosensor," in IEEE Transactions on Electron Devices, vol. 64, no. 9, pp. 3841–3848, 2017, doi: 10.1109/TED.2017.2732820.
3. S. Yadav, D. Sharma and D. Soni et al., "Controlling Ambipolarity With Improved RF Performance by drain/gate Work Function Engineering and Using High κ Dielectric Material in Electrically Doped TFET: Proposal and Optimization," in The Journal of Computational Electronics, vol. 16, pp. 721–731, 2017, doi: 10.1007/s10825-017-1019-2.
4. M. K. Anvarifard, Z. Ramezani and I. S. Amiri, "High Ability of a Reliable Novel TFET-Based Device in Detection of Biomolecule Specifies—A Comprehensive Analysis on Sensing Performance," in IEEE Sensors Journal, vol. 21, no. 5, pp. 6880–6887, 2021, doi: 10.1109/JSEN.2020.3044056.
5. W. Li and J. C. S. Woo, "Optimization and Scaling of Ge-Pocket TFET," in IEEE Transactions on Electron Devices, vol. 65, no. 12, pp. 5289–5294, 2018, doi: 10.1109/TED.2018.2874047.
6. A. Bhattacharyya, M. Chanda and D. De, "Analysis of Noise-Immune Dopingless Heterojunction Bio-TFET Considering Partial Hybridization Issue," in IEEE

Transactions on Nanotechnology, vol. 19, pp. 769–777, 2020, doi: 10.1109/ TNANO.2020.3033966.

7. M. K. Anvarifard and A. A. Orouji, "Energy Band Adjustment in a Reliable Novel Charge Plasma SiGe Source TFET to Intensify the BTBT Rate," in IEEE Transactions on Electron Devices, vol. 68, no. 10, pp. 5284–5290, 2021, doi: 10.1109/ TED.2021.3106891.

8. K. Vanlalawmpuia and B. Bhowmick, "Analysis of Hetero-Stacked Source TFET and Heterostructure Vertical TFET as Dielectrically Modulated Label-Free Biosensors," in IEEE Sensors Journal, vol. 22, no. 1, pp. 939–947, 1 Jan. 2022, doi: 10.1109/ JSEN.2021.3128473.

9. D. Mohata et al., "Barrier-Engineered Arsenide–Antimonide Heterojunction Tunnel FETs With Enhanced Drive Current," in IEEE Electron Device Letters, vol. 33, no. 11, pp. 1568–1570, Nov. 2012, doi: 10.1109/LED.2012.2213333.

10. T. Krishnamohan, D. Kim, S. Raghunathan and K. Saraswat, "Double-gate strained-Ge heterostructure tunneling FET (TFET) with record high drive currents and ≪60mV/ dec subthreshold slope", IEDM Tech. Dig., pp. 1–3, Dec. 2008.

11. G. Dewey et al., "Fabrication, characterization, and physics of III–V heterojunction tunneling Field Effect Transistors (H-TFET) for steep sub-threshold swing," 2011 International Electron Devices Meeting, Washington, DC, USA, 2011, pp. 33.6.1– 33.6.4, doi: 10.1109/IEDM.2011.6131666.

12. A. M. Walke et al., "Fabrication and Analysis of a $Si/Si_{0.55}Ge_{0.45}$ Heterojunction Line Tunnel FET," in IEEE Transactions on Electron Devices, vol. 61, no. 3, pp. 707–715, March 2014, doi: 10.1109/TED.2014.2299337.

13. W. Cheng et al., "Fabrication and Characterization of a Novel Si Line Tunneling TFET With High Drive Current," in IEEE Journal of the Electron Devices Society, vol. 8, pp. 336–340, 2020, doi: 10.1109/JEDS.2020.2981974.

14. C. Chen, Q. Huang, J. Zhu, Y. Zhao, L. Guo and R. Huang, "New Understanding of Random Telegraph Noise Amplitude in Tunnel FETs," in IEEE Transaction on Electron Devices, vol. 64, no. 8, pp. 3324–3330, 2017.

15. J. Wan, C. Le Royer, A. Zaslavsky and S. Cristoloveanu, "Low-Frequency Noise Behavior of Tunneling Field Effect Transistors," in Applied Physics Letters, vol. 97, no. 24, 2010

16. Q. Huang, R. Huang, C. Chen, C. Wu, J. Wang, C. Wang and Y. Wang, Deep insights into low frequency noise behavior of tunnel FETs with source junction engineering, in: Symposium on VLSI Technology (VLSI-Technology): Digest of Technical Papers, IEEE, 2014, pp. 1–2.

17. G. V. Luong, S. Strangio, A. T. Tiedemann, P. Bernardy, S. Trellenkamp, P. Palestri, S. Mantl and Q. T. Zhao, Experimental Characterization of the Static Noise Margins of Strained Silicon Complementary Tunnel-FET SRAM, European Solid-State Device Research Conference IEEE, 2017, pp. 42–45.

18. R. Pandey, B. Rajamohanan, H. Liu, V. Narayanan and S. Datta, "Electrical Noise in Heterojunction Interband Tunnel FETs," in IEEE Transaction on Electron Devices, vol. 61, no. 2, pp. 552–560, Feb. 2014, doi: 10.1109/TED.2013.2293497.

19. H. Im, X. J. Huang, B. Gu and Y. K. Choi, "A Dielectric-Modulated Field Effect Transistor for Biosensing," in Nature Nanotechnology, vol. 2, no. 7, pp. 430–434, 2007.

20. S. Kanungo, S. Chattopadhyay, P. S. Gupta, K. Sinha and H. Rahaman, "Study and Analysis of the Effects of SiGe Source and Pocket-Doped Channel on Sensing Performance of Dielectrically Modulated Tunnel FET-Based Biosensors," in IEEE Transactions on Electron Devices, vol. 63, no. 6, pp. 2589–2596, 2016, doi: 10.1109/ JSEN.2021.3122582.

21. P. Bergveld, "Thirty Years of ISFETOLOGY: What Happened in the Past 30 Years and What May Happen in the Next 30 Years," in Sensors & Actuators, B: Chemical, vol. 88, no. 1, pp. 1–20, 2003.

22. H. D. Sehgal, Y. Pratap, M. Gupta and S. Kabra, "Performance Analysis and Optimization of Under-Gate Dielectric Modulated Junctionless FinFET Biosensor," in IEEE Sensors Journal, vol. 21, no. 17, pp. 18897–18904, 2021, doi: 10.1109/JSEN.2021.3090263.

23. D. Sen, S. D. Patel and S. Sahay, "Dielectric Modulated Nanotube Tunnel Field-Effect Transistor as a Label Free Biosensor: Proposal and Investigation," in IEEE Transactions on NanoBioscience, doi: 10.1109/TNB.2022.3172553.

24. A. Gedam, B. Acharya and G. P. Mishra, "Design and Performance Assessment of Dielectrically Modulated Nanotube TFET Biosensor," in IEEE Sensors Journal, vol. 21, no. 15, pp. 16761–16769, 2021, doi: 10.1109/JSEN.2021.3080922.

25. F. I. Sakib, M. A. Hasan and M. Hossain, "Negative Capacitance Gate-All-Around Tunnel FETs for Highly Sensitive Label-Free Biosensors," in IEEE Transactions on Electron Devices, vol. 69, no. 1, pp. 311–317, 2022, doi: 10.1109/TED.2021.3129711.

26. P. Goma and A. K. Rana, "Analysis and Evaluation of $Si_{(1-x)}Ge_{(x)}$ Pocket on Sensitivity of a Dual-Material Double-Gate Dopingless Tunnel FET Label-Free Biosensor," Micro and Nanostructures, vol. 170, Oct. 2022, doi: 10.1016/j.micrna.2022.207393.

27. N. Kannan and M. J. Kumar, "Dielectric-Modulated Impact-Ionization MOS Transistor as a Label-Free Biosensor," in IEEE Electron Device Letters, vol. 34, no. 12, pp. 1575–1577, 2013, doi: 10.1109/LED.2013.2283858.

28. S. M. Sze, Y. Li and K. K. Ng, "Physics of semiconductor devices," John Wiley & Sons, 2021.

29. P. Venkatesh, K. Nigam, S. Pandey, D. Sharma and P. N. Kondekar, "A Dielectrically Modulated Electrically Doped Tunnel FET for Application of Label Free Biosensor," in Superlattices and Microstructures, vol. 109, pp. 470–479, 2017, doi: 10.1016/j.spmi.2017.05.035

30. S. Anand, A. Singh, S. I. Amin and A. S. Thool, "Design and Performance Analysis of Dielectrically Modulated Doping-Less Tunnel FET-Based Label Free Biosensor," in IEEE Sensors Journal, 19(12), 4369–4374, 2019, doi: 10.1109/JSEN.2019.2900092.

31. S. Kumar, Y. Singh, B. Singh and P. K. Tiwari, "Simulation Study of Dielectric Modulated Dual Channel Trench Gate TFET-Based Biosensor," in IEEE Sensors Journal, vol. 20, no. 21, pp. 12565–12573, 2020, doi: 10.1109/JSEN.2020.3001300.

32. P. Dwivedi and A. Kranti, "Overcoming Biomolecule Location-Dependent Sensitivity Degradation Through Point and Line Tunneling in Dielectric Modulated Biosensors," in IEEE Sensors Journal, vol. 18, no. 23, pp. 9604–9611, 1 Dec. 2018, doi: 10.1109/JSEN.2018.2872016.

33. S. Agarwal, G. Klimeck and M. Luisier, "Leakage-Reduction Design Concepts for Low-Power Vertical Tunneling Field-Effect Transistors," in IEEE Electron Device Letters, vol. 31, no. 6, pp. 621–623, 2010, doi: 10.1109/LED.2010.2046011.

34. K. Ganapathi, Y. Yoon and S. Salahuddin, "Analysis of InAs Vertical and Lateral Band-to-Band Tunneling Transistors: Leveraging Vertical Tunneling for Improved Performance," in Applied Physics Letters, vol. 97, no. 3, p. 033504, 2010, doi: 10.1063/1.3466908.

35. Q. Zhang, Y. Lu, C. A. Richter, D. Jena and A. Seabaugh, "Optimum Bandgap and Supply Voltage in Tunnel FETs," in IEEE Transactions on Electron Devices, vol. 61, no. 8, pp. 2719–2724, 2014, doi: 10.1109/TED.2014.2330805.

36. P. Kaushal and G. Khanna, "MoS_2 Based Thickness Engineered Tunnel Field-Effect Transistors for RF/analog Applications," in Materials Science in Semiconductor Processing, vol. 151, no. 107016, 15 Nov. 2022, doi: 10.1016/j.mssp.2022.107016.

6 Optimization of Hetero Buried Oxide Doped-Pocket Gate-Engineered Tunnel FET Structures

Sirisha Meriga and Brinda Bhowmick

6.1 INTRODUCTION

Future nanoelectronic devices have to operate at a sub-1.5 V supply voltage with low power dissipation. The downsizing of metal–oxide–semiconductor field-effect transistors (MOSFETs) is prevented by adverse short-channel effects (SCEs), in particular punch-through, drain-induced barrier narrowing, hot carrier, and also velocity saturation, which gradually reduce the performance of the devices [1–3]. Tunnel field-effect transistors (TFETs) are the successors of MOSFETs in future nanoelectronic devices that operate on the band-to-band tunneling (BTBT) methodology and have a subthreshold slope (SS) of less than 60 mV/dec subthreshold slope (SS) [4–6]. BTBT ceases the SCEs in the TFETs and also improves the SS of the device, compared to thermionic emission–operated MOSFETs. Hence, substantial research has been pursued in the field of TFET design [7–12]. The ambipolar conduction and low ON-state current are important issues to consider while designing the device. For silicon TFETs, ON-state current that is possibly boosted by placing an n+ pocket at the source terminal will uplift the internal electric field. Silicon TFETs can operate with low power supplies, exhibit steeper SS, and have enhanced reliability compared to conventional p-i-n TFETs, and they also possess great boosted ON-state current [13, 14]. For direct current (DC) circuit applications, the typical behavior of ambipolar conduction in TFETs limits their viability [15, 16]. Several types of device architecture—like large bandgap material at the output side, spacers (low-k), low drain doping, gate–drain underlap, and, later on, lateral heterostructures—have been reported [17, 18]. Despite that, low ambipolar current, a diminished ON-state current, and fabrication complexity instigate greater drain series resistance and lead to higher manufacturing costs [17–20]. Sahay et al. reported that, at the drain–channel interface, the width of the tunnelling barrier can be controlled by introducing a hetero buried oxide (HBOX) beneath the silicon substrate to suppress the ambipolarity [21].

In this chapter, a HBOX layer beneath the silicon surface, at the source terminal, a heavily doped n+ pocket is introduced. Along with these, gate-to-drain underlap and gate–source overlap have evolved to circumvent the drawbacks of conventional TFETs. The outline of this chapter is as follows: Section 6.2 narrates the stated HBOX-TFET structure and the simulation approach. DC analysis of the mentioned

DOI: 10.1201/9781003393542-6

TFET is illustrated in Section 6.3. Next, Section 6.4 elaborates the analog/radiofrequency (RF) performance interpretation of the stated TFET. Finally, the conclusion follows.

6.2   DEVICE STRUCTURE AND SIMULATION APPROACH

A HBOX doped-pocket gate-engineered TFET structure is shown in Figure 6.1. A low-k oxide layer that is 30 nm thick and 50 nm long is placed at the drain side of the TFET, and a high-k oxide with the same thickness is deposited at the source side, to inhibit ambipolar current and magnify the ON-state current, respectively [22]. A 5-nm-thick heavily doped n+ pocket doping concentration is presumed to be 5×10^{19}/cm^3 at the source terminal; it modulates the bandgap in the midst of the source and channel. It also produces an enhanced electric field. The source and drain concentrations are considered as 1×10^{20}/cm^3 and 5×10^{18}/cm^3, respectively, and their thickness is 20 nm. The channel thickness is 20 nm, and it is doped with a boron concentration of 1×10^{16}/cm^3.

The source, channel, and drain terminal lengths are taken as 40 nm, 30 nm, and 30 nm, respectively. In this structure, a HBOX comprising a low-k oxide (SiO$_2$) at the drain side and a high-k oxide (HfO$_2$) at the source side is contemplated. A 15 nm gate-to-drain underlap length is considered, and a gate-to-source overlap length of 5 nm to decrease the ambipolar current is included. A hafnium oxide layer of 2 nm thick at the gate stack and a 1-nm-thick TiN gate with 4.33 eV work function are preferred. The total length of this HBOX-TFET is chosen as 100 nm.

The simulation of the stated HBOX-TFET is performed by employing two-dimensional (2D) Synopsis TCAD software [23]. Shockley–Reade–Hall recombination is constituted for minority recombination effects. The full Schenk bandgap-narrowing model is encompassed, as high carrier concentrations injects

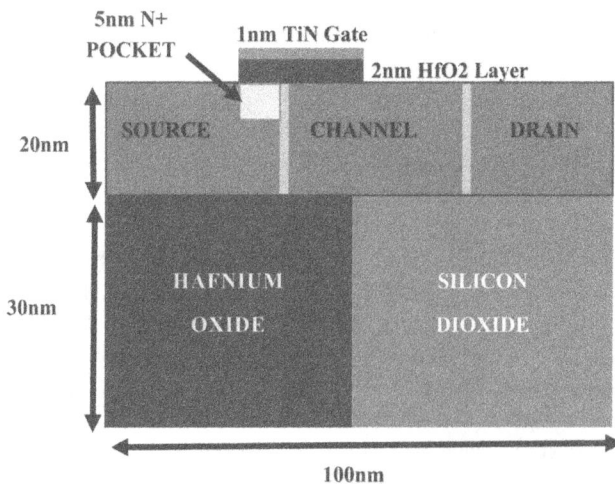

FIGURE 6.1   Hetero buried oxide–tunnel field-effect transistor (HBOX-TFET) structure.

uplifted electric field ensues bandgap narrowing. To mimic the tunneling effects, the Fowler–Nordheim method is initiated as the tunneling of electrons from the valence band ascends the input terminal. This current is crucial, while it finely illuminates the potential of the flying substrate for a SOI floating body at the same time. A heavily doped source, drain, and pocket ($>10^{19}$/cm^3) require a doping-dependent mobility model.

6.3 DC PARAMETER ANALYSIS

The performance of the device under DC conditions is evaluated. The transfer characteristics of this HBOX-TFET are delineated by (1) altering the thickness of the silicon substrate and HBOX but keeping the total device thickness constant, (2) optimizing HfO$_2$ length in the HBOX, and (3) altering the silicon thickness, which are explored in this section. From the characteristics, it is clear that the device's ambipolar current is minor (i.e., 7×10^{-13}A/μm); an enhanced ON-state current of 1.123×10^{-3}A/μm with a SS of 41.4 mV/dec is noticeable. This HBOX-TFET exhibited an ON-OFF current ratio of 0.16×10^{10} with a threshold voltage of 0.67 V by altering the silicon thickness. Hence, the proposed device operation is absolutely good for DC applications.

In silicon substrate TFETs, the introduction of pocket-doping engineering at the source terminal is reported [13, 24] to create the local minima at the conduction band and generate a large breakthrough to decrease the width of the tunneling region. To obtain gate control, the pocket thickness should not be too long because it doesn't get fully depleted. Hence, the pocket thickness is considered as only 5 nm. An intensified ON-state current results because of the high electric field in this region, and it is depicted in Figure 6.2(a) and (b). An increase in electric field is examined; this is caused by the origination of an n+ pocket and can be seen in Figure 6.2(a). The silicon thickness variation also changed the electric field in this stated hetero TFET, as revealed in Figure 6.2(b).

Another exceptional factor that will change the output current is the surface potential, and this is portrayed in Figure 6.3. The BOX thickness effect on the surface potential is less, and the n+ pocket (source terminal) uplifts the potential at the surface, which is shown in Figure 6.3. The surface potential φs relies on the height of the barrier and its dependency on the gate oxide and depletion capacitances C_{ox} and C_{dep}, respectively. This is given by Equation (6.1):

$$\varphi s = \frac{C_{ox}}{C_{ox} + C_{dep}} V_{gs} \tag{6.1}$$

An increase in the input voltage will affect the potential at the surface of the TFET. The SS of the stated device depends on this potential and is given by Equation (6.2):

$$SS = \frac{\partial V_{gs}}{\partial \log_{10} I_d} = \frac{\partial V_{gs}}{\partial \varphi_s} \times \frac{\partial \varphi_s}{\partial \log_{10} I_d} = \frac{\partial V_{gs}}{\partial \varphi_s} \times (2.3) V_t \tag{6.2}$$

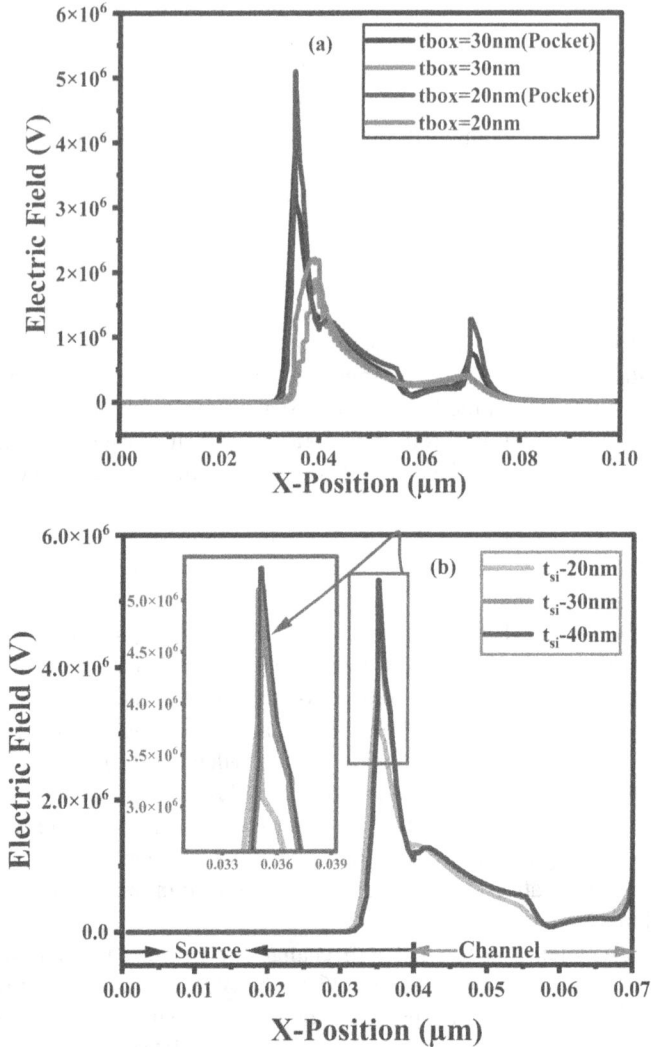

FIGURE 6.2 Electric field (a) along the full length of the geometry, and (b) at the source and channel regions.

Hence, from Equation (6.2), it is obvious that the SS alters with the surface potential of the device. A SS of 41.4 mV/dec has been exposed by the device with this improved surface potential by inserting an n+ pocket.

Energy band diagrams of a HBOX-TFET operated at $V_{gs} = 1$ V and $V_{ds} = 0.7$ V (ON-state) and differentiated in the OFF-state are depicted in Figure 6.4. In a TFET, the current flows primarily due to the source–channel tunnel junction. The introduction of an n+ pocket narrows the bandgap, and this leads to decreased barrier height. If $V_{gs} < 0$, the device operates in the OFF-state, leading to zero current under ideal conditions. With a gradual increase in the input voltage (i.e., $V_{gs} \geq V_{th}$), the device

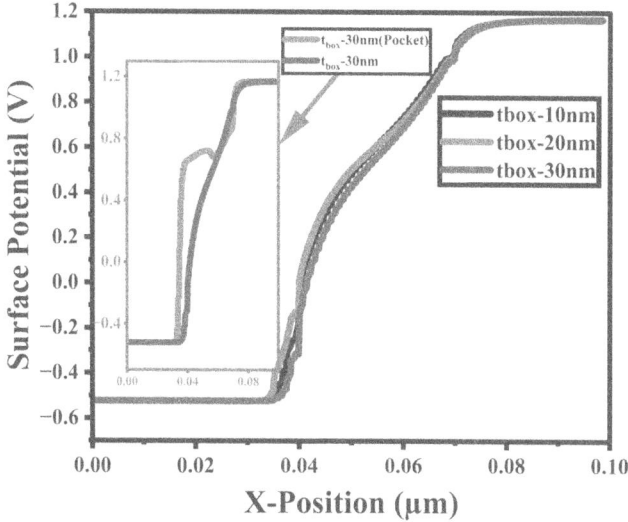

FIGURE 6.3 Surface potential of the HBOX-TFET.

enters the ON-state as the tunneling of electrons evolves from one band to another band of the source and the channel regions. Output current flowing in the device increments, with applied bias voltage in the subthreshold region by virtue of the energy band bending, is shown in Figure 6.4.

The input (transfer) characteristics are depicted in Figure 6.5, and these can be obtained by adjusting the HBOX thickness of the device mentioned. Initially, when the input voltage is negative, it operates in the OFF-state as BTBT is inhibited.

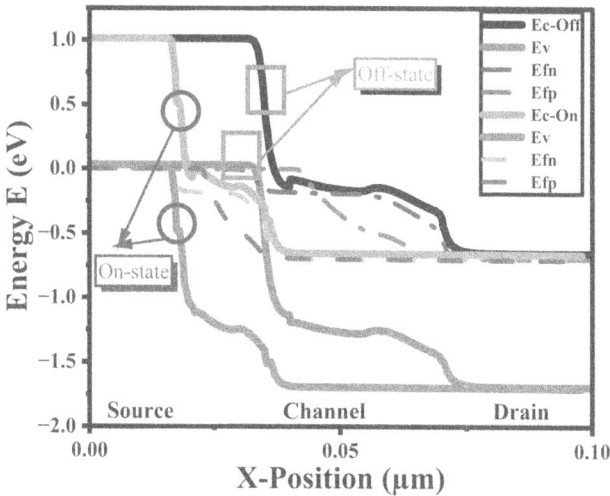

FIGURE 6.4 Energy band diagram showing the conduction and valence bands and their band bending in the ON-state and OFF-state.

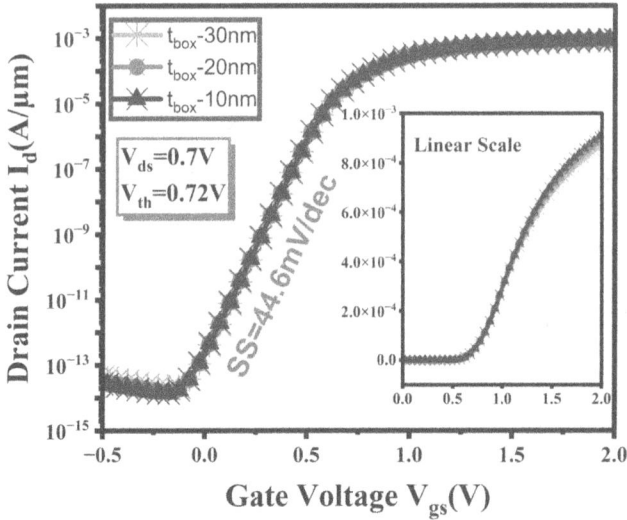

FIGURE 6.5 I_d-V_{gs} transfer characteristics when changing the HBOX thickness.

The source valence band approaches the conduction band of the channel with an intensification of the biased input voltage. Then, the current flows in the device as electrons move from one band to the other. This is known as BTBT. Because of the heavily doped n+ pocket at the source terminal, line tunnelling enhances SS than that of the point tunnelling [25]. It also enhances the electric field, leading to boosted ON-state current. The device characteristics are verified by sweeping the input voltage V_{gs} and is shown in Figure 6.5 by optimizing the HBOX thickness. The device showed a threshold voltage of 0.7 V, OFF-state current of 4.14×10^{-13}A/μm, and ON-state current of 3.1×10^{-4} A/μm, and it exhibited a SS of 44.6 mV/dec.

Figure 6.6 depicts the device transfer characteristics by changing the silicon thickness to obtain better OFF- and ON-state currents. It is noticeable from Figure 6.6 that the ON-state current is increased to 1.12×10^{-3} A/μm, and there is an improved SS of 41.4 mV/dec. This is as a consequence of the change in the viable bandgap narrowing with an applied voltage at the gate and source terminals. The effective tunneling between the bands increases with an increment in the silicon thickness. The device will switch quickly from on to off, and vice versa as SS lessens.

Homojunction TFETs exposed an inferior I_{ON}/I_{OFF} ratio. Due to energy bandgap considerations, simultaneously obtaining a high ON-state current and lower OFF-state current is unviable. IC fabrication engineers use silicon material, as it is abundantly available and able to emanate a silicon dioxide layer smoothly over it; although, as silicon dioxide has a low-k (3.9) value, the ON-state current lessens. As a means to increase the ON-state current, hafnium oxide is placed in the BOX at the source side, as it has a high-k (22) value. The length of this oxide layer is altered to obtain the high ON-state current.

The device characteristics with respect to the alteration in the length of hafnium oxide present in the HBOX are exhibited in Figure 6.7. With an increase in the HfO₂

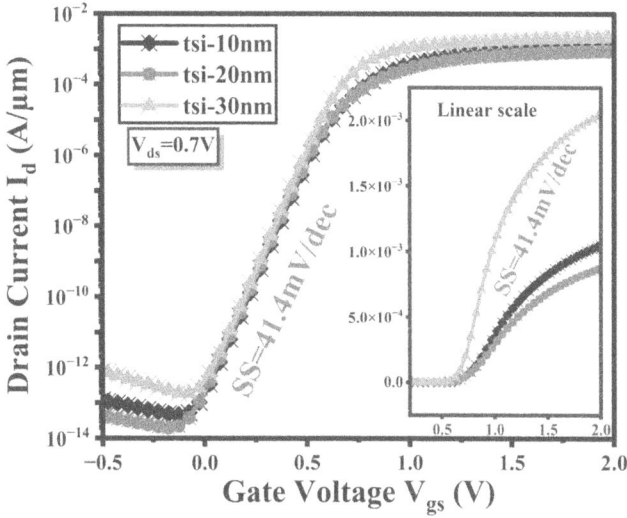

FIGURE 6.6 I_d-V_{gs} with respect to silicon thickness variation.

layer length, a reduction in ON-state current is observed as it penetrates the channel and drain regions. A slight improvement in ON- and OFF-state currents has been obtained from the device, and it is depicted in Figure 6.7. Here, the threshold voltage and SS remain constant at 0.7 V and 44.6 mV/dec, respectively. The ambipolar current is reduced to 10^{-13}A/μm, and the ON-OFF current ratio is 10^9.

The extracted DC parameters of the proposed HBOX-TFET from the Synopsis TCAD software are tabulated in Table 6.1.

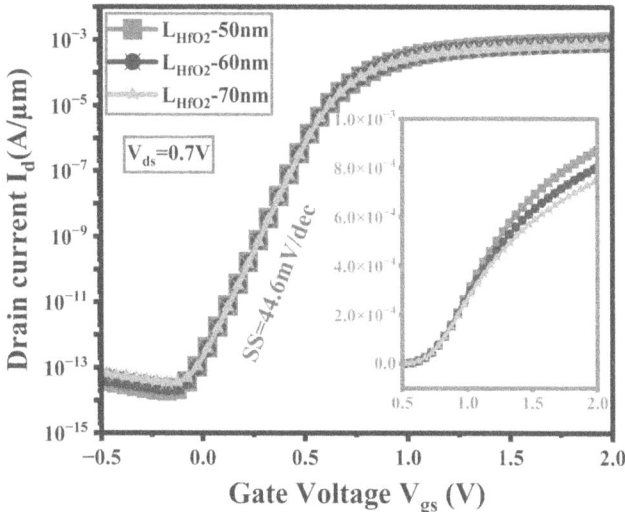

FIGURE 6.7 I_d-V_{gs} characteristics when varying the HfO$_2$ length in a HBOX.

TABLE 6.1

Extracted DC Parameters of the Proposed TFET (Varying Silicon Thickness but Keeping BOX Thickness at 30 nm)

Buried Oxide Thickness (nm)	Silicon Thickness (nm)	V_{th} (V)	I_{off} (A/µm)	I_{on} (A/µm)	I_{on}/I_{off}	SS (mV/dec)
30	10	0.73	3.39×10^{-13}	3.91×10^{-4}	1.15×10^{9}	48.6
30	20	0.72	4.41×10^{-13}	3.15×10^{-4}	0.71×10^{9}	44.6
30	30	0.67	7.01×10^{-13}	1.12×10^{-4}	0.16×10^{9}	41.4

6.4 ANALOG/RF PERFORMANCE ANALYSIS

For logic applications, conventional MOSFETs can be replaced by TFETs because of the latter's unidirectional conduction and lesser SS. Moreover, recently, researchers analyzed the advantages of employing TFETs in various analog circuits [26–31]. In this section, the performance of the HBOX-TFET is analyzed and evaluated for analog/RF applications.

Parasitic capacitances exhibited by the proposed device are vital for ascertaining the frequency of operation of the device and its propagation delay. The gate-to-source capacitance C_{gs} is depicted in Figure 6.8. It depends up on the doping concentration of the source terminal. The coupling among gate and source diminishes with high input voltage V_{gs} as the region of inversion increases. As seen in the figure, the reduction in silicon thickness has reduced the capacitance C_{gs}. Generally, C_{gs} is less than C_{gd}. The gate–drain capacitance C_{gd} is shown in Figure 6.9. It has a similar manner to C_{gs} with the variation in the silicon thickness. These two capacitance values are in

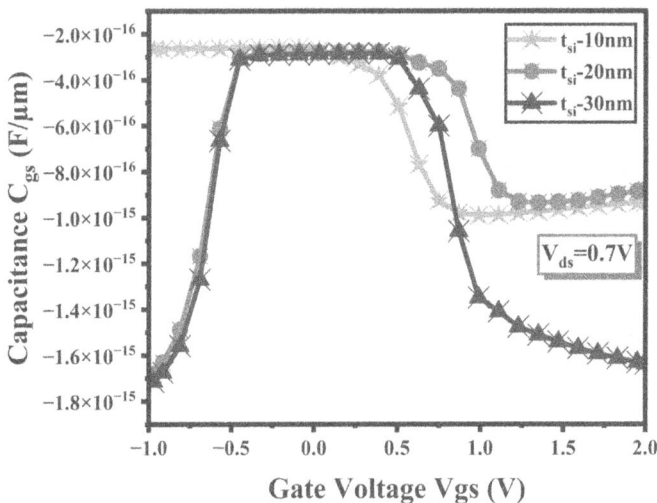

FIGURE 6.8 Gate–source capacitance C_{gs}.

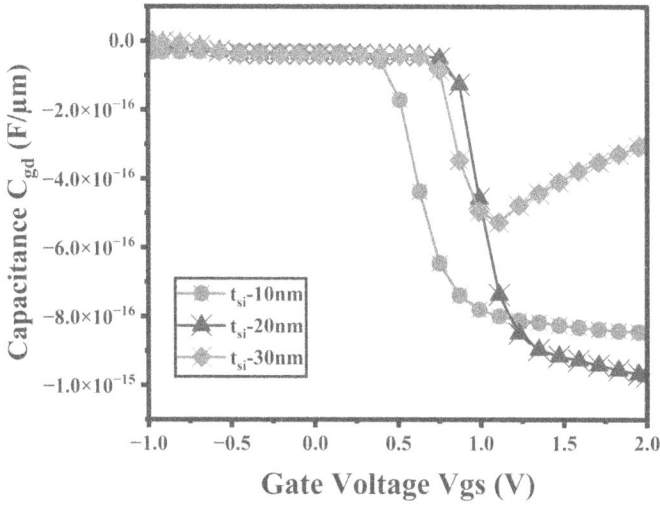

FIGURE 6.9   Gate–drain capacitance C_{gd}.

the range of 10^{-16} F/μmm with a silicon thickness of 30 nm and HBOX thickness of 20 nm. The total capacitance C_{gg} is shown in Figure 6.10. It is on the order of 10^{-15} F/μm.

First and foremost, an important factor to operate the device well in analog/RF applications is the transconductance g_m. It is depicted in Figure 6.11 and is given by:

$$g_m = \frac{\partial I_d}{\partial V_{gs}} : Vds \rightarrow \text{Constant} \tag{6.3}$$

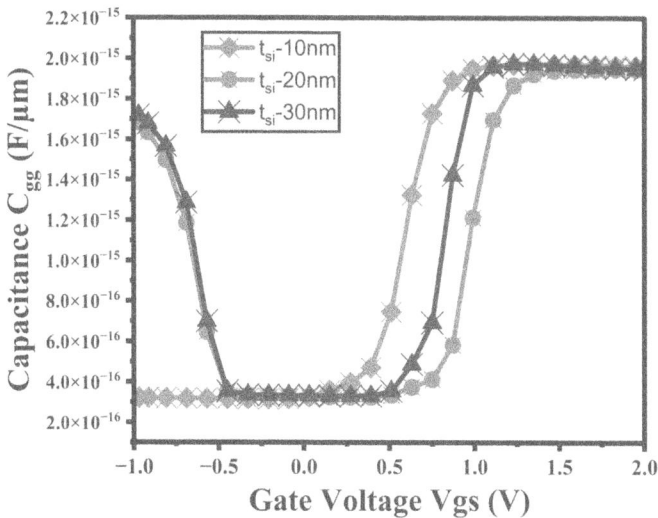

FIGURE 6.10   Total gate capacitance C_{gg}.

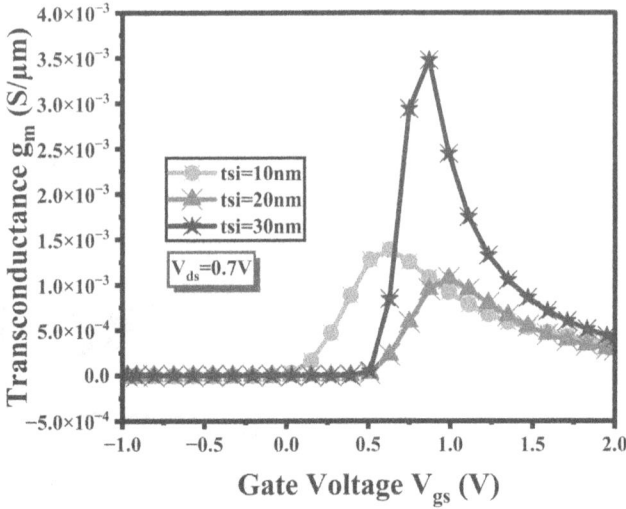

FIGURE 6.11   Transconductance (g_m).

At low-input voltages, g_m increases because of the mobility degradation. An increase in the input voltage leads to a decrease in transconductance value. It can be observed that an increase in silicon thickness gives rise to transconductance.

Another figure of merit is the output conductance g_{ds}. It is given by:

$$g_{ds} = \frac{\partial I_d}{\partial V_{ds}} : Vgs \rightarrow \text{Constant} \tag{6.4}$$

In the linear region, the output conductance is maximum, and in the saturation region it is minimum; this is shown in Figure 6.12. The proposed HBOX-TFET is exceptionally suitable for analog/RF applications as the transconductance, and the output conductance values are good.

The ability of the device to amplify signals over a range of frequencies possibly inferred by the cutoff frequency f_t is given as Equation (6.5):

$$f_t = \frac{g_m}{2\pi C_{gg}} \tag{6.5}$$

The plot of cutoff frequency is shown in Figure 6.13. It is recognized that the device can operate well in the 1–100 MHz frequency range without any attenuation because of the lower capacitance values and higher transconductance value. Similarly, another essential factor for analog/RF applications is the transit time τ; this factor is inversely proportional to f_t, and it is represented as Equation (6.6):

$$\tau = \frac{1}{20 \times \pi \times f_t} \tag{6.6}$$

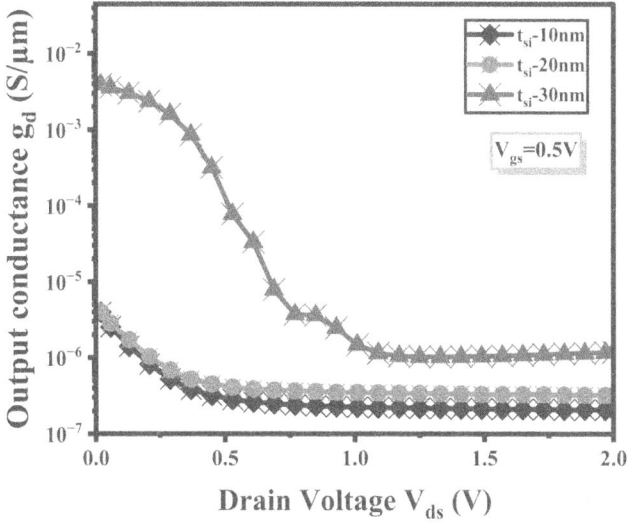

FIGURE 6.12   Output conductance (g_{ds}).

The decay in transit time corresponds to a high speed of operation. Figure 6.14 portrays the transit time. This factor decreases on account of decreased barrier tunneling width with the input voltage. This is on the order of 10^{-14}sec, even by changing the thickness of the silicon wafer.

Intrinsic gain (×1,000) of the stated HBOX-TFET is shown in Figure 6.15. The gain of the device is on the order of 6×10^3, elevating the applicability of the device

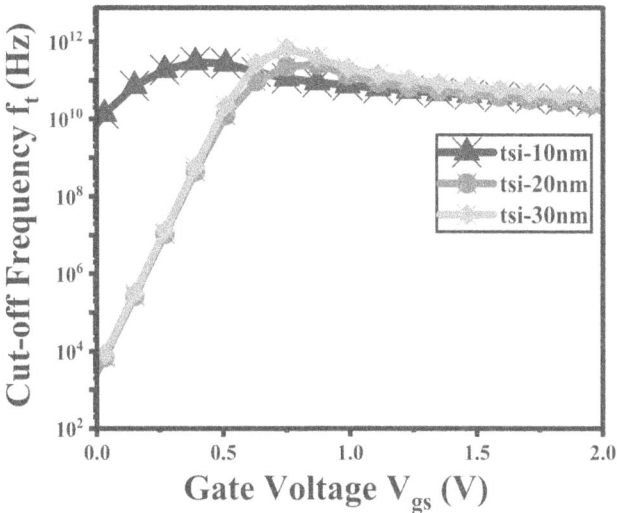

FIGURE 6.13   Cutoff frequency (f_t).

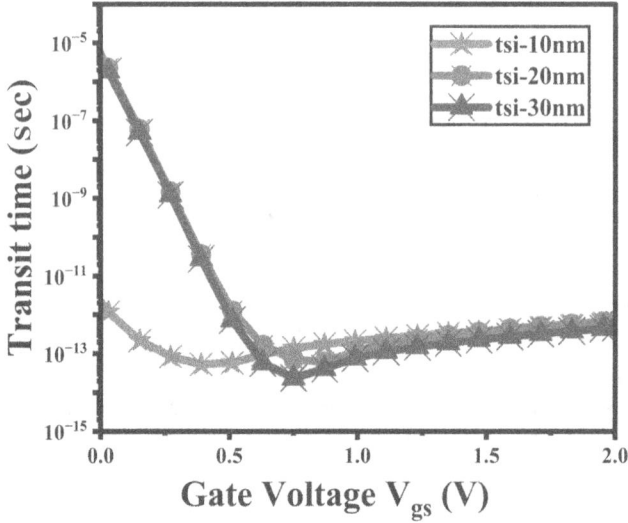

FIGURE 6.14   Transit time (τ).

in analog/RF applications. The amplification provided by the device is very high, which indicates that the device works good as an amplifier even at high frequencies.

At a high-frequency region of operation of the device, the transistor frequency product (TFP) is also an important factor for evaluating the analog/RF performance analysis of TFETs. It is expressed as [24, 32–35]:

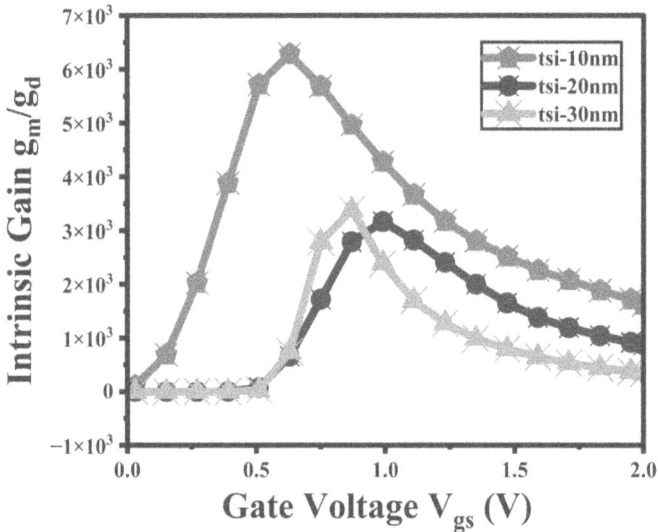

FIGURE 6.15   Intrinsic gain (g_m/g_d).

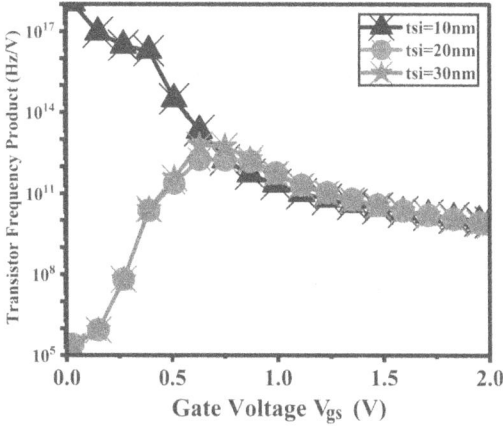

FIGURE 6.16 Transistor frequency product (TFP).

$$TFP = f_t \left(\frac{g_m}{I_d} \right) \tag{6.7}$$

This factor is directly proportional to the transconductance and inversely proportional to the output current of the device. At low input voltages, the TFP increases in the subthreshold region, although it continues to be stable in the above-subthreshold region; its plot is shown in Figure 6.16.

From Figure 6.16, the TFP exhibited by the proposed HBOX-TFET is 10 THz/V. Hence, it is observed that the device exhibited good TFP.

The gain bandwidth product (GBP) is the agreement among the bandwidth and gain. This factor is expressed as Equation (6.8) [27]:

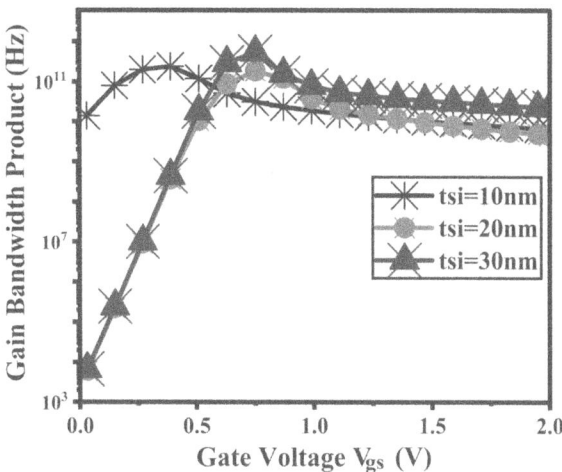

FIGURE 6.17 Gain bandwidth product (GBP).

TABLE 6.2

Extracted Analog/RF Parameters of a Proposed HBOX-TFET from Synopsis TCAD

Silicon Thickness (nm)	C_{gs} (Ff/μm)	C_{gd} (Ff/μm)	C_{gg} (Ff/μm)	g_m (mS/μm)	Cutoff Frequency (GHz)	Transit Time (Sec)	Gain (×10³)	GBP (GHz)	TFP (GHz/V)
10	0.98	0.78	1.9	0.92	75	2.1×10^{-13}	4.3	19	220
20	0.69	0.45	1.2	1	140	1.1×10^{-13}	3.2	37	580
30	0.13	0.49	1.8	2.4	200	7.6×10^{-14}	2.4	78	510

$$GBP = \frac{g_m}{20\pi X C_{gd}} \tag{6.8}$$

This factor changes with respect to the transconductance and drain capacitance of the device. In the subthreshold region, this factor slowly improves, and in the above-subthreshold region it reaches a constant value, as shown in Figure 6.17. A dominant value of this factor exhibited by the device is 10^{11} Hz. The proposed HBOX-TFET parameters extracted from Synopsis TCAD are tabulated in Table 6.2.

Henceforth, from Tables 6.1 and 6.2, it is remarkable that the proposed HBOX-TFET DC and analog/RF performance is prominent in terms of all the characteristics. Therefore, the applicability of the stated TFET can be considered in the design of electronic circuits to operate at low-power voltages.

6.5　CONCLUSION

The doped-pocket gate-engineered HBOX-TFET structure described in this chapter is perfectly suitable for both DC and analog/RF applications. The device exhibited a SS of 41.4 mV/dec, an OFF-state current of 7.01×10^{-13} A/μm, an ON-state current of 1.12×10^{-3} A/μm, and an ON-OFF current ratio of 0.16×10^{10}. The reduction in the parasitic capacitances (1.6×10^{-15} F/μm) results in a high speed of operation. The conductance exhibited by the device is effective for analog/RF performance evaluation. The cutoff frequency, intrinsic gain, transit time, GBP, and TFP illustrate the highlighted efficiency of the device and its suitability at high frequencies (1–100 MHz), with perfect amplification (10^4) at high speed. Henceforth, the device is feasible for ultra-low-power analog/RF and DC applications.

REFERENCES

1. D. J. Frank, et al, 'Device scaling limits of Si MOSFETs and their application dependencies'. Proceedings of the IEEE, vol. 89, no. 3, pp. 259–288, 2001.
2. A. M. Ionescu and H. Riel, 'Tunnel field-effect transistors as energy-efficient electronic switches'. Nature, vol. 479, no. 7373, pp. 329–337, 2011.

3. K. Boucart and A. M. Ionescu, 'Double-gate tunnel FET with high-k gate dielectric'. IEEE Transactions on Electron Devices, vol. 54, no. 7, pp. 1725–1733, 2007.

4. S. Chander, S. K. Sinha, S. Kumar, P. K. Singh, K. Baral, K. Singh and S. Jit, 'Temperature analysis of Ge/Si heterojunction SOI-tunnel FET'. Superlattices and Microstructures, vol. 110, pp. 162–170, 2017.

5. A. Chattopadhyay and A. Mallik, 'Impact of a spacer dielectric and a gate overlap/underlap on the device performance of a tunnel field-effect transistor'. IEEE Transactions on Electron Devices, vol. 58, no. 3, pp. 677–683, 2011.

6. S. Chen, H. Liu, S. Wang, W. Li, X. Wang and L. Zhao, 'Analog/RF performance of T-shape gate dual-source tunnel field-effect transistor'. Nanoscale Research Letters, vol. 13, no. 1, 321, 2018.

7. Y. Taur, et al, 'CMOS scaling into the nanometer regime'. Proceedings of the IEEE, vol. 85, no. 4, pp. 486–504, 1997.

8. R. Asra, M. Shrivastava, K. V. Murali, R. K. Pandey, H. Gossner and V. R. Rao, 'A tunnel FET for VDD scaling below 0.6 V with a CMOS-comparable performance'. IEEE Transactions on Electron Devices, vol. 58, no. 7, pp. 1855–1863, 2011.

9. R. H. Dennard, F. H. Gaensslen, V. L. Rideout, E. Bassous and A. R. LeBlanc, 'Design of ion-implanted MOSFET's with very small physical dimensions'. IEEE Journal of Solid-State Circuits, vol. 9, no. 5, pp. 256–268, 1974.

10. M. Kobayashi, K. Jang, N. Ueyama and T. Hiramoto, 'Negative capacitance for boosting tunnel FET performance'. IEEE Transactions on Nanotechnology, vol. 16, no. 2, pp. 253–258, 2017.

11. P. Ghosh and B. Bhowmick, 'Analysis of kink reduction and reliability issues in low-voltage DTD-based SOI TFET'. Micro & Nano Letters, vol. 15, no. 3, pp. 130–135, 2020.

12. S. Singh and P. N. Kondekar, 'A novel electrostatically doped ferroelectric Schottky barrier tunnel FET: Process resilient design'. Journal of Computational Electronics, vol. 16, pp. 685–695, 2017.

13. V. Nagavarapu, R. Jhaveri and J. C. S. Woo, 'The tunnel source (PNPN) n-MOSFET: A novel high-performance transistor'. IEEE Transactions on Electron Devices, vol. 55, no. 4, pp. 1013–1019, 2008.

14. W. Cao, C. J. Yao, G. F. Jiao, D. Huang, H. Y. Yu and M.-F. Li, 'Improvement in reliability of tunneling field-effect transistor with p-n-i-n structure'. IEEE Transactions on Electron Devices, vol. 58, no. 7, pp. 2122–2126, 2011.

15. D. B. Abdi and M. J. Kumar, 'In-built N+ pocket p-n-p-n tunnel field-effect transistor'. IEEE Electron Device Letters, vol. 35, no. 12, pp. 1170–1172, 2014.

16. T. Krishnamohan, D. Kim, S. Raghunathan and K. Saraswat, 'Doublegate strained-Ge heterostructure tunneling FET (TFET) with record high drive currents and <60 mV/dec subthreshold slope'. IEDM Tech. Dig., San Francisco, CA, USA, Dec. 2008, pp. 1–3.

17. A. Hraziia, A. Vladimirescu, C. Amara and Anghel, 'An analysis on the ambipolar current in Si double-gate tunnel FETs'. Solid-State Electron, vol. 70, pp. 67–72, 2012.

18. J. Wan, C. Le Royer, A. Zaslavsky and S. Cristoloveanu, 'Tunneling FETs on SOI: Suppression of ambipolar leakage, low-frequency noise behavior, and modeling'. Solid-State Electron, vol. 65–66, pp. 226–233, 2011.

19. C. Anghel, H. Hraziia, A. Gupta, A. Amara and A. Vladimirescu, '30-nm tunnel FET with improved performance and reduced ambipolar current'. IEEE Transactions on Electron Devices, vol. 58, no. 6, pp. 1649–1654, 2011.

20. A. S. Verhulst, W. G. Vandenberghe, K. Maex and G. Groeseneken, 'Tunnel field-effect transistor without gate-drain overlap'. Applied Physics Letters, vol. 91, no. 5, pp. 053102–053103, 2007.

21. S. Sahay and M. J. Kumar, 'Controlling the drain side tunnelling width to reduce ambipolar current in tunnel FETs using hetero-dielectric box'. IEEE Transactions on Electron Devices, vol. 62, no. 11, November 2015.

22. G. D. Das, G. P. Mishra and S. Dash, 'Impact of source-pocket engineering on device performance of dielectric modulated tunnel FET'. Superlattices and Microstructures, vol. 124, pp. 131–138, 2018.

23. Synopsys, TCAD Sentaurus Device User's Manual, Mountain View, CA, USA, 2010.

24. S. Meriga and B. Bhowmick, 'Investigation of a dual gate pocket-doped drain engineered tunnel FET and its reliability issue'. Applied Physics A, vol. 129, no. 2, pp. 1–11, 2023.

25. D. Leonelli, A. Vandooren, R. Rooyackers, A. Verhulst, C. Huyghebaert, S. De Gendt, M. Heyns and G. Groeseneken, 'Novel architecture to boost the vertical tunneling in tunnel field effect transistors'. SOI Conference (SOI), 2011 IEEE International, pp. 1–2, 2011.

26. B. Senale-Rodriguez, Y. Lu, P. Fay, D. Jena, A. Seabaugh, H. G. Xing, L. Barboni and F. Silveira, 'Perspectives of TFETs for low power analog ICs'. IEEE Subthreshold Microelectronics Conference (SubVT), pp. 1–3, IEEE, 2012.

27. A. R. Trivedi, S. Carlo and S. Mukhopadhyay, 'Exploring tunnel-FET for ultra-low power analog applications: a case study on operational transconductance amplifier'. Proceedings of the 50th Annual Design Automation Conference, 109, IEEE, June 2013.

28. K. Ghosh and U. Singisetti, 'RF performance and avalanche breakdown analysis of InN tunnel FETs'. IEEE Transactions on Electron Devices, vol. 61, pp. 3405–3410, 2014.

29. P. G. D. Agopian, D. V. Martino, S. D. dos Santos, F. S. Neves, J. A. Martino, R. Rooyackers, A. Vandooren, E. Simoen, A.-Y. Thean and C. Claeys, 'Influence of the source composition on the analog performance parameters of vertical nanowire-TFETs'. IEEE Transactions on Electron Devices, vol. 61, pp. 16–22, 2015.

30. V. Vijayvargiya, B. Reniwal, P. Singh and S. Vishvakarma, 'Analogue/RF performance attributes of underlap tunnel field effect transistor for low power applications'. Electronics Letters, vol. 52(7), pp. 559–560, Feb. 2016.

31. K. Vanlalawmpuia and B. Bhowmick, 'Analysis of temperature dependent effects on DC, Analog/RF and linearity parameters for a delta doped heterojunction vertical tunnel FET'. Silico, vol. 14(13), pp. 7517–7529, 2022.

32. B. Das and B. Bhowmick, 'Effect of curie temperature on ferroelectric tunnel FET and its RF/analog performance'. IEEE Transactions on Ultrasonics, Ferroelectrics, and Frequency Control, vol. 68, no. 4, pp. 1437–1441, 2020.

33. S. K. Vishvakarma, A. Beohar, V. Vijayvargiya and P. Trivedi, 'Analysis of DC and analog/RF performance on Cyl-GAA-TFET using distinct device geometry'. Journal of Semiconductors, vol. 38, no. 7, 074003, 2017.

34. W. Li, H. Liu, S. Wang and S. Chen, 'Analog/RF performance of four different tunneling FETs with the recessed channels'. Superlattices and Microstructures, vol. 100, pp. 1238–48, 2016.

35. M. Saravanan and E. Parthasarathy, 'Investigation of RF/Analog performance of Lg = 16 nm Planner In0.80Ga0.20As TFET'. 2021 Fourth International Conference on Electrical, Computer and Communication Technologies (ICECCT), pp. 1–4. IEEE, 2021.

7 Comprehensive Analysis of NC-L-TFETs

Yuvraj Kadale, Prabhat Singh, and Dharmendra Singh Yadav

7.1 INTRODUCTION

Transistors serve as the basic building blocks of most electronic gadgets such as radios, televisions, computers, and cellphones. Transistors can control the flow of current with a minimal input signal, making them popular switches and amplifiers in electronic circuits. Technological innovation in a variety of sectors has been facilitated by downsizing the size and improving the performance. The capacity to manufacture transistors at ever-smaller scales is a crucial component of transistor technology. Transistors have shrunk in size over time, allowing for a huge rise in the number of transistors that can be crammed into a given space on a computer chip. This phenomenon, called Moore's law [1], has been largely responsible for the tremendous increase in computing power that has occurred over the past few decades. However, when transistors go closer to the nanoscale, they start to run into several physical restrictions that can make it more challenging to design and produce them.

Another area of active study in transistor technology is the development of novel transistor designs that can offer superior functionality or performance compared to traditional designs. As an illustration, there has recently been a great deal of interest in creating transistors with unique properties using two-dimensional (2D) materials like graphene or transition metal dichalcogenides [2]. Adding these materials gives them advantages that include high carrier mobility, low power consumption, and flexibility. New transistor types that can operate at low voltages or that are simple to integrate with other electrical components have been the focus of various research projects. Transistors are used in conventional electrical devices as well as in emerging fields like bioelectronics and neuromorphic computing. Bioelectronic devices can be utilized for purposes like health monitoring or delivering personalized medicinal therapies since they connect biological processes with electronic components. In contrast, neuromorphic computing makes use of electronic components to mimic how the human brain works.

Transistors are an essential part of both new sectors and are expected to become more crucial in the development of electronics and computers in the future [3–8].

7.1.1 INTRODUCTION TO NANOELECTRONICS

The study and use of electronic systems and devices at the nanoscale level constitute the intriguing topic of nanoelectronics. This industry is built on the usage of components and materials with nanometer-scale dimensions. Structures that are typically

DOI: 10.1201/9781003393542-7

1–100 nm in size is referred to as "nanoscale." The ability to produce devices that are far smaller and more effective than conventional electrical devices is one of the key benefits of nanoelectronics. This is because materials' electronic characteristics alter at the nanoscale, enabling distinctive electronic behaviors that can be utilized in devices.

In nanoelectronics, for instance, the application of quantum mechanics enables the development of devices that are quicker, more energy-efficient, and more sensitive than their conventional equivalents. Nanoelectronics has several uses, including the creation of quicker and more potent computers, more effective solar cells, and cutting-edge medical equipment. Sensors and other devices that can detect and react to environmental changes, such as temperature, pressure, and radiation, are also being developed using nanoelectronics.

Despite the potential of nanoelectronics, the industry still faces difficulties. The difficulty of controlling and manipulating nanoscale structures, which necessitates the employment of specialized tools and methods, is one of the main difficulties. The possible threats to human health and the environment from using nanoparticles are another difficulty. To ensure the ethical and safe use of nanoelectronics, there is a need for ongoing research and development in the area [9].

7.1.2   INVENTION OF TRANSISTORS

Three American physicists—John Bardeen, Walter H. Brattain, and William B. Shockley—developed the transistor in 1947 at Bell Telephone Laboratories (Bell Labs) in Murray Hill, New Jersey. The invention of a working transistor was announced a year later in 1948. The transistors replaced electron tubes and supplanted them in many applications in the late 1950s. The transistor can control the flow of current with a minimal input signal, making it popular as switches and amplifiers in electronic circuits. Germanium was used to make the original transistors, but silicon eventually took its place as the most widely used semiconductor material in modern transistors because of its better qualities, including its abundance, low cost, and ease of manufacture. In the 1960s, integrated circuits (ICs) made it possible to mass-produce transistors, which resulted in the fabrication of smaller, more potent electronic devices [10].

7.1.3   EVOLUTION OF TRANSISTORS

The invention of vacuum tubes at the beginning of the 20th century served as the foundation for the creation of transistors. An important turning point in the history of electronics occurred when the very first transistor was created by John Bardeen and his team in 1947 at Bell Labs. Initially, transistors were relatively massive and constructed of germanium. The widespread usage of transistors in electronic devices like radios, televisions, and computers was made possible by the discovery of silicon-based transistors in the 1950s and 1960s. The ability to combine various numbers of transistors on a single substrate was made possible by the development of ICs in the 1960s, which resulted in the shrinking of electronic devices and the emergence of the microelectronics industry. The creation of transistors that are smaller, quicker, and more efficient than ever before has been made possible in recent years by the discovery of novel materials and fabrication techniques, advancing technology in industries like computing, telecommunications, and energy [2, 11, 12].

Here is the year-wise evolution of transistors in brief:

- *1947*: At Bell Labs, the first transistor was invented. It was made of germanium and had two contacts.
- *1951*: William B. Shockley invented the point-contact transistor, which used a metal point to contact a semiconductor material.
- *1952*: The junction transistor was invented by Shockley; it had a p-n junction to control the flow of electrons.
- *1953*: Development of the first commercial transistor by Texas Instruments.
- *1956*: Invention of the planar transistor by Jean A. Hoerni; this used a thin layer of oxide to isolate the transistor from other components on the same chip.
- *1959*: Invention of the mesa transistor by John Saby and James Early, which used a mesa structure to isolate the transistor from other components on the same chip.
- *1960*: The first IC was invented, which combined multiple transistors on a single chip.
- *1961*: At Bell Labs, the invention of a metal–oxide–semiconductor (MOS) transistor took place; this type of transistor used a thin layer of oxide for insulation of the gate from the channel.
- *1963*: Invention of the complementary MOS (CMOS) transistor by Frank Wanlass at Fairchild Semiconductor, which used both n- and p-type MOS transistors to reduce power consumption.
- *1969*: Invention of the Schottky transistor by Walter Schottky, which used a metal–semiconductor junction to increase the frequency response and decrease the switching time.
- *1971*: Introduction of the first microprocessor by Intel, which contained 2,300 transistors on a single chip.
- *1987*: Invention of the heterojunction bipolar transistor (HBT) by Takashi Mimura; for this type of transistor, the base and emitters are made up of different semiconductor materials to increase speed and reduce power consumption.
- *1998*: Invention of the fin field-effect transistor (FinFET) by Chenming Hu and his team at the University of California, Berkeley, which used a three-dimensional (3D) structure to reduce leakage current and increase performance.
- *2011*: Introduction of the first 3D transistor by Intel, which used a tri-gate structure to improve performance and reduce power consumption.
- *2021*: Invention of the gate-all-around (GAA) transistor by the Taiwan Semiconductor Manufacturing Company (TSMC), which uses a vertical nanowire structure to improve performance and reduce power consumption.

7.1.4 TYPES OF TRANSISTORS

Here is some brief information about different types of transistors that are currently in use:

- *Bipolar junction transistor (BJT)*: This is a semiconductor device that is widely used as an amplifier or as a switch. Most of the other transistors use a single type of charge carrier, but a BJT uses both holes and electrons. This three-terminal device consists of p-n junctions, which are of two types, PNP and NPN [13–15].
- *Field-effect transistor (FET)*: In this type of transistor, the flow of current is governed by an electric field. A FET is a three-terminal device made up of a source, drain, and gate. It is a unipolar transistor, as it uses only one type of charge carrier. FETs are utilized in a variety of applications, including amplifiers, switches, oscillators, and in ICs for digital logic gates, amplifiers, and other circuits [16].
- *Heterojunction bipolar transistor (HBT)*: This is a special type of BJT whose base and emitter regions are made up of different semiconductor materials, producing a heterojunction. This design grants the HBT advantages over conventional BJTs, such as higher electron mobility and lower base resistance, allowing it to operate at much faster speeds with increased efficiency. As a result, HBTs are utilized in high-speed digital circuits, wireless communication systems, and optoelectronic devices, among other applications.
- *Metal–oxide–semiconductor field-effect transistor (MOSFET)*: This is a widely used electronic device for switching and amplification of electronic signals. It consists of gate, drain, body (substrate), and source terminals. This device operates by changing the width of the channel between the drain and source terminal. Voltage applied at the gate terminal regulates the channel width of the device and repels or attracts the charge carriers by creating an electric field. This property makes the MOSFET a versatile device, capable of functioning as either a switch or an amplifier [17, 18].
- *Junction field-effect transistor (JFET)*: The flow of current in a JFET is controlled by an electric field. It is a three-terminal device. JFETs are often used as voltage-controlled resistors, amplifiers, or switches, and they work in depletion mode, which means that when negative voltage is applied to JFETs, they are turned off [19, 20].
- *Insulated-gate bipolar transistor (IGBT)*: This transistor is used for electronic switches for high-power applications. It is designed to achieve the combined objectives of MOSFETs' high current and bipolar transistors' low-saturation-voltage capabilities. IGBTs are popular in various uses in power supplies, electric vehicles, home appliances, and renewable energy systems, all due to their capabilities of high efficiency and faster switching [21, 22].
- *Fin field-effect transistor (FinFET)*: This is a 3D device that uses a multi-gate structure to improve its performance. FinFETs are constructed on

substrate and place the gate on two, three, or four sides of the channel, forming a double- or even multi-gate structure. This unique design allows the gate to fully "wrap" around the channel on three sides, providing better control of the electric state and reducing leakage current. As the surface area is more between the source and the gate, the output and the performance of the device are best suited for the electronics industry [23, 24].

* *Tunneling field-effect transistor (TFET)*: This is a unique type of transistor that utilizes a tunneling junction in place of the conventional p-n junction to facilitate the injection of charge carriers into a channel region. This tunneling junction is made up of two heavily doped regions of opposite polarity. A TFET is a good option for low-power electronics and has recently been supplemented by negative-capacitance ferroelectric field-effect transistors (NC-FETs) as another potential alternative [25, 26].

* *Nanowire transistor*: A nanowire transistor is a FET that uses a nanowire as the channel for its current. The nanowire is made of semiconductor materials like silicon, germanium, or III-V semiconductors, and is controlled by a gate that regulates the flow of electrons through the channel. Nanowire transistors can overcome a few limitations of conventional transistors, such as their size and power consumption, and are being studied for various uses in sensors, memory devices, and logic circuits [27].

* *Carbon nanotube transistor*: This device replaces the traditional transistor in the channel region with carbon nanotubes as its channel material. Carbon nanotubes are cylindrical in structures and are arranged in a hexagonal lattice. They are desirable for use in electronic devices, due to their distinctive mechanical and electrical qualities. Transistors made of carbon nanotubes have the potential to outperform silicon-based transistors in terms of speed, size, and energy efficiency. Although they are still at the research and development stage, they have great potential for electrical devices in the future [28].

* *Spin field-effect transistor (SpinFET)*: SpinFETs control the flow of current by spinning electrons. These are based on the principle of spintronics, which is a field of electronics that deals with the spin of electrons rather than their charge. SpinFETs have several advantages over conventional transistors, including lower power consumption, faster switching speeds, and higher integration density. SpinFETs work by using a ferromagnetic material as the gate electrode instead of a dielectric material. Only electrons with a particular spin orientation are permitted to travel through the ferromagnetic material, which serves as a spin filter. The magnetic field direction decides how the current flows in this device. SpinFETs are still in the experimental stage and are not yet widely used in commercial applications. However, they have shown great promise for use in low-power electronics and other applications where high integration density and fast switching speeds are required [29].

In this chapter, we will study TFETs, specifically L-shaped TFETs.

7.2 L-TFETs

TFETs that are designed in an L-shape are called L-TFETs. One of them is shown in Figure 7.1. The L-TFET is a promising device for the coming generation of electronic gadgets because it operates at very low power and has high-performance features. It is a type of transistor that operates on the tunneling of the electrons through a potential barrier. Compared to conventional transistors, L-TFETs can achieve superior performance with lower energy consumption, which is crucial for the development of low-power electronic devices [30–32].

Ferroelectric materials have special properties such as spontaneous polarization, hysteresis, and piezoelectricity. These properties make them useful for various types of applications, including electronic devices, memory devices, and sensors. Ferroelectric materials have been investigated extensively for applications in electronic devices due to their ability to consume less power and enhance device performance. The use of ferroelectric materials in L-TFETs has the potential to enhance device performance by utilizing the negative-capacitance effect. The negative-capacitance effect is a phenomenon where the effective capacitance of a ferroelectric material is negative. This effect can be used to overcome the limitations of conventional MOSFETs, which suffer from the subthreshold slope (SS) problem [33–35].

In this chapter, we will discuss L-TFETs with ferroelectric materials, as well as their structure, working principle, and significance in the field of electronics. We will also provide a brief overview of ferroelectric materials and their properties, as well as the effect of negative capacitance in L-TFETs. In addition, we will review the latest research on L-TFETs with ferroelectric materials and their potential applications [36].

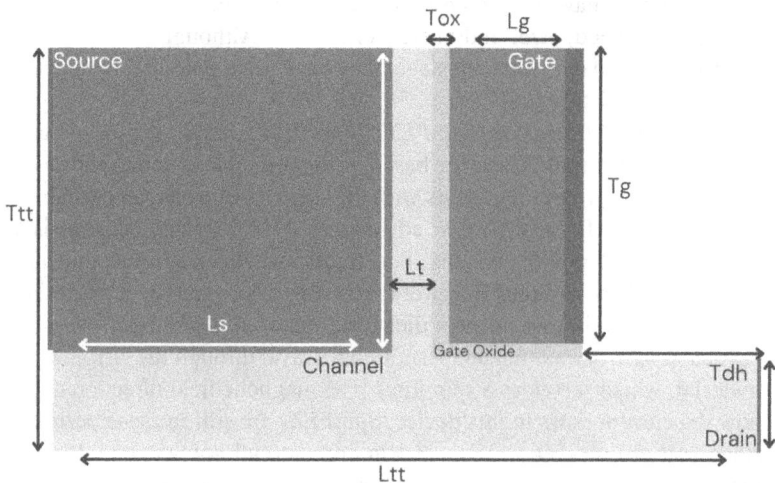

FIGURE 7.1 Cross-sectional view of an L-TFET (conventional device). $L_g = 16$ nm, $L_t = 4$ nm, $T_{ox} = 1$ nm, $T_g = 35$ nm, $T_s = 36$ nm, $T_{dh} = 6$ nm, $L_s = 67$ nm, $L_d = 67$ nm, $N_s = 1 \times 10^{20}$ cm^{-3}, $N_d = 1 \times 10^{17}$ cm^{-3}, $N_c = 1 \times 10^{15}$ cm^{-3}, $WF_g = 4.8$ eV, $T_{tt} = 42$ nm, $L_{tt} = 160$ nm, and SiO$_2$ as the gate oxide.

7.2.1 FERROELECTRIC MATERIALS

Ferroelectric materials in the presence of an external electric field have reversible spontaneous polarization. Nonvolatile memory, piezoelectric sensors and actuators, electro-optic modulators, and pyroelectric detectors are just a few of the many devices that can use ok them [37]. Ferroelectric materials are used in NC-FETs to improve the subthreshold swing of conventional FETs [38]. Recently, there has been much ongoing research on combining ferroelectric materials with novel device concepts for transistors, to examine their capability for high-performance, low-power, next-generation electronic devices. The primary topics of discussion in this chapter will be the integration of ferroelectric materials with L-TFETs and the impact of ferroelectricity on device performance. Due to ferroelectric materials' potential use in nonvolatile memory systems, such as ferroelectric random-access memory (FeRAM), and their capacity to display negative capacitance, which helps to improve the performance of electronic devices, they have been intensively explored.

7.2.2 PROPERTIES OF FERROELECTRIC MATERIALS

- *Spontaneous polarization*: In ferroelectric materials, spontaneous polarization is a property that arises due to the alignment of electric dipoles in a particular direction without any external electric field [39]. Ferroelectric materials exhibit spontaneous polarization because of their crystal structure, and they are characterized by a hysteresis loop in their polarization–electric field curve [40]. The magnitude and direction of spontaneous polarization depend on the crystal structure of the material [41].
- *Switchable polarization*: Unlike other dielectric materials, ferroelectric materials have the capacity to change the direction of their polarization when an external electric field is present. This characteristic is essential for their use in FeRAM and other nonvolatile memory systems.
- *Hysteresis*: The polarization of a ferroelectric material exhibits hysteresis behavior, meaning that the history of the electric field will decide the behavior of the material. This characteristic is crucial for their use in capacitors because it enables the material to store charge even after the electric field has been removed.
- *Piezoelectricity*: Piezoelectricity is a property of generating an electric field when a mechanical stress is applied or any deformation takes place. This property is important for their application in sensors and actuators.
- *Nonlinear dielectric response*: Ferroelectric materials have a nonlinear dielectric response, which means that the displacement is inversely proportional to the electric field applied. This property is important for their application in frequency doubling and other nonlinear optical devices.
- *Temperature dependence*: The properties of ferroelectric materials are strongly temperature dependent. Their Curie temperature (the temperature at which they lose their ferroelectric properties) is a critical parameter that limits their application in high-temperature environments.

Overall, these properties make ferroelectric materials attractive for various applications, including memory devices, capacitors, sensors, actuators, and nonlinear optical devices [42–45].

7.2.3 TYPES OF FERROELECTRIC MATERIALS

- *Perovskite-based ferroelectrics*: These are the most extensively studied ferroelectric materials, and their unique properties are mainly attributed to their structure. Examples of perovskite-based ferroelectrics include lead zirconate titanate ($PbZrTiO_3$) and barium titanate ($BaTiO_3$).
- *Organic ferroelectrics*: These are relatively new ferroelectric materials and are composed of organic molecules or polymers. Examples for organic ferroelectric materials include polyvinylidene fluoride (PVDF) and its copolymers.
- *Relaxor ferroelectrics*: These are a class of disordered ferroelectric materials that exhibit high piezoelectricity and electromechanical coupling coefficients. Examples of relaxor ferroelectric materials include lead scandium tantalate (PST) and lead magnesium niobate–lead titanate (PMN-PT).
- *Biomorphic ferroelectrics*: These are biomimetic ferroelectric materials that are synthesized to replicate the structure and function of natural materials. An example of a biomorphic ferroelectric is hydroxyapatite ($Ca_{10}(PO_4)_6(OH)_2$) with ferroelectric polarization along the *x*-axis.
- *Hybrid ferroelectrics*: These are a class of ferroelectric materials that combine inorganic and organic components, resulting in unique properties. An example of a hybrid ferroelectric is the metal–organic framework (MOF) Zn(2,5-dimethoxybenzene dicarboxylate).
- *Thin-film ferroelectrics*: These are thin ferroelectric material films that are typically created by physical or chemical deposition processes. Lead titanate ($PbTiO_3$) and barium strontium titanate ($Ba_{0.5}Sr_{0.5}TiO_3$) are examples of thin-film ferroelectrics.

7.3 L-TFETs WITH FERROELECTRIC MATERIALS

The L-TFET is a valuable device for lower power consumption in high-speed electronics. Ferroelectric materials provide a negative-capacitance effect that can reduce effective gate voltage, thereby improving the SS and reducing the OFF-current of the device [46, 47]. L-TFET devices can be simulated using a variety of software tools, including commercial tools like Silvaco ATLAS and Synopsys' QuantumATK, Sentaurus TCAD, and Sentaurus Device, as well as open-source tools. These simulation tools employ numerical methods to solve the fundamental equations of charge and current transport, including Poisson's equation, Schrödinger's equation, and current continuity equations. Researchers can gain insights into a device's performance from studying the simulation results, which also show the electric field, band energy diagrams, I_d-V_g characteristics, and a few other static properties. To simulate an L-TFET device, these steps can be followed:

1. Define the device geometry, including the gate length, gate oxide thickness, and ferroelectric material thickness.

2. Specify the material properties of the various components, including the semiconductor material, gate oxide material, and ferroelectric material. These properties can include permittivity, mobility, doping concentration, and other relevant parameters.
3. Define the bias conditions for the device, including the source and drain voltages, gate voltage, and temperature.
4. Select the appropriate numerical methods and models for solving the charge and current transport equations, including the quantum mechanical models for tunneling and the models for the ferroelectric material.
5. Run the simulation and analyze the results, including the current–voltage characteristics, SS, ON-current, and OFF-current.

Overall, L-TFETs with ferroelectric material have shown promising results in reducing the subthreshold swing and improving the performance of the device compared to conventional MOSFETs. The simulation tools can provide necessary information on the device's behavior and can help guide the design and optimization of L-TFETs for future electronics applications.

7.3.1 ABOUT THE SIMULATION SOFTWARE

For the simulation of material-based devices like solar cells, photodiodes, transistors, and sensors, Silvaco's ATLAS software is a powerful tool. It can simulate the physical processes and electrical properties of germanium, silicon, and a variety of other device structures and materials. ATLAS can also perform optimization and analysis of device performance using various methods such as parameter extraction, curve fitting, optimization algorithms, and statistical techniques. ATLAS is part of the TCAD suite of products by Silvaco, which also includes process simulation, mesh generation, visualization, and characterization tools. ATLAS is a modular and flexible platform that allows you to create and analyze realistic models of any type of solar cell, transistor, diode, sensor, or other device. ATLAS software can handle complex geometries, materials, doping profiles, contacts, interfaces, and physical phenomena such as quantum effects, tunneling, recombination, mobility models, temperature effects, stress effects, and more [48]. ATLAS consists of several modules to perform different tasks, including the following:

Device: The Poisson and continuity equations for electrons and holes in 1D, 2D, or 3D systems are solved in this fundamental module. Devices such as MOSFETs, BJTs, HBTs, TFETs, light-emitting diodes (LEDs), and lasers can all be simulated.

Blaze: This is a module that extends a device to simulate optoelectronic devices such as solar cells, photodetectors, and organic light-emitting diodes (OLEDs). Blaze can calculate optical generation rates using rigorous coupled-wave analysis (RCWA) or ray-tracing techniques. The optical losses brought on by reflection, transmission, absorption, scattering, and so on can also be modeled using Blaze.

Luminous: This is a module that extends Blaze to simulate advanced optoelectronic devices such as quantum well lasers, vertical-cavity surface-emitting lasers (VCSELs), and quantum dot devices. Luminous can account for quantum confinement effects using effective mass or k·p methods. Luminous can also model carrier transport and recombination in quantum structures using drift diffusion or Schrödinger–Poisson methods.

TonyPlot: This is a module that provides visualization and analysis capabilities for ATLAS. A TonyPlot can display 1D, 2D, or 3D data from TCAD simulations or parasitic extraction tools. A TonyPlot can also perform various operations on the data, such as scaling, averaging, integration, differentiation, and interpolation.

7.4   DEVICE STRUCTURE AND PARAMETERS

The designed L-TFET is presented in Figure 7.1. The source region of the device is doped with germanium (Ge) with a source doping concentration (N_s) of 1×10^{20} cm^{-3}, keeping the source length (L_s) as 67 nm and thickness as 36 nm. The length of the tunneling region (L_t) is set to 4 nm, followed by a gate oxide (T_{ox}) of 1 nm thickness made of SiO$_2$. The gate length (L_g) and thickness (T_g) are set as 16 nm and 35 nm, respectively. The channel doping concentration (N_c) is 1×10^{15} cm^{-3}, and the drain doping concentration (N_d) is 1×10^{17} cm^{-30}. The length (L_d) and thickness (T_{dh}) of the drain have set values of 67 nm and 6 nm, respectively. Thus, the overall length and thickness of the device sum to 160 nm length and 42 nm thickness. The work function for the device is set to 4.8 eV for the complete simulation.

As discussed above regarding the types and properties of ferroelectric materials, there we find the scope of improvements in the current L-TFET structure (e.g., adding ferroelectric material to the gate stack region can improve the performance and efficiency of the device). Figure 7.2 is the new proposed device that has a ferroelectric

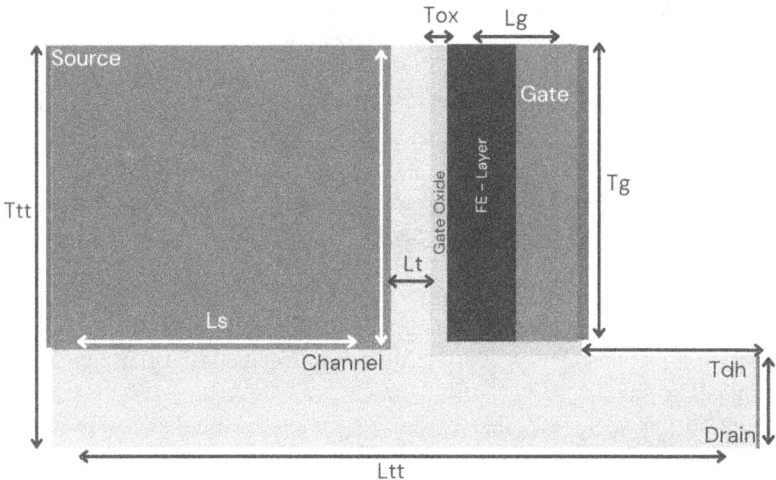

FIGURE 7.2   Cross-sectional view of a NC-L-TFET (proposed device). L_g = 16 nm, L_t = 4 nm, T_{ox} = 1 nm, T_g = 35 nm, T_s = 36 nm, T_{dh} = 6 nm, L_s = 67 nm, T_{FE} = 35 nm, L_{FE} = 5 nm, L_d = 67 nm, N_s = 1×10^{20} cm^{-3}, N_d = 1×10^{17} cm^{-3}, N_c = 1×10^{15} cm^{-3}, WF_g = 4.8 eV, T_{tt} = 42 nm, L_{tt} = 160 nm, and SiO$_2$ as the gate oxide.

TABLE 7.1

Parameters and Values Used for Simulation of the Device

Parameters	Abbreviations	Set Values
Gate length	L_g	16 nm
Length of tunneling region	L_t	4 nm
Thickness of gate oxide	T_{ox}	1 nm
Thickness of gate region	T_g	35 nm
Source thickness	T_s	36 nm
Drain thickness	T_{dh}	6 nm
Source length	L_s	67 nm
Ferroelectric layer thickness	T_{FE}	35 nm
Ferroelectric layer length	L_{FE}	5 nm
Drain length	L_d	67 nm
Source doping concentration	N_s	1×10^{20} cm^{-3}
Drain doping concentration	N_d	1×10^{17} cm^{-3}
Channel doping concentration	N_c	1×10^{15} cm^{-3}
Gate work function	WF_g	4.8 eV
Total thickness of device	T_{tt}	42 nm
Total length of device	L_{tt}	160 nm

layer of 5 nm length and 35 nm thickness stacked between the gate and oxide layer [49, 50]. All the device parameters and values used for simulation is listed in Table 7.1.

7.5   SIMULATION RESULTS AND PERFORMANCE ANALYSIS

Here, we will observe, analyze, and compare the performance of a NC-L-TFET, the newly proposed device, with respect to an L-TFET (the conventional device). We will observe the quasi-Fermi level (QFL) and energy band diagrams, followed by the electric field, recombination rate, potential, I_d-V_g characteristics, and analysis of capacitance from gate to drain and gate to source. From here onward, analysis along *AB* and *XY* will be referred to as performing the analysis of the device by taking the respective cutlines, as shown in Figure 7.3.

7.5.1   Electron Quasi-Fermi Level

The electron QFL is a concept used in semiconductor physics to describe the energy level of electrons in a semiconductor material. It is a measure of the energy distribution of electrons in a material and is used to describe the behavior of electrons in a semiconductor device. The QFL is defined as the energy level at which there is an equal probability of finding an electron or hole [51]. In other words, it is a measure of the energy level at which there are equal numbers of electrons and holes in a material. The electron QFL can be used to describe the behavior of electrons in a semiconductor device such as a solar cell or transistor [52]. From Figure 7.4(a), we

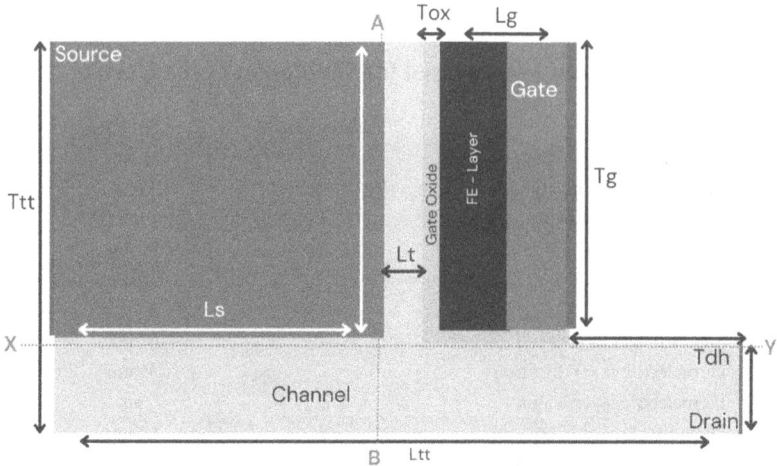

FIGURE 7.3 Cross-sectional view of a NC-L-TFET with vertical line *AB* and horizontal line *XY* as cutlines.

observe the electron QFL in electronvolts. The plot (which follow linear relation from 0.06 to 0.16 um) QFL represents the electron QFL of a NC-L-TFET, whereas blue represents the QFL of an L-TFET.

7.5.1.1   QFL along the *AB* Cutline

From Figure 7.4(a), we find that the electrons present in the L-TFET have relatively more energy compared to the NC-T-TFET. However, from Figure 7.4(b), we see that the holes' energy is almost zero throughout the device except for the initial position.

7.5.1.2   QFL along the *XY* Cutline

Similarly, we look at Figure 7.5(a) and (b), which depicts the electrons and holes QFL when the horizontal cutline is taken along *XY*.

7.5.2   Effect of Ferroelectric Materials on Energy Band Diagrams

A pictorial representation of the energy levels of an atom, molecule, or solid is called an "energy band diagram." It serves as a visual representation of how electrons are distributed across various energy states. Energy band diagrams are frequently used in solid-state physics to depict the electronic structure of materials like semiconductors and insulators. The permitted and prohibited energy levels in a substance are displayed on the energy band diagram. The diagram's horizontal axis denotes position or distance, while the vertical axis represents energy. The permissible energy levels are depicted as bands, and the energy levels themselves are portrayed as horizontal lines. In semiconductors, the conduction band is present over the valence band; at the initial stage, all the electrons are accumulated in the valence band, whereas the conduction band is just partially or sparsely populated [53]. When voltage is applied to the device, the electrons from the valence band gain energy and are excited to the

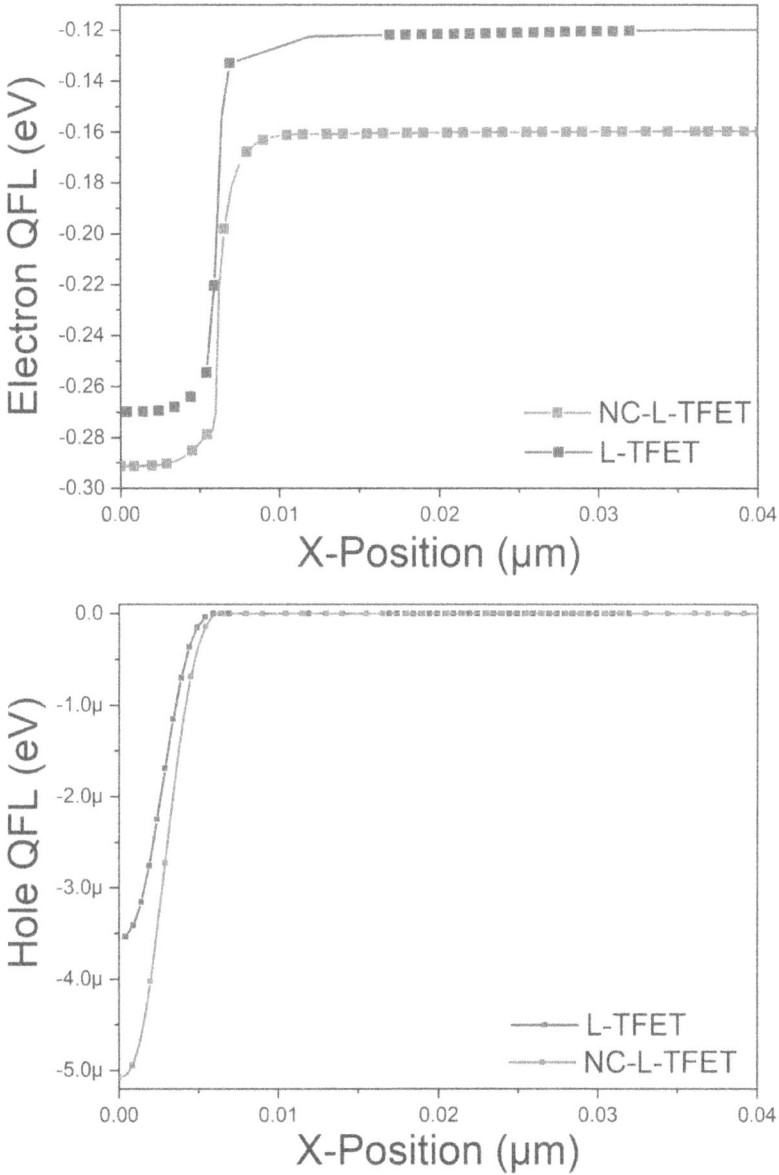

FIGURE 7.4 Structure in accordance with Figure 7.3. Quasi-Fermi-level (QFL) analysis of (a) electrons and (b) hole charge carriers along the vertical cutline *AB*.

conduction band, which makes the flow of current possible in the device. The energy band diagram can be used to build electronic components like transistors and diodes as well as to forecast the electrical properties of materials like conductivity. Here, we will study energy bands at two different cutlines, the vertical cutline (*AB*) and the horizontal cutline (*XY*). Figure 7.6 shows the newly proposed NC-L-TFET with

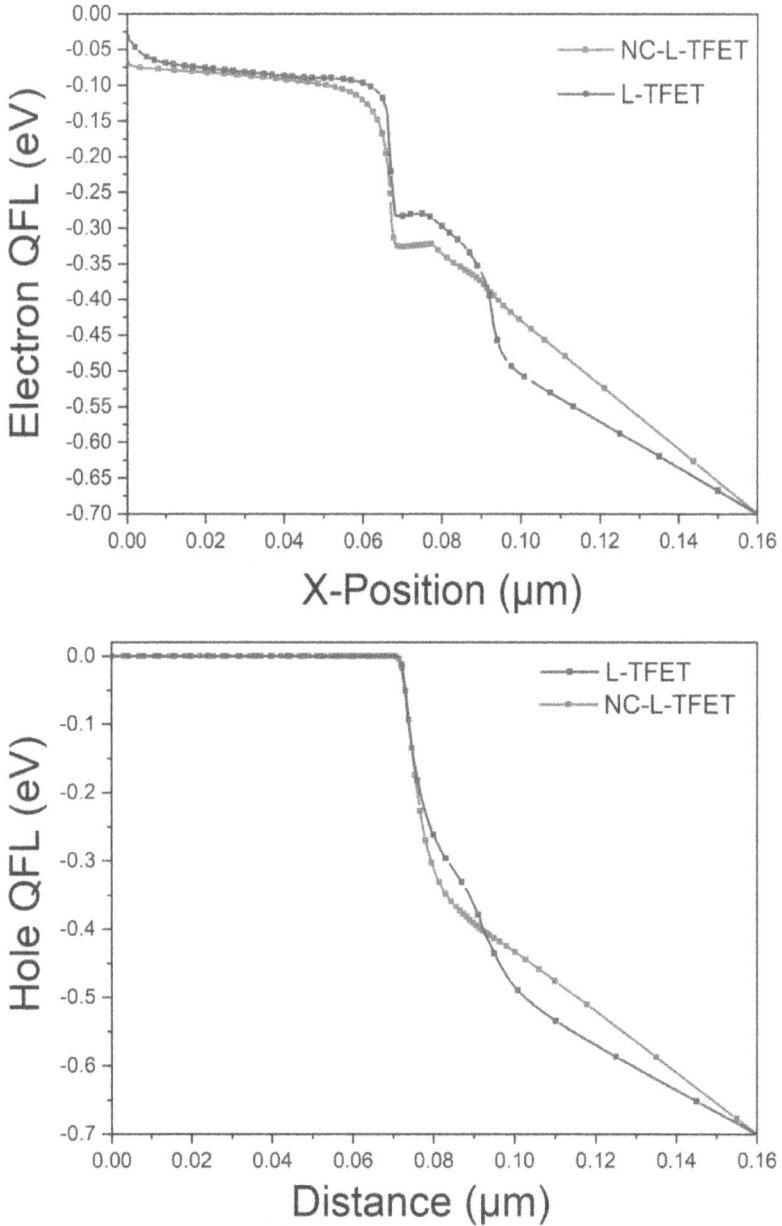

FIGURE 7.5 Structure in accordance with Figure 7.3. Quasi-Fermi-level (QFL) analysis of (a) electrons and (b) hole charge carriers along the horizontal cutline *XY*.

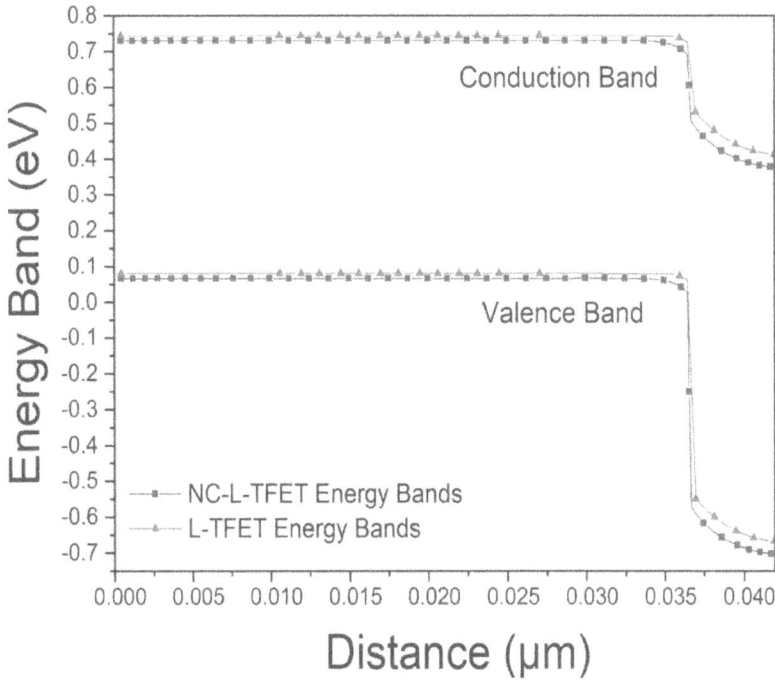

FIGURE 7.6 Energy band diagram of an L-TFET with rectangle symbol markings and NC-L-TFET with tringle symbol markings along the *AB* cutline. The dotted line shows the band energy of the L-TFET, whereas plane lines indicate the band energy of the NC-L-TFET in the source–channel region.

AB and *XY* cutlines for analysis. Here, we will observe and compare the energy band difference between the L-TFET and NC-L-TFET.

7.5.2.1 Analysis along the *AB* Cutline

Figure 7.6 represents the valence and conduction band of L-TFETs both with and without ferroelectric material. We observe no major difference in the energy band diagrams.

7.5.2.2 Analysis along the *XY* Cutline

In Figure 7.7, the dotted lines represent the bands of the L-TFET, and the sharp thick line with dots represents the bands of the NC-L-TFET. The rectangle symbol stands for the conduction band, whereas the black current represents valence bands. If we observe in Figure 7.7, the NC-L-TFET shows relatively better tunneling compared to the L-TFET. Due to the negative-capacitance effect, the NC-L-TFET shows better tunneling. The NC-L-TFET's gate ferroelectric material displays a negative-capacitance response, which significantly lowers the device's threshold voltage. The ferroelectric material acts as a negative capacitor when the voltage is applied at the gate, storing a charge with the polarity opposite to that of the applied voltage. By lowering the

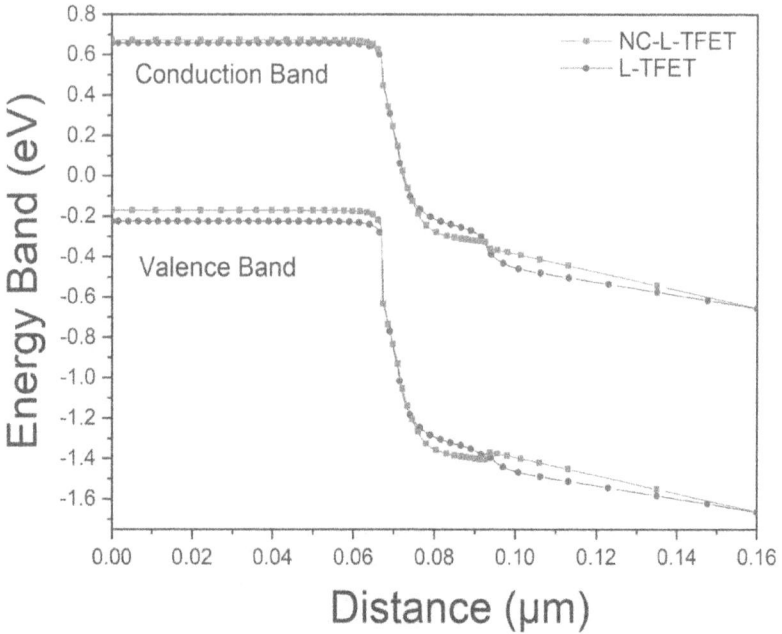

FIGURE 7.7   Energy band diagram of an L-TFET (with circle symbol) and NC-L-TFET (with rectangle symbol) along the *XY* cutline.

voltage drop across the tunneling zone, this negative-capacitance effect can also narrow the effective tunneling barrier and increase the tunnel current.

7.5.3   Effect of Ferroelectric Materials on the Electric Field

In an L-TFET, the electric field distribution in the channel region of the device is referred to as the "electric field." The electric field in an L-TFET is essential for allowing electrons to tunnel through the barrier created by the gate and channel region. The potential barrier in the channel region is shaped by the electric field and controls how likely it is that electrons will tunnel through it. In an L-TFET, the channel's source side has the strongest electric field, which gradually diminishes while moving toward the drain side. This is because of the presence of the gate terminal, and when a voltage is applied to it, it generates an electric field in the channel region. The ferroelectric substance, which can strengthen the electric field in the channel region, has an impact on the electric field as well.

7.5.3.1   Analysis along the *AB* Cutline

Figure 7.8(c) and (d) represents the electric fields of an L-TFET and NC-L-TFET observed on the *AB* cutline, as shown in Figure 7.3. Between 0.035 and 0.040 μm, we find the maximum electric field and a relatively higher electric field in the case of the NC-L-TFET. Due to the existence of the ferroelectric layer, the electric field in the source–channel region of the NC-L-TFET is higher than that of the L-TFET.

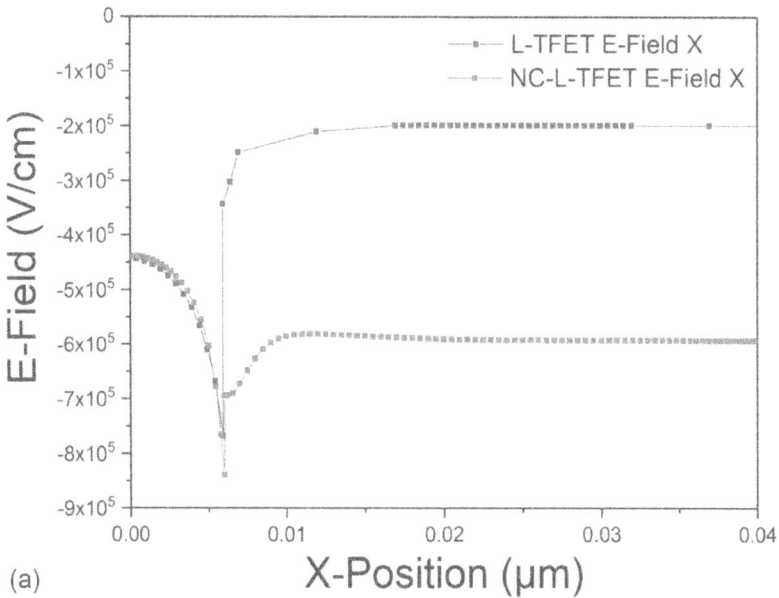

(a)

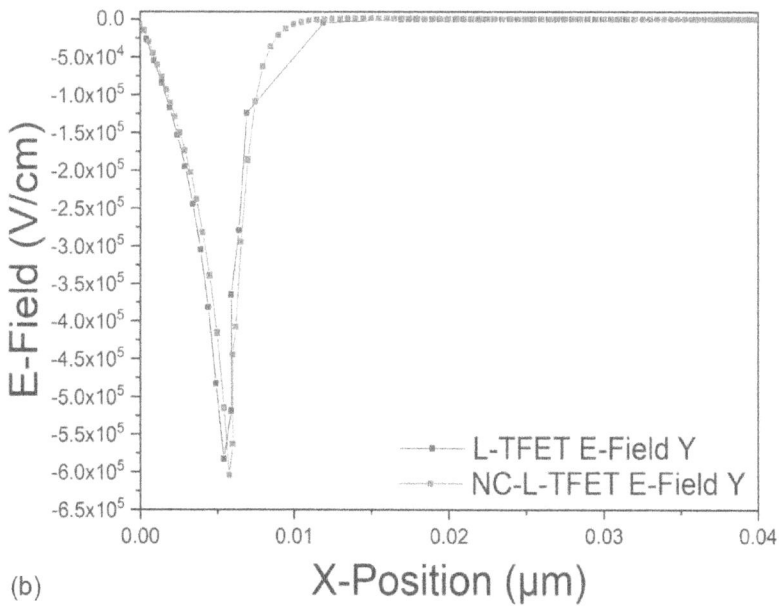

(b)

FIGURE 7.8   Structure in accordance with Figure 7.3. (a) E_{Field} intensity analysis in horizontal (x-direction) and (b) vertical (y-direction). *(Continued)*

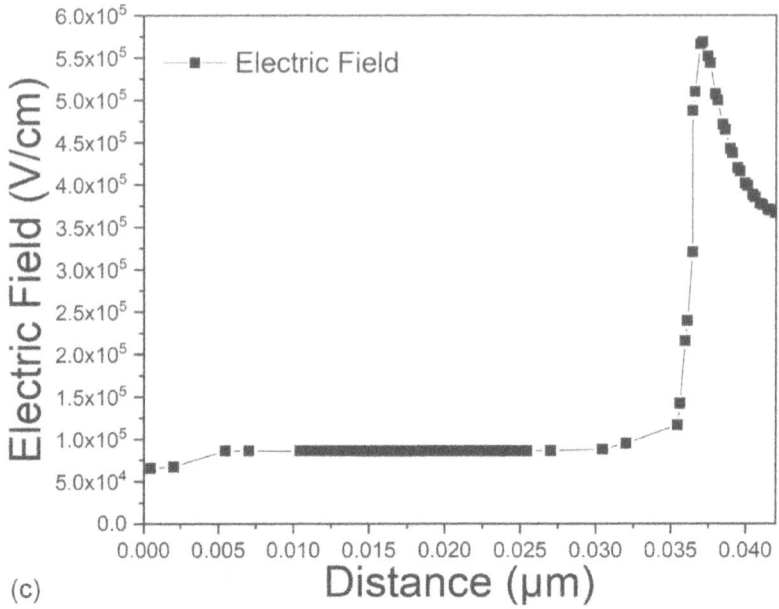

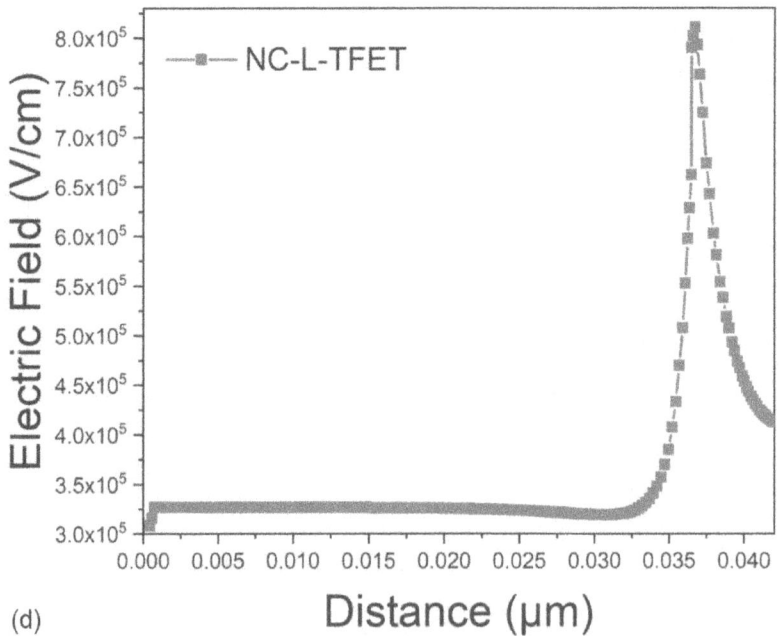

FIGURE 7.8 *(Continued)* Structure in accordance with Figure 7.3. (c) Total electric field of the L-TFET, (d) total electric field of the NC-L-TFET.

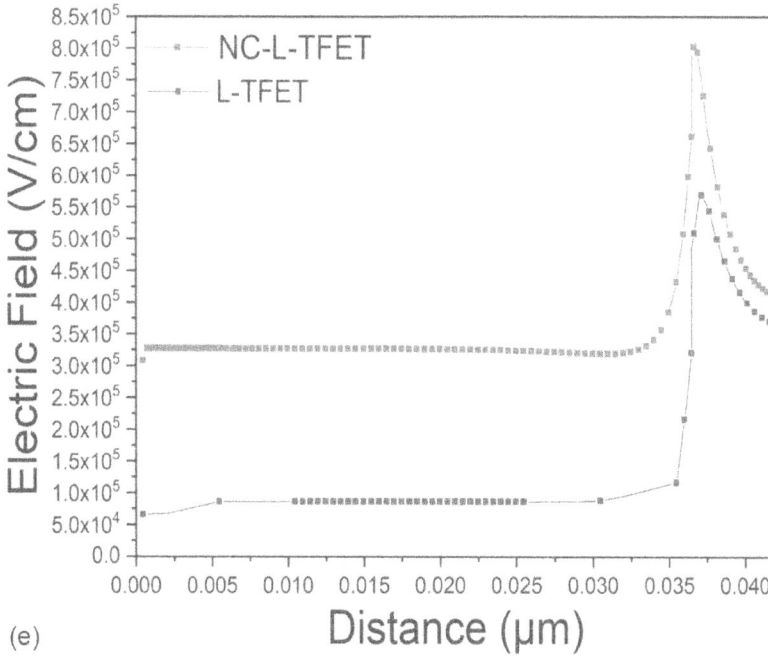

(e)

FIGURE 7.8 *(Continued)* Structure in accordance with Figure 7.3. (e) Overlay of (c) and (d) observed along the *AB* cutline.

A negative-capacitance effect that the ferroelectric layer can produce can improve gate control on the channel and increase the electric field in the source–channel region. The L-TFET geometry also has an impact on the distribution of the electric field, which is later modified by the presence of the ferroelectric layer.

7.5.3.2 Analysis along the *XY* Cutline

Figure 7.9 represents the electric field observed on taking the XY cutline in the drain–channel interface. Figure 7.9(c) represents the electric field in the L-TFET, whereas Figure 7.9(d) represents the electric field in the NC-L-TFET. We observe that the electric field remains the same in both devices at 0.07 μm and decreases as we move close to the drain. However, between 0.09 μm and 0.10 μm, we find a huge drop in the electric field in the NC-L-TFET. The NC-L-TFET's drain–channel area has experienced a sudden drop in the electric field because of the ferroelectric material's negative-capacitance effect. A ferroelectric material's polarization shifts in response to an applied voltage, which raises the effective gate voltage and lowers the subthreshold swing. This effect results in a lower voltage drop across the channel region, leading to a reduced electric field in that region. In contrast, L-TFETs do not have a ferroelectric layer, so they do not exhibit negative capacitance, and the electric field does not drop as much in the drain–channel region. Other factors, such as interface trap charge and material parameters, can also affect the electric field distribution in the device, but negative capacitance is a possible explanation for the observed behavior in the NC-L-TFET.

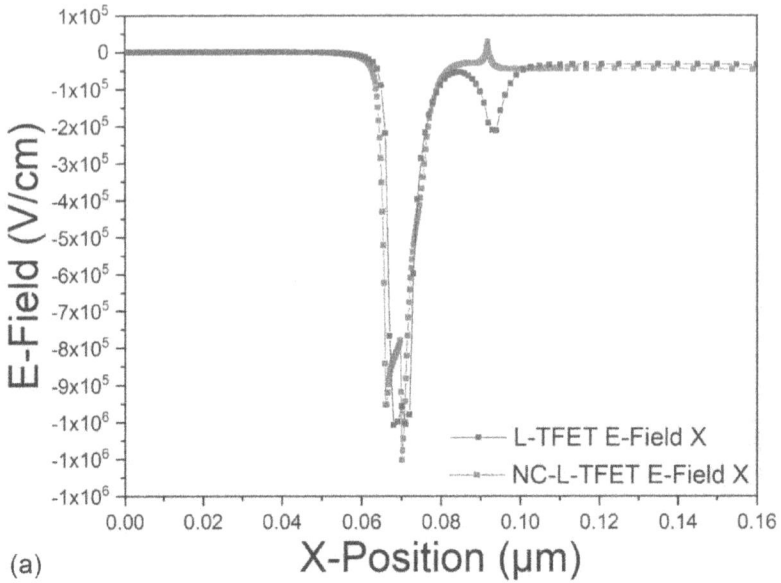

(a)

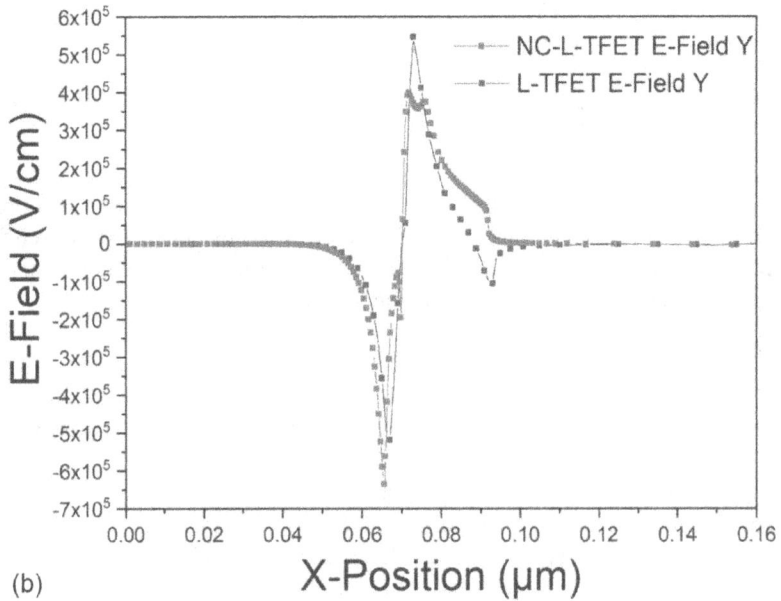

(b)

FIGURE 7.9   Structure in accordance with Figure 7.3. (a) E_{Field} intensity analysis in horizontal (x-direction) and (b) vertical (y-direction) *(Continued)*

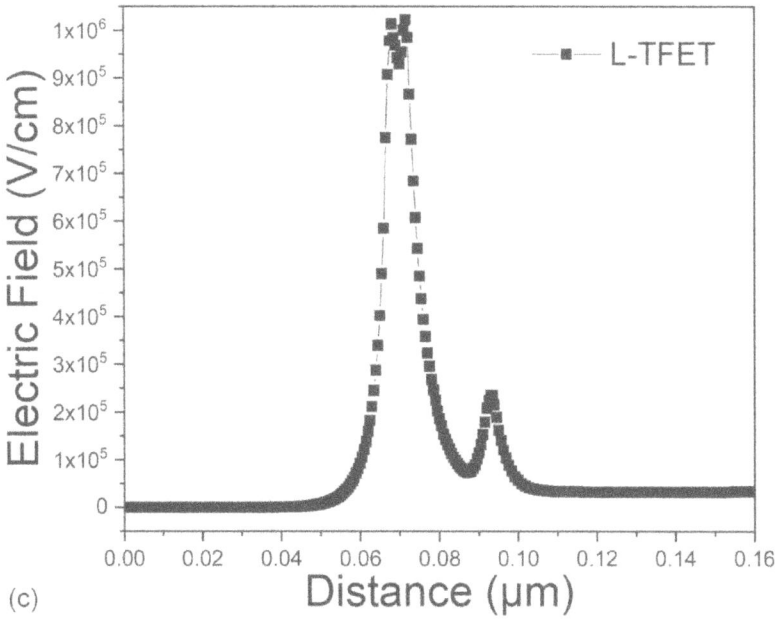

(c)

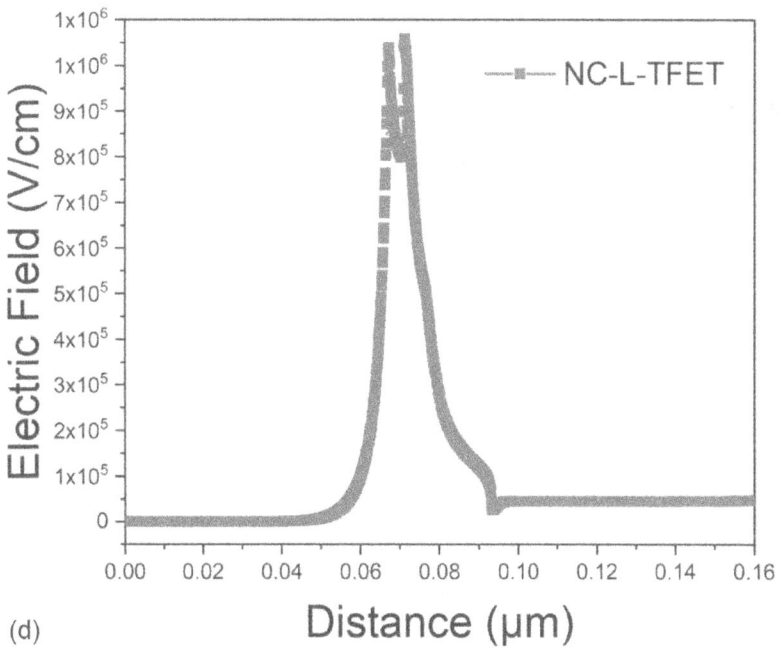

(d)

FIGURE 7.9 *(Continued)* Structure in accordance with Figure 7.3. (c) Total electric field of the L-TFET, (d) total electric field of the NC-L-TFET.

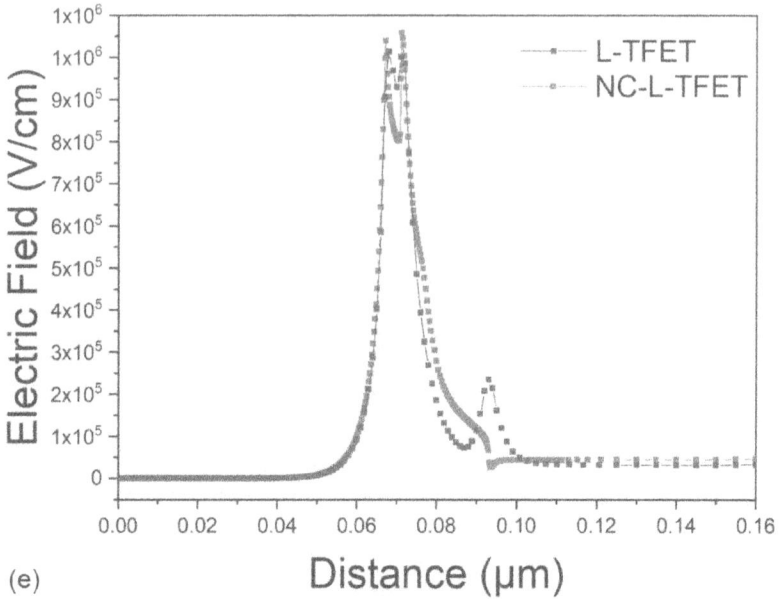

(e)

FIGURE 7.9 *(Continued)* Structure in accordance with Figure 7.3. (e) Overlay of (c) and (d) observed along the *XY* cutline.

7.5.4 POTENTIAL OF THE DEVICE

7.5.4.1 Analysis along the *AB* and *XY* Cutlines

Figures 7.10 and 7.11 reveal that NC-L-TFETs have higher potential values compared to conventional L-TFETs. However, if we move horizontally along the *XY* cutline, we find that the L-TFET's potential increases after the intersection point at 0.09 µm.

7.5.5 EFFECT OF FERROELECTRIC MATERIALS ON THE RECOMBINATION RATE

The recombination rate in transistors refers to the rate at which electrons and holes recombine in the device, leading to a reduction in the number of charge carriers and a decrease in the transistor's performance. Recombination can occur through a variety of mechanisms, including surface recombination, bulk recombination, and Shockley–Read–Hall recombination [54, 55]. The recombination rate is influenced by factors such as the doping concentration, temperature, and radiation exposure. In bipolar transistors, recombination can occur in the base–emitter depletion region, leading to hole losses [56].

7.5.5.1 Analysis along the *AB* and *XY* Cutlines

Observing Figures 7.12 and 7.13, we can comment that there is no major difference in the recombination rate of the conventional L-TFET and the proposed NC-L-TFET. The recombination rate of both the devices is observed to be -7×10^{16} cm^{-3}s^{-1}.

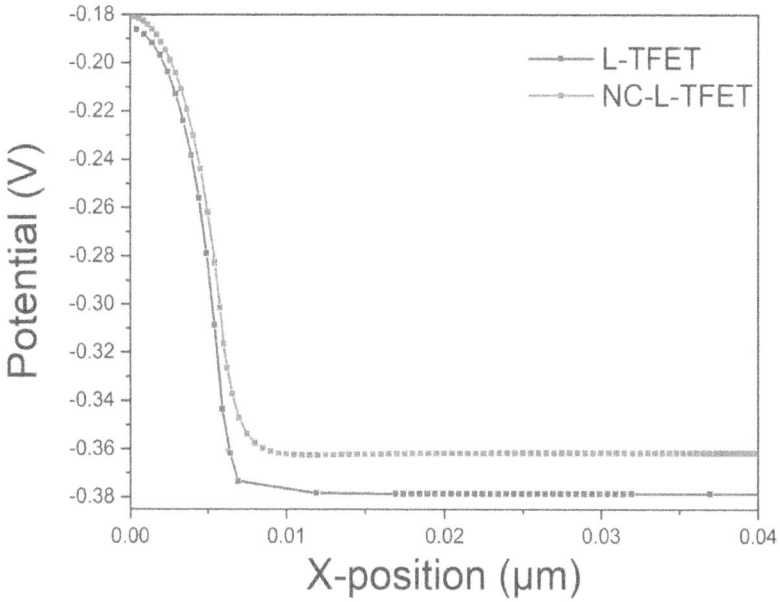

FIGURE 7.10 Potential of an L-TFET (lowest plot) and NC-L-TFET (higher value data plot) along the vertical *AB* cutline.

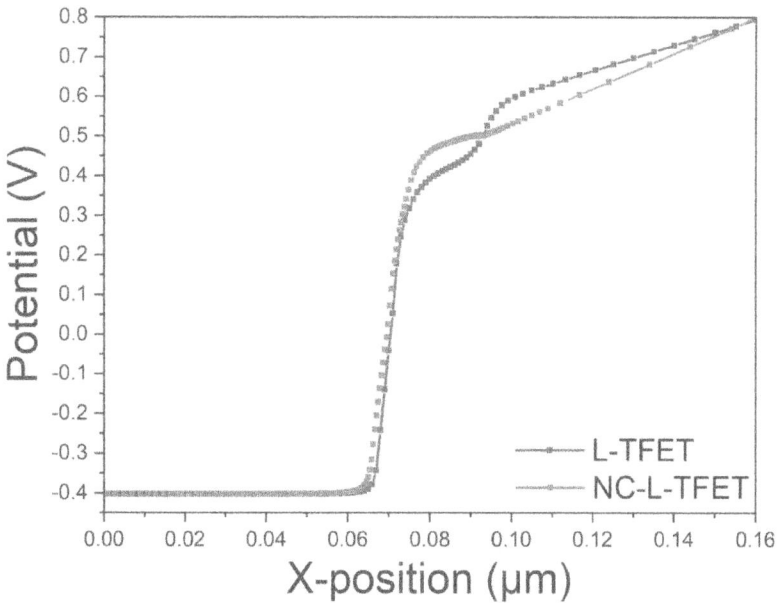

FIGURE 7.11 Potential of an L-TFET (higher plot between 0.10 um to 0.16 um) and NC-L-TFET (lowest and linear plot between 0.10 um to 0.16 um) along the horizontal *XY* cutline.

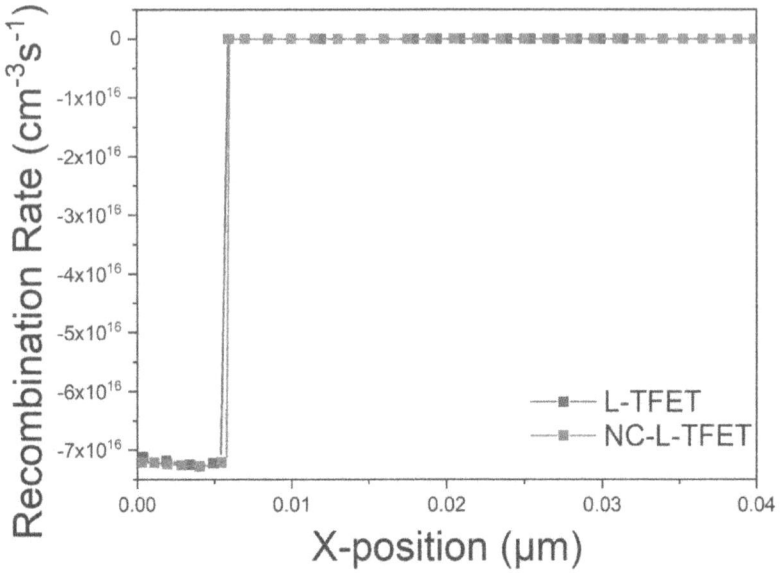

FIGURE 7.12 Recombination rate of an L-TFET and NC-L-TFET along the vertical cutline *AB (approximately both are same).*

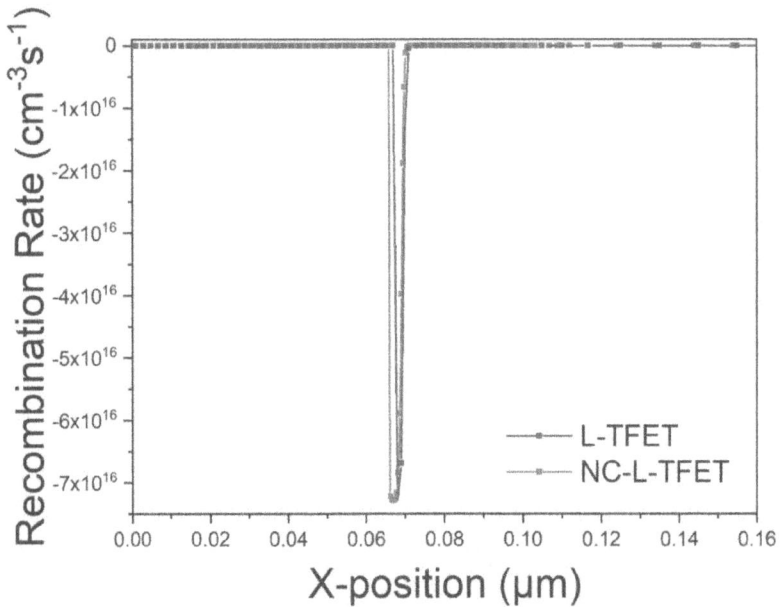

FIGURE 7.13 Recombination rate of an L-TFET and NC-L-TFET along the horizontal cutline *XY (approximately both are same).*

7.5.6   Effect of Ferroelectric Materials on I_D-V_G Characteristics

For L-TFETs, the I_D-V_G characteristics demonstrate the correlation between I_D and V_G, which stand for current at drain terminal and voltage at gate terminal, respectively. L-TFETs, in contrast to ordinary MOSFETs, use band-to-band tunneling to conduct current. As a result, the SS is smaller than the 60 mV/decade that is the theoretical upper limit for ordinary MOSFETs [57], making L-TFETs potential candidates for low-power applications. The I_D-V_G characteristics for L-TFETs typically exhibit a very low drain current at low gate voltages, which is referred to as the "off-state" region. With an increase in the voltage at the gate, the tunneling probability increases, resulting in a gradual increase in the drain current. However, the current at the drain in L-TFETs is typically lower than that of conventional MOSFETs at the same gate voltage due to the lower SS. The steepness of the I_D-V_G curve in the ON-state region is determined by the effective tunneling barrier height and the gate voltage. The gate voltage also affects the turn-on voltage and the maximum achievable drain current. In addition, the I_D-V_G characteristics may be affected by other factors such as the doping profile, gate oxide thickness, and temperature.

7.5.6.1   Subthreshold Swing

In subthreshold operation, when gate voltage is lower than the threshold voltage, subthreshold swing is a crucial parameter for assessing the performance of transistors. It is defined as the change in gate voltage required to modify the drain current by one decade, and is expressed in volts per decade (V/dec). Due to their extremely low power consumption during subthreshold operation, TFETs place a special emphasis on subthreshold swing. The ON-OFF current ratio of the device and, consequently, its performance are directly impacted by the steepness of the SS. The Boltzmann limit for subthreshold swing for ordinary MOSFETs is generally accepted to be 60 mV/dec at room temperature. However, because of the tunneling current's sharp turn-on property, subthreshold swing in TFETs can be much smaller than this limit [58]. The subthreshold swing in TFETs has been improved by several design and material optimizations. The subthreshold swing, for instance, can be improved by using heterojunctions made of Si-Ge or III-V materials, which can also shorten the effective tunneling distance. Like this, by changing the band structure of the device, the use of strained materials or bandgap engineering can likewise result in improved subthreshold swing [59]. Ferroelectric L-TFETs have been a promising candidate in recent years for obtaining even lower subthreshold swing. Negative capacitance is made possible by using ferroelectric materials, which can effectively balance out the device's positive capacitance and lessen subthreshold swing. In addition, the geometry of L-TFETs has the ability to increase tunneling probability while decreasing effective tunneling distance, resulting in even lower subthreshold swing. Subthreshold swing is a metric used to describe the steepness of a transistor's transfer characteristic curve in the subthreshold region [60, 61].

The equation for subthreshold swing is:

$$\text{Subthreshold swing} = \left(\partial(log Id_{10})/\partial V_{GS}\right)^{-1} \qquad (7.1)$$

where V_{GS} is the gate-to-source voltage.

In practical terms, a lower subthreshold swing indicates a more efficient transistor that can switch its state from on to off or off to on more quickly and with less energy consumption. If we observe the characteristics represented in Figure 7.14(a), we see more drain current in the NC-L-TFET compared to the L-TFET; also, there is good improvement in SS in the case of the NC-L-TFET, making it the best device for faster switching. The reason for improvement is the negative-capacitance effect in the NC-L-TFET that causes the SS to be less than the theoretical limit of 60 mV/dec, which leads to improved ON-state current and lower OFF-state leakage current [62].

The negative capacitance in the gate oxide layer helps to overcome the potential barrier and lowers the required gate voltage to achieve a given current level. In addition, the incorporation of a ferroelectric layer in the gate oxide results in improved gate control of the channel charge and a higher current drive. This leads to better drain current and good subthreshold swing in NC-L-TFETs as compared to L-TFETs.

7.5.7 ANALYSIS AND COMPARISON OF GATE-TO-DRAIN CAPACITANCE (C_{gd})

C_{gd} is the capacitance between the gate and drain terminals of a FET, which represents the coupling between the gate and drain terminals. The C_{gd} capacitance affects the device's performance in high-frequency applications, such as radiofrequency (RF) circuits, as it can impact the gain and the stability of the device [63]. In Figure 7.15, the circle symbolrepresents the gate-to-drain capacitance value of a NC-L-TFET with respect to the voltage applied to the device, whereas the black lines indicate the gate-to-drain capacitance value for a conventional L-TFET. In this figure, we observe that the C_{gd} value of the NC-L-TFET is relatively higher than that of the conventional L-TFET.

7.5.8 ANALYSIS AND COMPARISON OF GATE-TO-SOURCE CAPACITANCE (C_{gs})

The C_{gs} of a transistor is defined as the capacitance between the gate and source terminals of a transistor when the drain voltage is zero. In Figure 7.16, we observe that NC-L-TFETs have lower C_{gs} compared to a conventional L-TFET. As the voltage increases from 0 to 1 V, the rate at which the C_{gs} decreases is higher for NC-L-TFETs compared to L-TFETs.

7.5.9 ANALYSIS AND COMPARISON OF TRANSCONDUCTANCE (gm)

"Transconductance" in TFETs refers to the ratio of the change in the I_D with respect to the V_{GS} of the device [64]. It is a measure of the ability of the device to amplify an input signal and is an important parameter in the design of low-power, high-frequency analog ICs [65]. Transconductance can be enhanced in TFETs by optimizing the device's geometry, doping concentration, and gate oxide thickness [66]. TFETs have a large ratio of transconductance, which allows them to work at low power and low OFF-currents, but ON-currents are highly reduced due to low transconductance [64]. Conduction band, valence band, and electron-and-hole QFL profiles along the channel at various bias conditions can be used to illustrate how the device functions [67].

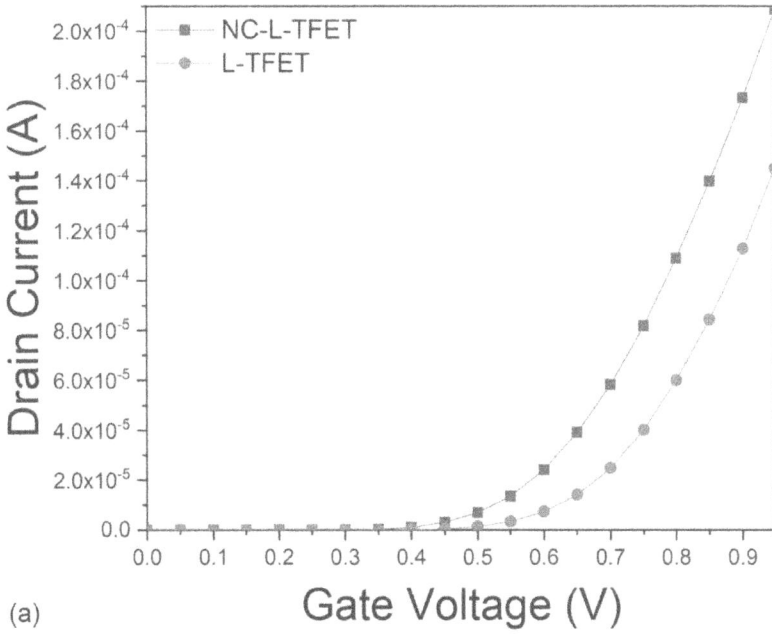

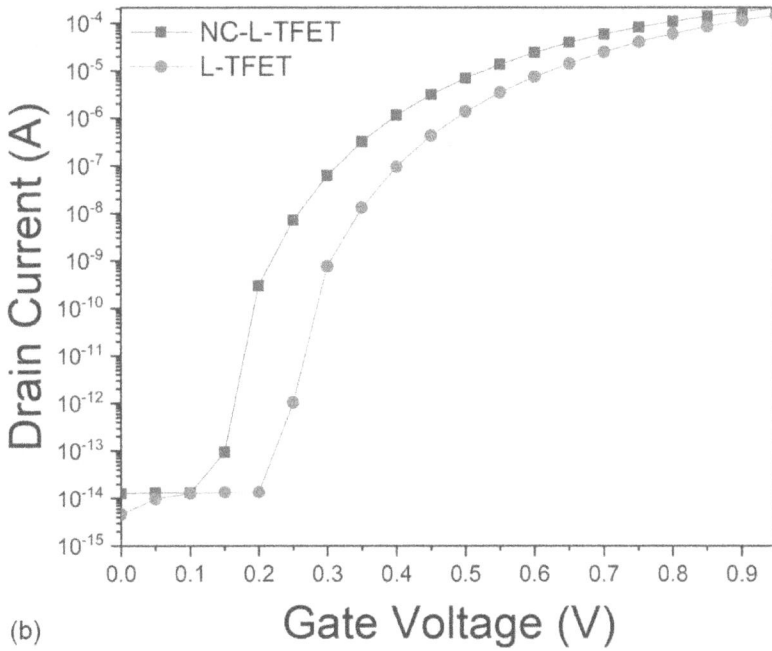

FIGURE 7.14 (a) I_d-V_g characteristics on a linear scale and (b) I_d-V_g characteristics on a log scale.

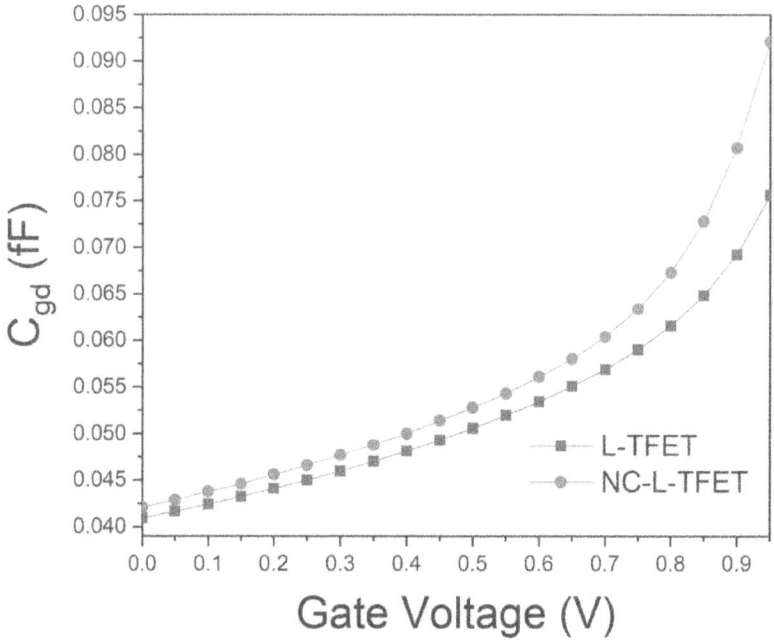

FIGURE 7.15 Overlay plot of the gate-to-drain capacitance of an L-TFET (rectangle symbol) and NC-L-TFET (circle symbol).

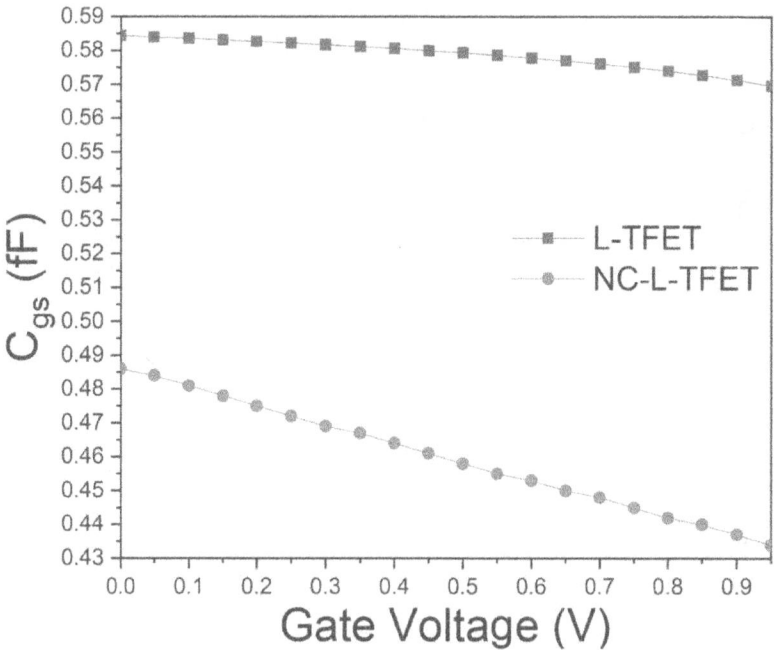

FIGURE 7.16 Overlay plot of the gate-to-source capacitance of an L-TFET and NC-L-TFET.

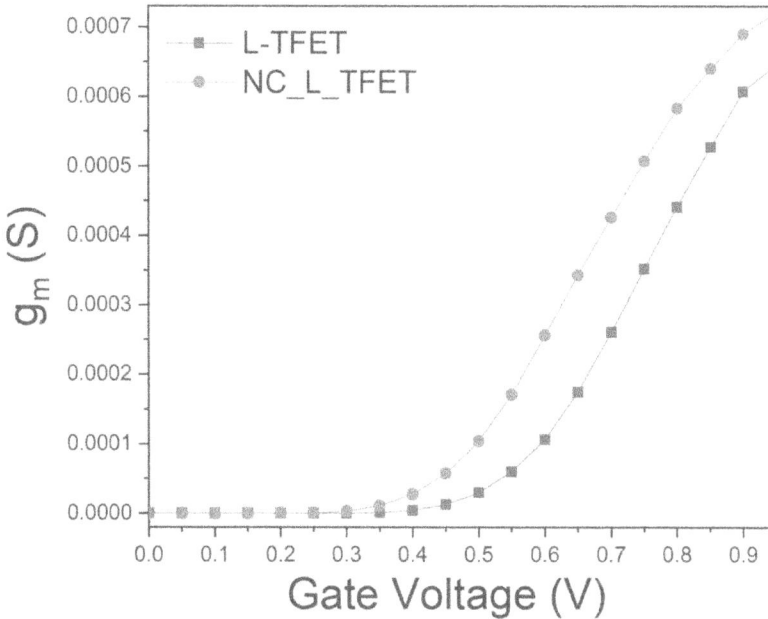

FIGURE 7.17 Overlay plot of transconductance (gm) of an L-TFET and NC-L-TFET.

Transconductance is the change in drain current with respect to voltage. Transconductance can be calculated by taking the differentiation of drain current and gate voltage. From Figure 7.17, we see that the transconductance of NC-L-TFETs is higher than that of L-TFETs, meaning that the proposed device can be a good amplifier for electronic signals as it has the capability to produce a higher difference at a particular applied voltage.

7.6 DISCUSSION ON FERROELECTRIC MODELS USED

7.6.1 LANDAU'S THEORY OF FERROELECTRICITY

Landau's theory of ferroelectricity is a mathematical framework used to describe the behavior of ferroelectric materials. It was developed by the Russian physicist Lev Landau in the 1930s. The theory describes how ferroelectric materials, with the help of an external electric field, can change the spontaneous polarization. Landau's theory uses a mathematical expression called a "Landau potential" to describe the energy of the material as a function of the polarization. The Landau potential has terms that represent the energy associated with the spontaneous polarization, the energy associated with an external electric field, and the energy associated with the interaction between neighboring atoms or molecules in the material. By minimizing the Landau potential, one can determine the equilibrium polarization of the material as a function of temperature, pressure, and other external parameters. Landau's theory has been widely used to study the properties of ferroelectric materials and to design new ferroelectric devices [68].

7.6.2 THE FERROELECTRIC PERMITTIVITY MODEL

The behavior of ferroelectric materials in an applied electric field is described by the ferroelectric permittivity model. It considers the hysteresis effect, which is the dependence of the polarization of the material on the history of the electric field, and the nonlinear nature of the material's response to the electric field. The model is based on Landau's theory of ferroelectricity and includes parameters such as the coercive field, remnant polarization, and saturation polarization. The ferroelectric permittivity model is commonly used in the simulation of ferroelectric devices, like FeRAMs and ferroelectric field-effect transistors (FeFETs). Landau's theory of ferroelectricity provides a framework for understanding the behavior of ferroelectric materials and their response to external fields. It defines temperature, external field, and internal strain as functions of ferroelectric polarization [69, 70].

The ferroelectric permittivity model is a mathematical model that uses the Landau theory to describe the polarization and permittivity of ferroelectric materials. It considers the effects of hysteresis and nonlinearity and links polarization to the electric field and permittivity to temperature and polarization. In other words, the ferroelectric permittivity model is a practical implementation of Landau's theory, which allows for the calculation of the permittivity and polarization of ferroelectric materials in specific situations. Poisson's equation relates the space charge density and electrostatic potential by:

$$div(\varepsilon \nabla \psi) = -\rho \qquad (7.2)$$

where ε is the local permittivity, ρ is the local space charge density, and ψ is the electrostatic potential.

In the MODELS statement, add the *FERRO* parameter to activate the ferroelectric model. The following functional form is given to the permittivity in this model, which is employed in Poisson's equation:

$$\varepsilon(E) = FERRO.EPSF + FERRO.PS * 2\delta * sech \ 2\left[(E - FERRO.EC)/2\delta\right] \quad (7.3)$$

where E stands for the electric field, *FERRO.EPSF* is the permittivity, and δ is defined as follows:

$$\delta = FERRO.EC \left[\log\left((1 + (FERRO.PR \ / \ FERRO.PS))\right) / 1 - (FERRO.PR \ / \ FERRO.PS)\right)\right]^{-1}$$
$$(7.4)$$

where the MATERIAL statement's parameters *FERRO.EPSF*, *FERRO.PS*, *FERRO. PR*, and *FERRO.EC* are variables, and their values can be altered.

With respect to the Silvaco ATLAS simulation, *FERRO.EPSF*, *FERRO.PR*, *FERRO.PS*, and *FERRO.EC* are commonly used parameters that relate to the ferroelectric material properties. Here is a brief explanation of each parameter [48]:

- *FERRO.EPSF*: This parameter represents the static or zero-frequency relative permittivity of the ferroelectric material. It determines the amount of electrical energy stored per unit volume in the ferroelectric material when the external electric field is zero. The unit of *FERRO.EPSF* is farads per meter (F/m).

- *FERRO.PS*: The ferroelectric material's spontaneous polarization is represented by this parameter. That means it is the polarization that exists in ferroelectric materials when the external electric field is zero. The unit of *FERRO.PS* is coulombs per square meter (C/m^2).

- *FERRO.PR*: This parameter represents the piezoelectric coefficient of the ferroelectric material. It describes the relationship between the mechanical strain and the resulting polarization in the material. The unit of *FERRO.PR* is coulombs per square meter per strain (C/m2·m).

- *FERRO.EC*: This parameter represents the critical electric field of the ferroelectric material. To switch the direction of polarization, the amount of electric field is given by this parameter. The unit of *FERRO.EC* is volts per meter (V/m).

These variables are used to characterize how ferroelectric materials behave in electrical devices, including transistors, sensors, and ferroelectric memories. They are important for accurately modeling the electrical properties of ferroelectric materials in simulation software like Silvaco ATLAS [48, 71].

Landau's theory of ferroelectricity gives the information for understanding the behavior of ferroelectric materials and their response to external fields. It describes how temperature, external field, and internal strain affect ferroelectric polarization. The ferroelectric permittivity model is a mathematical model that uses the Landau theory to describe the polarization and permittivity of ferroelectric materials. With the effects of hysteresis and nonlinearity considered, it links polarization to the electric field and permittivity to temperature and polarization. In other words, the ferroelectric permittivity model is a practical implementation of Landau's theory, which allows for the calculation of the permittivity and polarization of ferroelectric materials in specific situations.

7.7 CONCLUSION

This chapter thoroughly examined negative capacitance on L-TFETs. We have analyzed the electrical characteristics and performance of NC-L-TFETs and compared them with the conventional L-TFETs. Our results show that NC-L-TFETs have better electrical characteristics, such as higher ON-current and lower subthreshold swing. The negative-capacitance effect has lowered the effective threshold voltage and boosted the subthreshold swing, along with further improvements in the electric field, band bending, and the capacitance values (both C_{GS} and C_{GD}). In addition, the research into the ferroelectric permittivity model and Landau's theory of ferroelectricity has aided in our understanding of the fundamental processes that underlie the effects of negative capacitance in ferroelectric materials. Enhancing the performance of nanoelectronics devices through the incorporation of ferroelectric materials into NC-L-TFETs has proven to be a promising strategy. The use of advanced simulation tools like Silvaco ATLAS has enabled researchers to study the behavior of L-TFETs in detail and optimize their performance. With the continuous development of new materials and fabrication techniques, there is immense potential for the L-TFET with ferroelectric material to become a leading material for future nanoelectronics devices. This chapter has provided a view on the L-TFET with ferroelectric material and its properties, various types of transistors, ferroelectric materials, and simulation tools.

REFERENCES

1. R. S. Williams, "What's next?," *Comput Sci Eng*, vol. 19, no. 2, pp. 7–13, 2017, doi: 10.1109/MCSE.2017.31.
2. W. Choi, N. Choudhary, G. H. Han, J. Park, D. Akinwande and Y. H. Lee, "Recent development of two-dimensional transition metal dichalcogenides and their applications," *Materials Today*, vol. 20, no. 3, pp. 116–130, 2017, doi: 10.1016/j.mattod.2016.10.002.
3. A. Romeo and S. P. Lacour, "Stretchable metal oxide thin film transistors on engineered substrate for electronic skin applications," in *Proceedings of the Annual International Conference of the IEEE Engineering in Medicine and Biology Society, EMBS*, Institute of Electrical and Electronics Engineers Inc., Nov. 2015, pp. 8014–8017, doi: 10.1109/EMBC.2015.7320252.
4. S. Priyanka, M. Singh and Panchore, "Dopingless-TFET leaky-integrated-fire (LIF) neuron for high-speed energy efficient applications," *IEEE Trans Nanotechnol*, vol. 21, pp. 110–117, 2022, doi: 10.1109/TNANO.2022.3151241
5. J. Luo, C. Chen, Q. Huang and R. Huang, "A biomimetic tunnel FET-based spiking neuron for energy-efficient neuromorphic computing with reduced hardware cost," *IEEE Trans Electron Devices*, vol. 69, no. 2, pp. 882–886, 2022, doi: 10.1109/TED.2021.3131633.
6. M. C. Chen *et al*, "A 3-D stackable maskless embedded metal-gate thin-film-transistor nanowire for use in bioelectronic probing," *IEEE Trans Electron Devices*, vol. 61, no. 3, pp. 897–901, 2014, doi: 10.1109/TED.2014.2298462.
7. M. K. Sarma, P. K. Sharma and J. C. Dutta, "Enzyme modified field effect transistors for applications in bioelectronic sensors: Modelling and technology," in *International Conference on Electrical, Electronics, and Optimization Techniques, ICEEOT 2016*, Institute of Electrical and Electronics Engineers Inc., Nov. 2016, pp. 409–414, doi: 10.1109/ICEEOT.2016.7755471.
8. S. Salvi, S. Pahar and Y. Kadale, "Smart glass using IoT and machine learning technologies to aid the blind, dumb and deaf," *Journal of Physics: Conference Series*, 2021, doi: 10.1088/1742-6596/1804/1/012181.
9. V. Mishra *et al.*, "Indian nanoelectronics users' program: An outreach vehicle to expedite nanoelectronics research in India," in *Biennial University/Government/Industry Microelectronics Symposium - Proceedings*, 2010, doi: 10.1109/UGIM.2010.5508858.
10. "The_Transistor_at_75_The_past_present_and_future_of_the_modern_worlds_most_important_invention", https://eds.ieee.org/images/files/newsletters/Newsletter_July22.pdf.
11. R. A. Street *et al*, "From printed transistors to printed smart systems," *Proceedings of the IEEE*, vol. 103, no. 4, pp. 607–618, 2015, doi: 10.1109/JPROC.2015.2408552.
12. S. Datta, "Invited tutorial: Advanced CMOS transistor technology: Past, present and future," Institute of Electrical and Electronics Engineers (IEEE), May 2011, pp. 1–1, doi: 10.1109/wmed.2011.5767289.
13. A. Van Der Ziel, A. H. Pawlikiewicz, X. Zhang and X. Zhang, "Location of 1/f noise sources in BJT's and HBJT's—I. Theory," *IEEE Trans Electron Devices*, vol. 33, no. 9, pp. 1371–1376, 1986, doi: 10.1109/T-ED.1986.22672.
14. "Bipolar Junction Transistor: Definition, Construction, Types, Function, Application, and FAQs." https://byjus.com/physics/bipolar-junction-transistor/ (accessed Apr. 16, 2023).

15. J. T. Maupin, "The tetrode power transistor," *IRE Transactions on Electron Devices*, vol. 4, no. 1, pp. 1–5, 1957, doi: 10.1109/T-ED.1957.14192.

16. "Field Effect Transistors." https://www.tutorialspoint.com/semiconductor_devices/semiconductor_devices_field_effect_transistors.htm (accessed Apr. 16, 2023).

17. C. S. P. Chuang, K. Y. G. Chen, Y. R. R. Hung, T. C. Kuo and C. C. T. Huang, "Forward-voltage-tunable schottky-integrated trench MOSFETs," in *Proceedings of the International Symposium on Power Semiconductor Devices and ICs*, Institute of Electrical and Electronics Engineers Inc., 2014, pp. 159–162, doi: 10.1109/ISPSD.2014.6856000.

18. "MOSFET - Construction, Signals, Types, Working & Uses." https://testbook.com/learn/physics-mosfet/ (accessed Apr. 16, 2023).

19. N. M. Biju and R. Komaragiri, "Dual gate enhancement-mode JFET (DG-JFET) for ultra low power applications," in *2012 IEEE Students' Conference on Electrical, Electronics and Computer Science: Innovation for Humanity, SCEECS 2012*, 2012, doi: 10.1109/SCEECS.2012.6184726.

20. "Junction Field Effect Transistor or JFET Tutorial." https://www.electronics-tutorials.ws/transistor/tran_5.html (accessed Apr. 16, 2023).

21. S. Dong *et al.*, "Threshold voltage improvement scheme for high-voltage IGBT," in *2019 IEEE 3rd International Conference on Electronic Information Technology and Computer Engineering, EITCE 2019*, Institute of Electrical and Electronics Engineers Inc., Oct. 2019, pp. 390–393, doi: 10.1109/EITCE47263.2019.9094813.

22. S. Abedinpour and K. Shenai, "Insulated gate bipolar transistor," *Power Electronics Handbook: Devices, Circuits, and Applications, Third Edition*, pp. 73–90, 2010, doi: 10.1016/B978-0-12-382036-5.00005-7.

23. W.-T. Chang, S.-W. Lin, C.-T. Shih and W.-K. Yeh, "Back Bias Modulation of UTBB FDSOI, Bulk FinFET, and SOI FinFET," Institute of Electrical and Electronics Engineers (IEEE), Oct. 2016, pp. 1–2, doi: 10.1109/inec.2016.7589260.

24. "What is a FinFET? – Benefits & How it Works | Synopsys." https://www.synopsys.com/glossary/what-is-a-finfet.html (accessed Apr. 16, 2023).

25. D. S. Yadav, D. Sharma, B. R. Raad and V. Bajaj, "Impactful study of dual work function, underlap and hetero gate dielectric on TFET with different drain doping profile for high frequency performance estimation and optimization," *Superlattices Microstruct*, vol. 96, pp. 36–46, 2016, doi: 10.1016/j.spmi.2016.04.027

26. "Tunneling Field Effect Transistors." http://large.stanford.edu/courses/2012/ph250/esfandyarpour1/ (accessed Apr. 16, 2023).

27. G. Larrieu and X. L. Han, "Vertical nanowire array-based field effect transistors for ultimate scaling," *Nanoscale*, vol. 5, no. 6, pp. 2437–2441, 2013, doi: 10.1039/C3NR33738C.

28. A. D. Franklin, M. C. Hersam and H. S. P. Wong, "Carbon nanotube transistors: Making electronics from molecules," *Science*, vol. 378, no. 6621, pp. 726–732, 2022, doi: 10.1126/SCIENCE.ABP8278.

29. "8 Spin Field-Effect Transistor." https://www.iue.tuwien.ac.at/phd/osintsev/disserch8.html (accessed Apr. 16, 2023).

30. A. I. Khan, *Negative capacitance for ultra-low power computing*. University of California, Berkeley. 2015.

31. G. Pahwa, A. D. Gaidhane, A. Agarwal and Y. S. Chauhan, "Assessing negative-capacitance drain-extended technology for high-voltage switching and analog applications," *IEEE Trans Electron Devices*, vol. 68, no. 2, pp. 679–687, 2021, doi: 10.1109/TED.2020.3044554.

32. P. Ghosh and B. Bhowmick, "Investigation of electrical characteristics in a ferroelectric l-patterned gate dual tunnel diode TFET," *IEEE Trans Ultrason Ferroelectr Freq Control*, vol. 67, no. 11, pp. 2440–2444, 2020, doi: 10.1109/TUFFC.2020.2999826.

33. A. Sharma, "Sub-10nm Transistors for Low Power Computing: Tunnel FETs Sub-10nm Transistors for Low Power Computing: Tunnel FETs and Negative Capacitance FETs and Negative Capacitance FETs," 2018. [Online]. Available: https://docs.lib.purdue.edu/open_access_dissertations/1884

34. P. Dasika, "Modeling and Simulation of Negative Capacitance MOSFETs," MS Thesis, 2018.

35. M. Kobayashi, K. Jang, N. Ueyama and T. Hiramoto, "Negative capacitance for boosting tunnel FET performance," *IEEE Trans Nanotechnol*, vol. 16, no. 2, pp. 253–258, 2017, doi: 10.1109/TNANO.2017.2658688.

36. P. Singh, D. P. Samajdar and D. S. Yadav, "A Low Power Single Gate L-shaped TFET for High Frequency Application," in *2021 6th International Conference for Convergence in Technology, I2CT 2021*, Institute of Electrical and Electronics Engineers Inc., Apr. 2021, doi: 10.1109/I2CT51068.2021.9418075.

37. S. M. Said, M. F. M. Sabri, F. Salleh and M. Ramadan, "Ferroelectrics and their applications," *Encyclopedia of Smart Materials*, vol. 3, pp. 495–506, 2021, doi: 10.1016/B978-0-12-815732-9.00105-4.

38. "Ferroelectric Materials: What Are They? (With Examples) | Electrical4U," *https://www.electrical4u.com/*, Accessed: Apr. 16, 2023. [Online]. Available: https://www.electrical4u.com/ferroelectric-materials/

39. M. Dawber, K. M. Rabe and J. F. Scott, "Physics of thin-film ferroelectric oxides," *Rev Mod Phys*, vol. 77, no. 4, 1083, 2005, doi: 10.1103/RevModPhys.77.1083.

40. "Ferroelectric Materials – Theory, Properties and applications." https://www.indiastudychannel.com/resources/117961-Ferroelectric-Materials-Theory-Properties-and-applications.aspx (accessed Apr. 16, 2023).

41. "What is Ferro-electricity and what are the properties of ferroelectric materials - Electrical - Industrial Automation, PLC Programming, scada & Pid Control System." https://forumautomation.com/t/what-is-ferro-electricity-and-what-are-the-properties-of-ferroelectric-materials/7412 (accessed Apr. 16, 2023).

42. I. H. Smaili, "Design and Simulation of Short Channel Si:HfO2 Ferroelectric Design and Simulation of Short Channel Si:HfO2 Ferroelectric Field Effect Transistor (FeFET) Field Effect Transistor (FeFET)." [Online]. Available: https://scholarworks.rit.edu/theses

43. R. Kawale, A. Pachouri, Y. Singh and L. Agarwal, "Design of TFET with Ferroelectric Gate Material for Low Power Applications," in *Proceedings of the 1st International Conference on Advances in Computing and Future Communication Technologies, ICACFCT 2021*, Institute of Electrical and Electronics Engineers Inc., 2021, pp. 110–113, doi: 10.1109/ICACFCT53978.2021.9837374.

44. Darmis, N., Alam, A. Z., & Hashim, M. M. (2021). Ferroelectric behavior and NCFETs-TCAD simulation. *Asian Journal of Electrical and Electronic Engineering*, *1*(1), 30–41.

45. P. Sengupta, R. Guo and A. Bhalla, "Modeling and simulation of novel ferroelectric gate stack in MOSFET for enhanced device performance," in *2018 IEEE 13th Nanotechnology Materials and Devices Conference, NMDC 2018*, Institute of Electrical and Electronics Engineers Inc., Jan. 2019, doi: 10.1109/NMDC.2018.8605823.

46. Y. S. Chauhan, "Fundamentals and Recent Progress in Negative Capacitance Transistors." [Online]. Available: http://home.iitk.ac.in/~chauhan/

47. G. Pahwa and Y. S. Chauhan, "Evaluation of Ferroelectric Negative Capacitance Technology for RF and High Voltage Applications Modeling of STT-MRAM devices View project ASM-GaN-HEMT: Industry Standard SPICE Model for GaN HEMTs View project." [Online]. Available: https://www.researchgate.net/publication/346771475

48. "ATLAS User's Manual. Device Simulation Software," 2004. [Online]. Available: www.silvaco.com

49. Y. Kadale, P. Singh and D. S. Yadav, "Comprehensive Analysis of DG-TFET with Ferro Electric Material," in *2023 7th International Conference on Computing Methodologies and Communication (ICCMC)*, IEEE, Feb. 2023, pp. 1474–1479, doi: 10.1109/ICCMC56507.2023.10083683.
50. M. Massarotto *et al.*, "Bridging Large-Signal and Small-Signal Responses of Hafnium-Based Ferroelectric Tunnel Junctions," in *2023 35th International Conference on Microelectronic Test Structure (ICMTS)*, IEEE, Mar. 2023, pp. 1–6, doi: 10.1109/ICMTS55420.2023.10094178.
51. "Quasi Fermi level - Wikipedia." https://en.wikipedia.org/wiki/Quasi_Fermi_level (accessed Apr. 19, 2023).
52. "What is quasi fermi level? | Technology Trends." https://www.primidi.com/what_is_quasi_fermi_level (accessed Apr. 19, 2023).
53. "Energy band diagram and layer structure for an n TFET consisting of an… | Download Scientific Diagram." https://www.researchgate.net/figure/Energy-band-diagram-and-layer-structure-for-an-n-TFET-consisting-of-an-n-th-source-S-p_fig1_228723526 (accessed Apr. 19, 2023).
54. Y. Shan, Y. Liu, H. Zheng and Z. Peng, "Simulation of ionizing/displacement synergistic effects on NPN bipolar transistors irradiated by mixed neutrons and gamma rays," *Science and Technology of Nuclear Installations*, vol. 2022, 2022, doi: 10.1155/2022/1283926.
55. H. P. D. Lanyon, "Shallow level recombination current dominance in transistor betas," *IEEE Electron Device Letters*, vol. 14, no. 2, pp. 49–50, 1993, doi: 10.1109/55.215102.
56. Spieler, H. Introduction to radiation detectors and electronics. *VI. Position-Sensitive Detectors.* 1998.
57. W. Y. Choi, B. G. Park, J. D. Lee and T. J. K. Liu, "Tunneling field-effect transistors (TFETs) with subthreshold swing (SS) less than 60 mV/dec," *IEEE Electron Device Letters*, vol. 28, no. 8, pp. 743–745, Aug. 2007, doi: 10.1109/LED.2007.901273.
58. A. Godoy, J. A. López-Villanueva, J. A. Jiménez-Tejada, A. Palma and F. Gámiz, "A simple subthreshold swing model for short channel MOSFETs," *Solid State Electron*, vol. 45, no. 3, pp. 391–397, 2001, doi: 10.1016/S0038-1101(01)00060-0.
59. "Subthreshold slope - Wikipedia." https://en.wikipedia.org/wiki/Subthreshold_slope (accessed Apr. 20, 2023).
60. G. H. Oh, J. G. An, S. Il Kim, J. C. Shin, J. Park and T. W. Kim, "Ultralow subthreshold swing 2D/2D heterostructure tunneling field-effect transistor with ion-gel gate dielectrics," *ACS Appl Electron Mater*, 2022, doi: 10.1021/ACSAELM.2C01277.
61. M. Sanaullah and M. H. Chowdhury, "Subthreshold swing characteristics of multilayer MoS2 tunnel FET," *Midwest Symposium on Circuits and Systems*, vol. 2015-September, 2015, doi: 10.1109/MWSCAS.2015.7282101.
62. C. W. Yeung, "Steep On/Off Transistors for Future Low Power Electronics," 2014. [Online]. Available: http://www.eecs.berkeley.edu/Pubs/TechRpts/2014/EECS-2014-226.html
63. N. Patin, "Case study – The variable speed drive," *Power Electronics Applied to Industrial Systems and Transports*, vol. 2, no. 2, pp. 215–251, 2015, doi: 10.1016/B978-1-78548-001-0.50006-X.
64. P. Long, Huang, J. Z., Povolotskyi, M., Verreck, D., Charles, J., Kubis, T., …, & Calhoun, B. H., A tunnel FET design for high-current, 120 mV operation. In *2016 IEEE International Electron Devices Meeting (IEDM)* (pp. 30–32). IEEE, December 2016.
65. L. Barboni, M. Siniscalchi and B. Sensale-Rodriguez, "TFET-based circuit design using the transconductance generation efficiency {g}-{m}/{I}-{d method," *IEEE Journal of the Electron Devices Society*, vol. 3, no. 3, pp. 208–216, May 2015, doi: 10.1109/JEDS.2015.2412118.

66. W. Y. Choi and J. W. Lee, "Design guideline of tunnel field-effect transistors (TFETs) considering negative differential transconductance (NDT)," *Solid State Electron*, vol. 163, 107659, 2020, doi: 10.1016/J.SSE.2019.107659

67. P. Chaturvedi and N. Goyal, "Effect of gate dielectric thickness on gate leakage in tunnel field effect transistor," In *2012 8th International Caribbean Conference on Devices, Circuits and Systems (ICCDCS)* (pp. 1–4). IEEE pp. 1–4, Mar. 2012.

68. J. Hattori, "Device Simulation of Ferroelectric Field-Effect Transistors Based on the Landau-Khalatnikov Equation," 2019.

69. V. O. Sherman, A. K. Tagantsev and N. Setter, "Model of a low-permittivity and high-tunability ferroelectric based composite," *Appl Phys Lett*, vol. 90, no. 16, 162901, Apr. 2007.

70. R. Placeres-Jiménez, J. P. Rino and J. A. Eiras, "Modeling ferroelectric permittivity dependence on electric field and estimation of the intrinsic and extrinsic contributions," *J Phys D Appl Phys*, vol. 48, no. 3, 035304, 2015.

71. V. Yerraguntla. *Simulation and experiments towards ferroelectric-gate heterojunction field effect transistors*, Doctoral dissertation, Chairperson of the Committee, 2010.

8 Thermal Behavior of Si-Doped MoS$_2$-Based Step-Structure Double-Gate TFETs

Priya Kaushal and Gargi Khanna

8.1 INTRODUCTION

Complementary metal–oxide–semiconductor (CMOS) field-effect transistors (FETs) have traditionally been aggressively downscaled to meet the rising need for integrated circuits' shrinking, operational speed, value, and varied functionality. A continuous decrease in transistor size is gravely affected by the high power dissipation present in CMOS technology, which has produced devices with dimensions much smaller than 100 nm [1, 2]. For metal–oxide–semiconductor field-effect transistors (MOSFETs), the cumulative impact of the increased leakage current and the wide range of supply voltage result in this increased power loss [1]. More leakage current was present because of the short-channel effects (SCEs), which can be mitigated to a certain extent by novel transistor design [3, 4]. Otherwise, a higher supply voltage must be used in order to provide a high drive current and, as a result, a faster operating speed. To lower a transistor's supply voltage without compromising its performance, its subthreshold swing must be as low as is practical. This will allow for a quick shifting between the off and on zones. In contrast, the lowest subthreshold swing of a semiconductor device is controlled with the restriction of 60 mV per decade at ambient temperature, which is substantially higher for operational nanotechnology transistors [2–4]. Research on substitute FETs with an incredibly tiny subthreshold swing is driven by these shortcomings in the current nanoscale MOSFET devices. The nanoscale MOSFET has been overtaken by the TFET [5, 6]. Due to the special carrier injection technique used in TFET designs, which is band-to-band tunneling (BTBT) of charge particles [7, 8], it is feasible to get a subthreshold swing of 60 mV per decade. The TFET is also capable of retaining a negligible leakage current flow in nanotechnology and is highly resistant to device downscaling. However, BTBT influences charge transfer in TFETs to generate an ON-state current (I_{on}) that is frequently several orders of magnitude lower than in traditional transistors. It is important to note that the possibility of BTBT, which is highly influenced by the electrostatics of the device and the characteristics of the semiconductor material [2], affects both the subthreshold swing and I_{on} of TFETs. In order to actually obtain smaller subthreshold swing and desirable I_{on}, the correct TFET device design and materials are both essential. As a result, growing research activity on TFET has been noticed over the previous decade, paving the way for thorough theoretical and design-level information. The selection of semiconductor material has always been one of the primary considerations for such techniques; it resulted in the development of several performance-improving strategies for TFETs.

Since the first production of graphene in 2004, the study of two-dimensional (2D) semiconducting materials including one or a few atomic layers has advanced rapidly. An increasing group of 2D materials has an unreported physical procedure that is produced by the materials' high quantum mechanical effects, fewer particle-scattering events, and expanded correlations [9, 10]. The exceptional electrical, optical, and magnetic properties of the 2D materials are notably distinct from those of their three-dimensional (3D) bulk materials [11, 12]. In addition, fast-evolving synthesis methods allow the continuous and controlled production of 2D materials of very excellent quality, with an atomic-thin size of layers on a range of substrates. In addition, there have been considerable improvements in relevant theoretical research using the computational nanomaterials approach, which has led to accrual of knowledge of the basic physics of those nanostructured materials. First-principles calculations utilizing density functional theory (DFT) are emerging as a major approach in investigating the characteristics of nanostructured materials and their related interactions with various external forces, successfully supplementing the present experimental findings [13–15]. These elements played a role in successful demonstrations of electrical, optoelectronic, and spintronic devices based on 2D materials that frequently display noticeable performance enhancements over the bulk materials [16].

As materials for FET channels, 2D materials are currently being extensively explored. These 2D transistors typically have benefits like good electrostatic control of the gate electrode because of a higher surface-to-volume proportion, the best electrical conductivity because of ballistic movement, and perfect surface areas that guarantee improved structural relationships with the insulators. In addition, they have tunable electrical properties that depend on layer and stacking and provide the transistor design with more flexibility [1, 17]. With the use of these innovative features, 2D nanomaterials for TFET creation might be developed that combine the benefits of increased electrostatic stability and tunneling barrier engineering. In conclusion, the use of TFETs with 2D material as a channel material has significantly increased in recent years. Particularly, 2D TFETs are regarded to be the most likely option for nanodevice size, in which channel lengths are approximately 40 nm, causing ballistic carrier movement in the channel. All nanoscale 2D TFETs must take specific design considerations because this results in significantly changed device physics. In consideration of the various 2D semiconductors' distinctive electrical properties, there has been a notable advancement in device and material co-optimization, involving a variety of unique techniques for designing devices as well as techniques for developing materials for 2D TFETs at the nanoscale. It is significant to emphasize that the primary source of this rise continues to be theoretical and simulation-based research. [18].

Ghosh et al. [19] have investigated the gate-controlled direct BTBT current in a single-layer TFET that utilizes a transition-metal dichalcogenide (TMD) channel. For this, the 2D sheet shapes of five TMD materials—MoS_2, $MoSe_2$, $MoTe_2$, WS_2, and WSe_2—are taken into consideration. DFT was first utilized to analyze the real and imaginary band structures of specific TMD materials. The Wentzel–Kramers-Brillouin (WKB) approximation was then used to compute the gate-controlled current. According to reports, all five TMDs allow direct BTBT in these configurations and offer an average I_{on} and subthreshold swing of 150 A/m (with $V_{dd} = 0.1$ V) and 4 mV/dec, respectively. In addition, it has been demonstrated how strain affects the complicated band structures and the behaviors of TFETs situated on TMDs. A particular tensile strain has been found to be advantageous for the enhancement of ON-current performances [20]. The performance of TMD (MoS$_2$, WSe$_2$, MoTe$_2$, and WTe$_2$) material-based TFETs has been examined using atomistic simulations. It has been demonstrated that an atomically thin channel alone is insufficient for high-performance TFET devices; the choice of channel material and design of the transistor are essential for achieving high I_{on}. Since the drain-induced barrier lowering (DIBL) and subthreshold swing values are substantially lower for TMD TFETs than for Si MOSFETs (by a ratio of one-third), TMD TFETs display reduced SCEs. According to our simulations, 2D materials with effective masses that are lower and bandgaps that are less extreme (0.5–0.7 eV) are better suited for high-performance TFETs [21]. By applying a nonequilibrium Green function (NEGF) in atomistic simulations, the ballistic transport feature of TMD TFETs is examined. The maximum limit of the I_{on}/I_{off} ratio in TMD TFETs depends on the crystal orientation, with a zigzag direction having a greater maximum limit than an armchair direction. This is the method to create a sheet of the material. The atomistic organization in the transport direction and the features of the subbands are related to the orientation-dependent transport. The current can be boosted at all V_{gs} tested using a greater source–drain dopant, while lowering the I_{on}/I_{off} relation, whereas a thinner sheet of dielectric material might simultaneously boost saturation current and diminish minimum current. Investigations on the scaling behavior of TMD TFETs revealed that the gate length widens the current gap between the two directions in the OFF-state. Due to a desirable combination of physical characteristics, such as bandgap tunability and moderately high electron mobility, semiconducting MoS$_2$ is one material that is attracting more interest. In terms of experiments, scientists have concentrated on the practical uses of 2D MoS$_2$, particularly the creation of FETs and the insignificant OFF-current. Monolayer MoS$_2$ is particularly appealing for transparent and flexible electronics due to its flexibility, stretchability, and optical transparency.

Khan et al. [22] have examined the current transport properties of TFETs based on MoS$_2$. We constructed an analytical model and verified the results with a computationally created NEGF model. The drain current I_{ds}, derived from an analytical model, is discovered to be higher than that obtained from the numerical computation. The subthreshold swing of both approaches, however, is essentially identical, with a minor deviation due to the WKB approximation. Scattering inside the channel has been researched to help explain the current mismatch between the linear and saturation areas. At 300 K, authors have determined the mobility as well as the ionized impurity. The results of the mathematical model with scattering are in satisfactory correlation with those of the NEGF simulator. Radisavljevic et al. [23] have developed single-layer MoS$_2$ transistors. They have observed I_{on}/I_{off} ratios of 10^8 and ultralow power loss at room temperature, as well as mobility of the single-layer MoS$_2$ not less than 200 cm^2 V^{-1} s^{-1}, which is comparable to the structure of graphene nanoribbons. As single-layer MoS$_2$ consists of a direct bandgap, it has the potential to cause BTBT transport in TFETs, which utilize very little power compared to conventional transistors. Monolayer MoS$_2$ could be used as a substitute for graphene in applications like optoelectronics and energy harvesting that call for thin, transparent semiconductors. Khan et al. [24] have proposed a double-gate (DG)-TFET based on MoS$_2$ with a current transport model that is derived by using analytical and numerical methods. The Landauer current transport mechanism has been used to solve and construct a 2D potential distribution model for a heterojunction with the appropriate energy level. On the basis of the NEGF formalism (approach or method), the output attributes and the quantum mechanical approach have been compared. At $V_{ds} = 0.5$ V, the subthreshold swing measured by NEGF and analytical procedures is 12.6 mV/decade and 10 mV/decade, respectively, with I_{on}/I_{off} ratios of 10^7 and 10^6, respectively. We incorporated the scattering phenomenon into our calculations and demonstrated that carrier movement is diffusive for 20 nm of channel length value. Furthermore, scattering displays strong concordance with the atomistic models and decreases the I_{on} by an order of magnitude. Finally, the mobility and ionized impurity concentrations were 27.2 cm^2/Vs and 1.17×10^{13}/cm^2, respectively.

One approach to change the structure is to enhance the functionality of TFETs. To boost the current and subthreshold swing, an L-shaped channel expands the source–channel terminal area. However, there is severe ambipolarity in the L-shaped TFET [22, 25]. Meshkin et al. [26] have presented a novel double Step Structure (SS)-TFET with a linear doping profile in the channel area. Since BTBT of carrier movement happens fairly close to the source–channel junction, the addition of a step-shaped structure to the transistor boosts the tunneling junction area significantly. The step-shaped design at the channel–drain junction decreases the distribution of the electric field near the drain intersection, which helps to inhibit the ambipolar tendency. Furthermore, by extending the effective length of the channel area, the double-step structure of the body restricts ambipolar conduction. The device properties are therefore significantly influenced by the height of the step in the presented transistor. In the channel region, the doping profile is thought to be linear. At the source side, the doping concentration is at its highest value and linearly falls to reach its lowest value down the x-axis. Numerical simulation results demonstrate that employing the linear doping distribution at the channel enhances device behavior by decreasing the tunneling barrier height at the source–channel interface. The ideal channel length for the linear-doped region is chosen to optimize the I_{on} without significantly reducing the I_{off}. A comparison of the proposed device with a standard silicon-on-insulator (SOI)-TFET has been shown. Dutta et al. [27] have presented a newly designed silicon-based TFET with a channel length of 21 nm. Analytical modeling and optimization of the suggested structure have been done. The TFET with a broken gate structure has extremely small ambipolar current, excellent subthreshold swing, and I_{on} that are analogous to recent architectures of comparable size. By resolving the 2D Poisson's equation, modeling is done to represent the device's surface potential and the electric field, and to model the I_{on} using Kane's generation rate. The analysis' findings are verified against simulations based on Synopsys Sentaurus TCAD's nonlocal tunneling model. Zhang et al. [28] have used a 2D simulation to examine the behaviors of double-gate TFETs with step channel thickness. The mismatch in between the source and drain is introduced by the step channel thickness; as a result, the ambipolar behavior is predicted to be reduced. According to the findings, the step channel TFET significantly reduces ambipolar current when compared to the traditional double-gate TFET. The physical understanding is explored with a full discussion of step channel TFET processes. In finding the ideal structure, the effects of the structural parameters on the onset voltage, subthreshold swing, I_{ds} in the ON-state, and ambipolar region of the drain current are demonstrated. Kaushal et al. [29] have proposed and investigated a novel MoS$_2$-based thickness-engineered TFET. The I_{on} has been improved in the device by utilizing the effect of channel layer thickness variations on the bandgap. The various channel lengths, including 7, 10, 14, and 22 nm, have been compared. Improved I_{on}/I_{off} ratios of 10^{13} have been achieved for transistors of 10 nm size, and the I_{on} has been increased at 1 A/m along with a subthreshold swing of 11.6 mV/decade. Furthermore, the radiofrequency (RF)/analog performance has been investigated.

The number of monolayers substantially influences the characteristics of MoS$_2$; hence, methods that allow for control over the number of monolayers deposited are highly desired. In this chapter, the step-structured DG-TFET is simulated to analyze the number of layer-dependent properties of MoS$_2$ material at different temperatures (i.e., 250 K, 300 K, 400 K, and 500 K).

8.2 DESIGN AND SIMULATION PARAMETERS

In Figure 8.1, as a channel material, a monolayer of MoS$_2$ has been utilized in the structure. In order to perform a first-principles calculation, we first built a monolayer using the Atomistix ToolKit's (ATK) DFT approach. The source and drain portions of the suggested design have varied widths and carrier concentrations. The suggested transistor has a channel width of 15 nm between the source side toward the channel's center and maintains a 5 nm width for the remaining channel in order to achieve thickness engineering along the channel. The source and drain areas are created by the silicon material using doping of $(1\times10^{20}$ cm$^{-3})$ and $(1\times10^{18}$ cm$^{-3})$, respectively, while the intrinsic channel is filled with doping of $(1\times10^{15}$ cm$^{-3})$. HfO$_2$ has been used as the dielectric material. A work function with a value of 4.2 eV has been used. Figure 8.1 shows a schematic for the Si-doped MoS$_2$-SS-DG-TFET. During the simulation phase, some fundamental models were also used for a thorough examination of the effect of temperature on the multiple performance features of the proposed device. The models that included the simulation time of the proposed device are as follows:

- Standard BTBT
- Bandgap narrowing (BGN)
- Trap-assisted tunneling (TAT)
- Shockley–Read–Hall (SRH)
- Concentration-dependent mobility (CONMOB)
- Field-dependent mobility (FLDMOB)
- Auger recombination, auger, and others are some often-used key models.

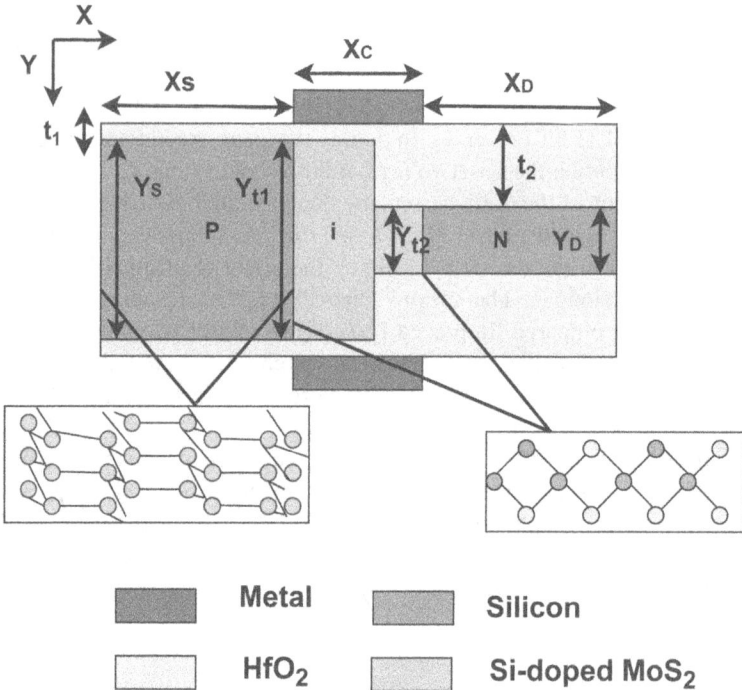

FIGURE 8.1 Two-dimensional (2D) schematic of a Si-doped MoS$_2$-based step-structure double-gate tunneling field-effect transistor (MoS$_2$-SS-DG-TFET).

In addition to these approaches and methods, Newton's numerical approach was also employed to establish the coupling between results driven for better current efficiency. Aside from that, the frequency is fixed to 1 MHz to investigate RF operation. Table 8.1 contains a list of design parameters for the simulated device. The design of the displayed device was created and implemented using a technology computer-aided design (TCAD) simulator.

8.3 RESULTS AND DISCUSSION

Temperature has a tremendous effect on transistor reliability. A transistor's electrical properties change with temperature. Here, we investigate the impact of temperature between 250 K and 500 K on the transfer properties, analog/RF characteristics, and linearity analysis of Si-doped MoS$_2$-SS-DG-TFETs.

The drain current (I_{ds}) of a transistor is determined by the mobility of charge particles, which is a temperature-dependent characteristic. Equation (8.1) provides the relationship between mobility and temperature.

$$\mu_{eff} = \mu_0 \left(\frac{T}{T_0} \right)^{-2} \tag{8.1}$$

TABLE 8.1
Design Parameters of the Proposed Device

Parameters	Silicon-Doped MoS$_2$-Based Step-Structure Double-Gate Tunneling Field-Effect Transistor
Source length (X_S)	20 nm
Channel length (X_C)	10 nm
Drain length (X_D)	20 nm
Source doping (N_S)	1×10^{20} cm^{-3}
Channel doping (N_C)	1×10^{12} cm^{-3}
Drain doping (N_D)	1×10^{18} cm^{-3}
Source width (Y_S)	15 nm
Channel width at source side (Y_{t1})	15 nm
Channel width at drain side (Y_{t2})	5 nm
Drain width (Y_D)	5 nm
Gate work function	4.2 eV
Oxide thickness (t_1)	1 nm
Oxide thickness (t_2)	6 nm

where μ_0 and μ_{eff} represent mobility at T_0 (T_0 is the temperature) and effective mobility, respectively. Due to the bandgap energy, which determines the voltage needed to cause an electron charge to move from the energy levels of valence to the conduction band, it also has an impact on the device's drain current. Equation (8.2) illustrates how the energy band is temperature dependent.

$$E_g(T) = E_g(300) - \frac{\alpha T^2}{T + \beta} \qquad (8.2)$$

where $E_g(300)$, α, and β denote the fitting factors. The transfer characteristics of the device in log and linear scale are shown in Figure 8.2. According to the linear characteristics, the current is low at lower temperatures and increases at higher temperatures. The current in a TFET will be substantially less temperature dependent, since the flow of electrons between one level of energy and another is not significantly affected by the temperature. However, as temperature rises, charge carriers become more mobile, which makes them more likely to collide with one another, increasing the diffusion component of I_{on}. This collision reduces their mobility, which causes the drift component of the I_{ds} to decrease, explaining why the I_{ds} varies slightly in the ON-state. At various temperatures, a significant difference in currents can be observed in the OFF-state.

The energy band controls the variance in I_{ds} at low gate voltages. As V_{gs} rises, the energy levels of the inversion layer begin to fall to coincide with the levels of the source region, allowing electrons to migrate from the source to the channel area, resulting in an increase in I_{ds}. Figure 8.3, in an evaluation of the device's performance at various temperatures, displays the device's energy band diagrams for

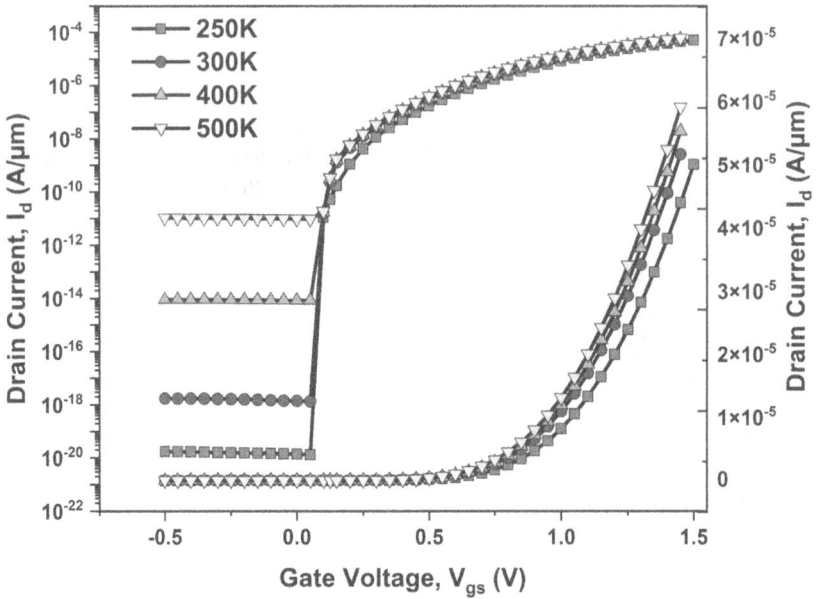

FIGURE 8.2   Transfer characteristics of Si-doped MoS_2-SS-DG-TFETs at various temperatures.

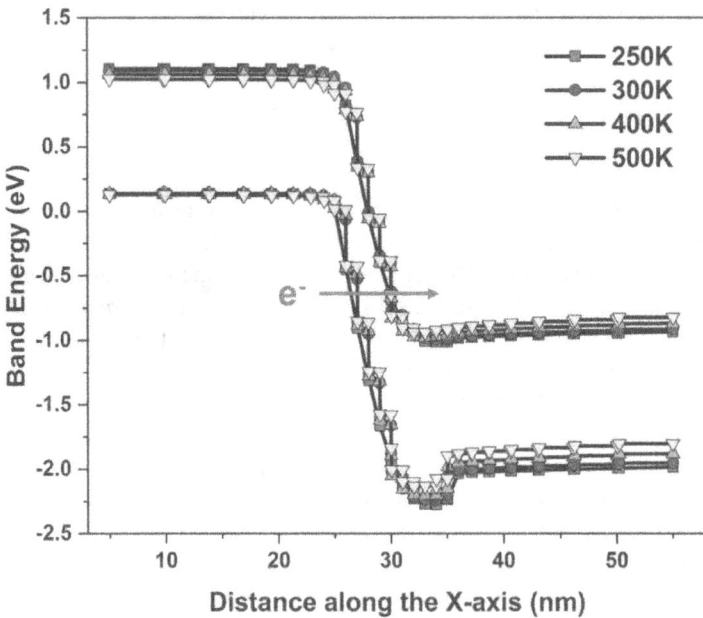

FIGURE 8.3   Impact of temperature on an energy band diagram of Si-doped MoS_2-SS-DG-TFETs.

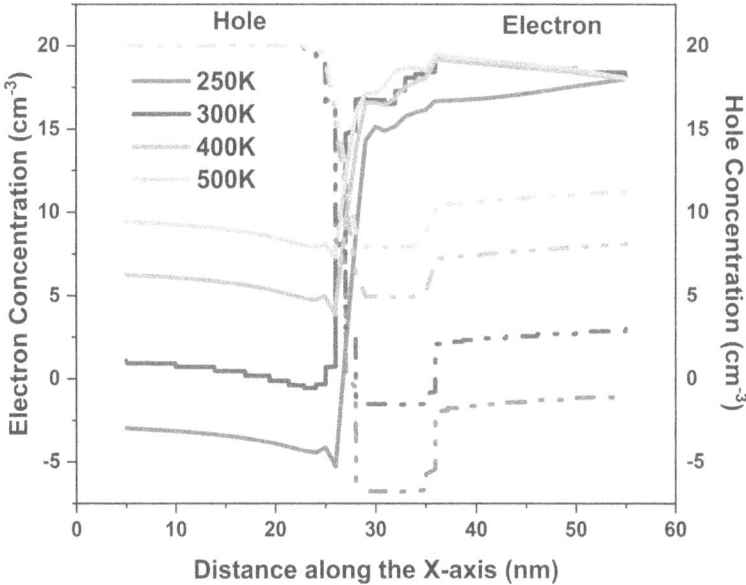

FIGURE 8.4 Impact of temperature on the electron–hole concentration of Si-doped MoS$_2$-SS-DG-TFETs.

temperatures ranging from 250 K to 500 K. As the temperature rises, the energy levels of the transistors' conduction band and valence band rise. The consequence is that when temperature increases, the energy bandgaps of Si-doped MoS$_2$-SS-DG-TFETs get smaller; as a result, the I_{off} increases. Furthermore, at high gate voltage, charge carrier mobility dominates and is inversely proportional to temperature change. As a result, the scattering effect causes the mobility of charge carriers to decrease with increasing temperature. When both components are present, the transfer properties degrade as the temperature rises.

Figure 8.4 depicts the charge carrier density of the simulated TFET at various temperatures. Since the donor atoms are bound with donor electrons in the area known as the "ionization region," at low temperature, no electron–hole pairs exist at this period. On the source side, the number of electrons is significantly larger at 500 K than at other temperatures, leading to a high I_{ds} in the ON-state. Thermal energy is produced by higher temperatures, which helps covalent bonds to break. This creates the separation of more donor particles from the donor states, which causes them to move into the conduction band and cause greater conduction.

8.3.1 ANALOG/RF PERFORMANCE

The transition to a nanoscale domain necessitates enhanced parameters of analog/RF performance for high-frequency operations. This subsection demonstrates how temperature affects the analog/RF properties of Si-doped MoS$_2$-SS-DG-TFETs. Device performance is evaluated at high frequency based on the examination of several parameters.

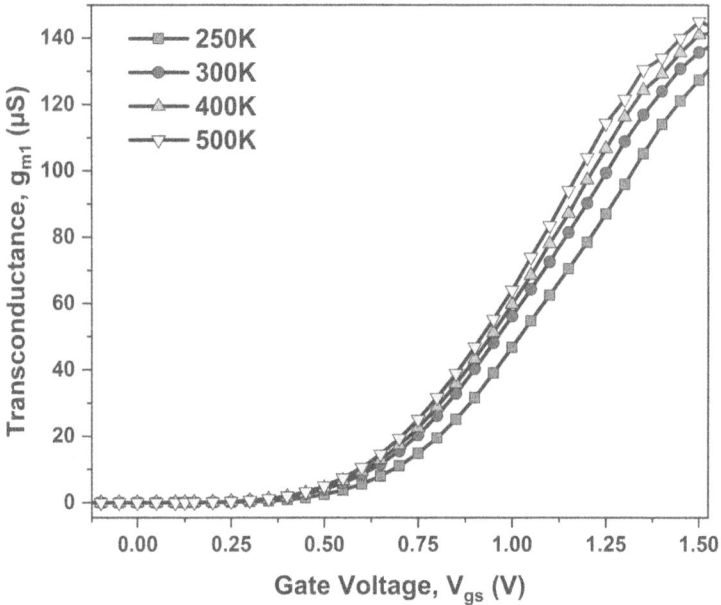

FIGURE 8.5 Transconductance (gm_1) at different temperatures.

Transconductance variability with temperature and V_{gs} can be shown in Figure 8.5. The cause for the g_{m1} increment is defined by the fluctuation in drain current. Equation (8.3) demonstrates the link between g_{m1} and I_{ds}.

$$g_{m1} = \frac{\partial I_{ds}}{\partial v_{gs}} \tag{8.3}$$

The g_{m1} rises proportionately as the drain current increases in response to V_{gs}. High g_{m1} values are preferred for a device to have a good amplifying response. The high work function at the source–channel junction that provides stability from the DIBL causes g_{m1} to rise with temperature at low V_{gs}. When the I_{ds} reaches its saturation point, the decrease in g_{m1} begins to occur, along with an increase in V_{gs}. At higher temperatures, g_{m1} exhibits a greater peak, similar to the behavior of the drain current. However, with larger V_{gs}, g_{m1} decreases as the temperature rises due to deterioration of the mobility of the charge particles. After $V_{gs} = 1.5$ V, the scattering effect causes a degradation in carrier mobility that results in drain current saturation.

According to Equation (8.4), output conductance (g_{ds}) represents the variation in I_{ds} brought on by a moderate change in V_{ds}.

$$g_{ds} = \frac{\partial I_{ds}}{\partial v_{ds}} \tag{8.4}$$

The g_{ds} is inversely related to the output resistance of the device. The device must have a high output resistance in order to have a strong amplification response; hence,

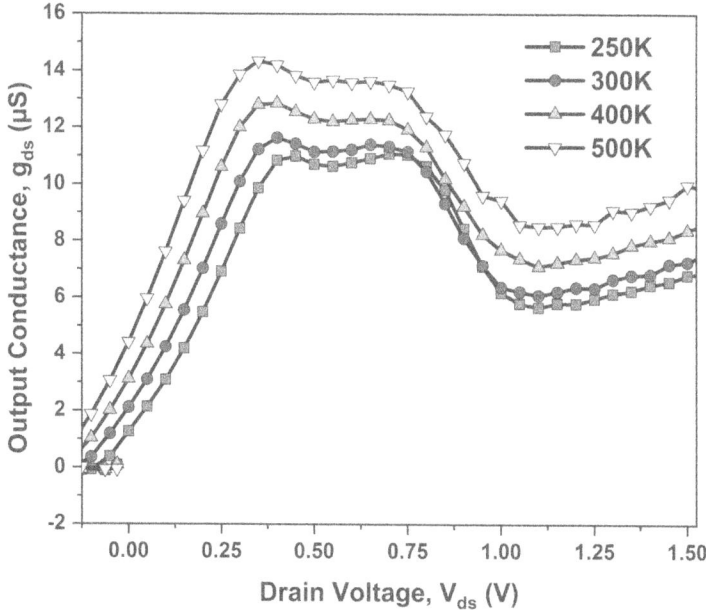

FIGURE 8.6 Output transconductance (g_{ds}) at different temperatures.

its g_{ds} should be as small as is feasible. Figure 8.6 illustrates how temperature affects the variability of g_{ds}. When V_{gs} seems low, g_{ds} is higher, and so as V_{gs} rises, g_{ds} begins to decrease rapidly. This causes the device to have a high current driving capacity when it is in the saturation zone. With a rise in temperature, the g_{ds} increases, showing an insufficient amplification response at high temperatures.

The highest voltage gain is known as "intrinsic gain," which is calculated using Equation (8.5) as the ratio between g_{m1} and g_{ds}. In this relationship, g_{m1} is the dominant factor, since g_{ds} is quite low in comparison to g_{m1}. As a conclusion, the impact of temperature on the intrinsic gain plot is comparable to that of a g_{m1} plot, and the intrinsic gain curve deviates downward with rising temperature, as shown in Figure 8.7.

$$Intrinsic\ Gain = \frac{g_{m1}}{g_{ds}} \tag{8.5}$$

Similarly, we must investigate the variability in cutoff frequency (f_t). According to Equation (8.6), the f_t has a direct relationship with g_{m1} and an inverse relationship with total gate capacitance ($C_{gg} = C_{gd} + C_{gs}$), where C_{gg} denotes the total capacitance, C_{gd} denotes the capacitance from gain to drain, and C_{gs} denotes the capacitance from gate to source. The g_{m1} is the dominant parameter for smaller V_{gs}, but when V_{gs} rises, C_{gg} takes over. High C_{gg} with higher V_{gs} causes the f_t curve to begin falling after reaching its peak, as shown in Figure 8.8.

$$f_t = \frac{g_{m1}}{2\Pi\left(C_{gs}+C_{gd}\right)} \tag{8.6}$$

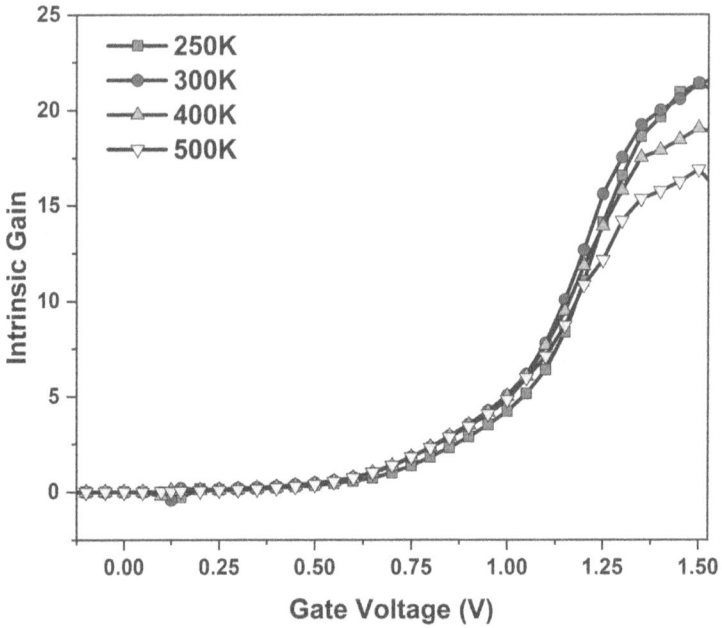

FIGURE 8.7 Intrinsic gain at different temperatures.

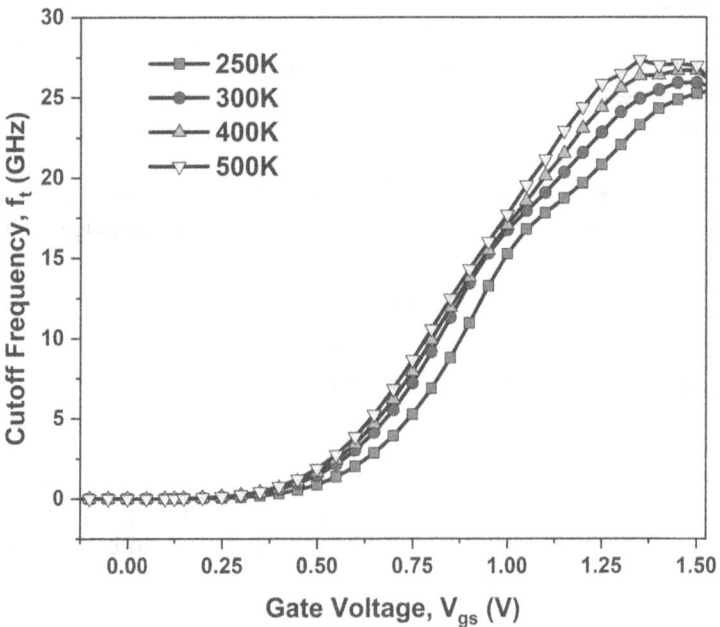

FIGURE 8.8 Cutoff frequency at different temperatures.

The term "gain bandwidth product" (GBP) refers to the bandwidth-gain product, which is used to calculate bandwidth. Equation (8.7) can be used to define GBP.

$$Gain\ Bandwidth\ Product\ (GBP) = \frac{g_{m1}}{2\Pi C_{gd}} \tag{8.7}$$

GBP and g_{m1} are directly proportional, as shown by Equation (8.7). GBP obtains the largest peak as g_{m1} with an increase in V_{gs}, and further increase in V_{gs} causes a decrease in the value of GBP. The cause of that occurrence is that the direct relation with g_{m1} and I_{ds} raises the GBP owing to a rise in itself, but at the point where the existing g_{m1} reaches saturation, g_{m1} starts to decrease, which has an impact on the GBP. In addition, C_{gd} has an inversely proportional impact on the GBP; as C_{gd} rises, the GBP falls. As shown in Figure 8.9, the GBP of the proposed device increases as the temperature rises.

A device metric that describes how well a device performs at its operating frequency is called the "transconductance frequency product" (TFP). The TFP represents a power–bandwidth trade-off. Equation (8.8) has been used to demonstrate how TFP relates to g_{m1}, f_t, and I_{ds}.

$$Transconductance\ Frequency\ Product,\ TFP = \frac{g_m f_t}{I_{ds}} \tag{8.8}$$

Equation (8.8) demonstrates how closely g_{m1}, f_t, and I_{ds} affect TFP; as a result, an increase in g_{m1} and f_t relative to V_{gs} will raise the TFP value, but after saturation,

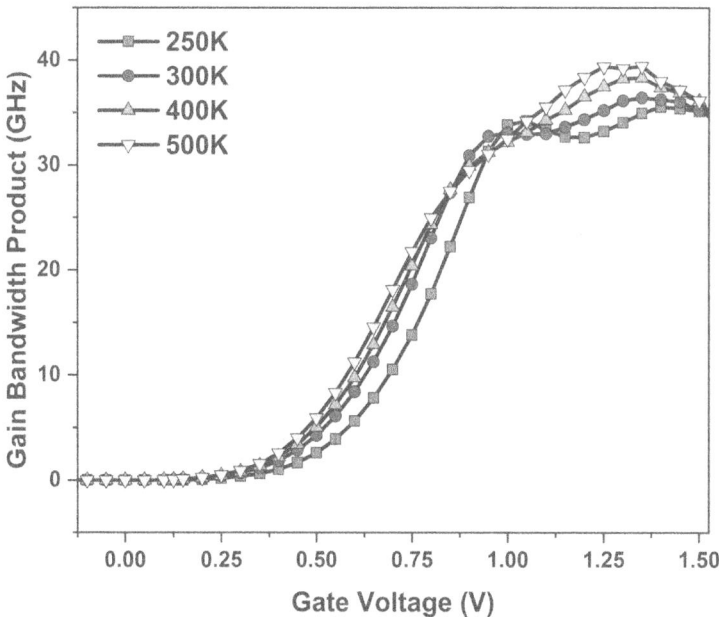

FIGURE 8.9 Gain bandwidth product (GBP) at different temperatures.

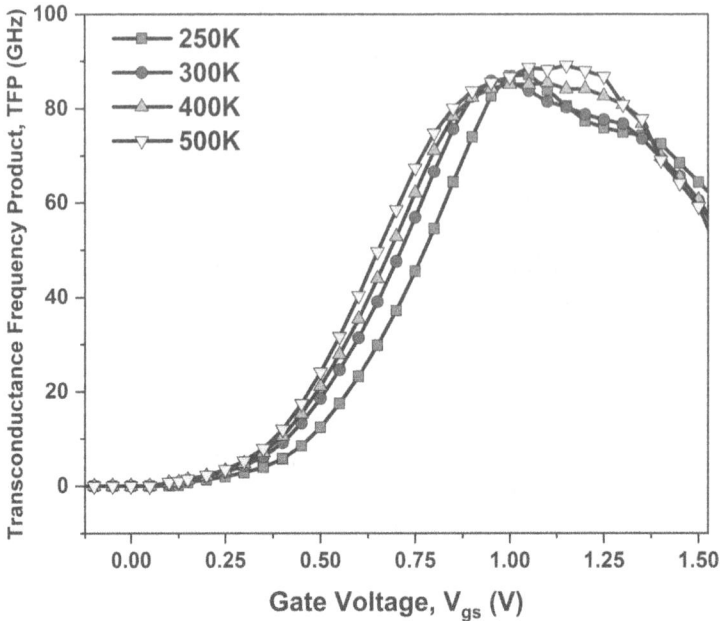

FIGURE 8.10   Transconductance frequency product (TFP) at different temperatures.

current will stop increasing and will no longer affect the TFP, while a decrease in g_{m1} and f_t will cause a decrease in the TFP's value. Figure 8.10 shows a higher TFP parameter value at higher temperatures.

A performance metric for devices that shows efficiency is the "transconductance generation factor" (TGF). The curve in Figure 8.11 depicts the connection of the TGF with V_{gs} and temperature. Equation (8.9) illustrates the direct and indirect proportionality of g_{m1} and I_{ds} with the TGF.

$$\text{Transconductance Generation Factor, } TGF = \frac{g_{m1}}{I_{ds}} \qquad (8.9)$$

According to the transfer characteristics, the I_{ds} for 500 K gets a high value, and in this case, the TGF has an indirect relation to I_{ds}, indicating that the TGF earned a lower value for 500 K and a greater value for 250 K. Figure 8.12 depicts a plot of the transit time (τ) of a Si-doped MoS_2-SS-DG-TFET in relation to V_{gs}. Transit time measures how quickly a device performs. Equation (8.10) can be used to calculate τ.

$$\text{Transit time, } \tau = \frac{1}{2\pi f_t} \qquad (8.10)$$

When V_{gs} is first applied, the device's enhanced τ for various temperatures gradually begins to diminish. This lowered τ reduces the delay of the device, which increases the speed of the device [30–32].

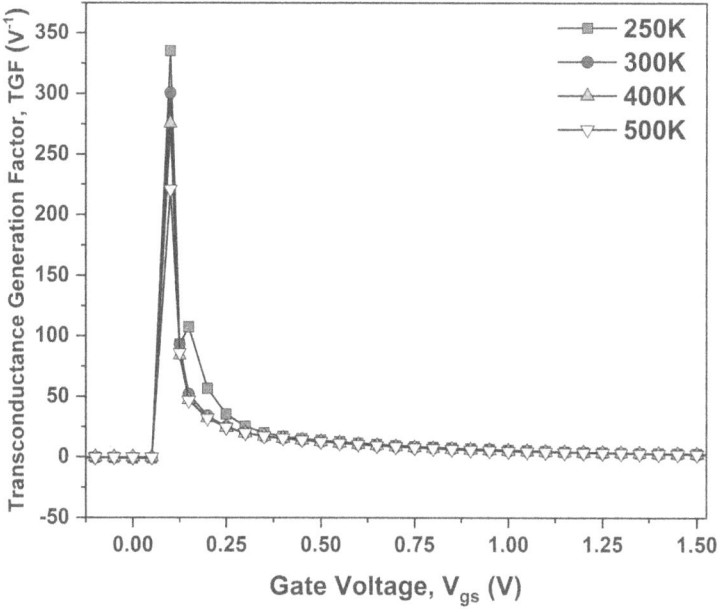

FIGURE 8.11 Transconductance generation factor (TGF) at different temperatures.

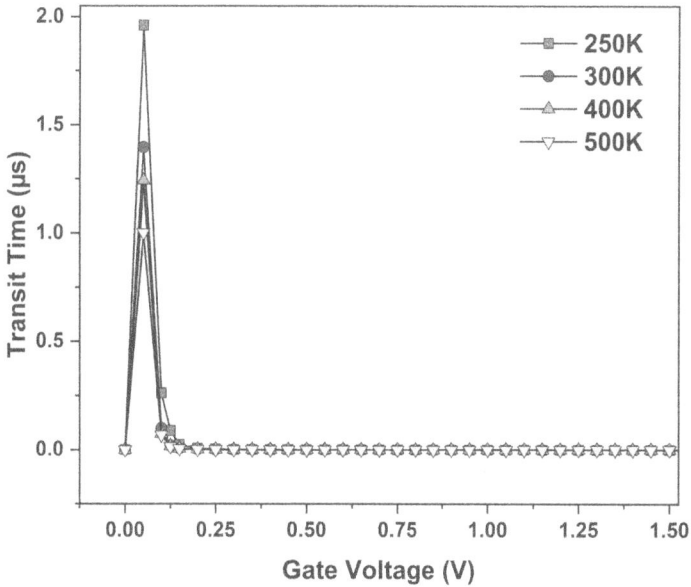

FIGURE 8.12 Transit time (τ) at different temperatures.

8.3.2 LINEARITY AND DISTORTION PERFORMANCE

Exploring linearity parameters for RF operations such as 3G and 4G technologies and for other operations of high-frequency purposes is an essential step. The term "linearity" refers to the measurement of distortion and noise in a technology. Devices with higher linearity cause less signal distortion and introduce less noise, which makes them suitable for low-noise applications. Linearity is controlled by the transconductance coefficients, which are g_{m1}, g_{m2}, and g_{m3}; voltage intercept points that include VIP_2 and VIP_3; input intercept points, such as IIP_2 and IIP_3; and intermodulation distortion, such as IMD_2 and IMD_3. The main factors that calculate the linearity of the device are the transconductance coefficients (g_{m1}, g_{m2}, and g_{m3}). All other linearity metrics are completely dependent on transconductance coefficients.

The values of second-order transconductance (g_{m2}) and third-order transconductance (g_{m3}) should be low, since the lower limit of distortion is controlled by these two parameters. Equation (8.11) can be used to describe g_{m2}, which is essentially the second derivative of I_{ds}. A low g_{m2} value improves linearity by lowering distortion.

$$\text{Second-Order Transconductance, } g_{m2} = \frac{\partial^2 I_{ds}}{\partial^2 v_{gs}} \tag{8.11}$$

The graph in Figure 8.13 can be used to study the modification of g_{m2} with V_{gs}. The value of g_{m2} initially rises; however, after reaching a maximum value, it begins to fall. This occurs while the impact of the I_{ds} on g_{m2} starts to decrease when it

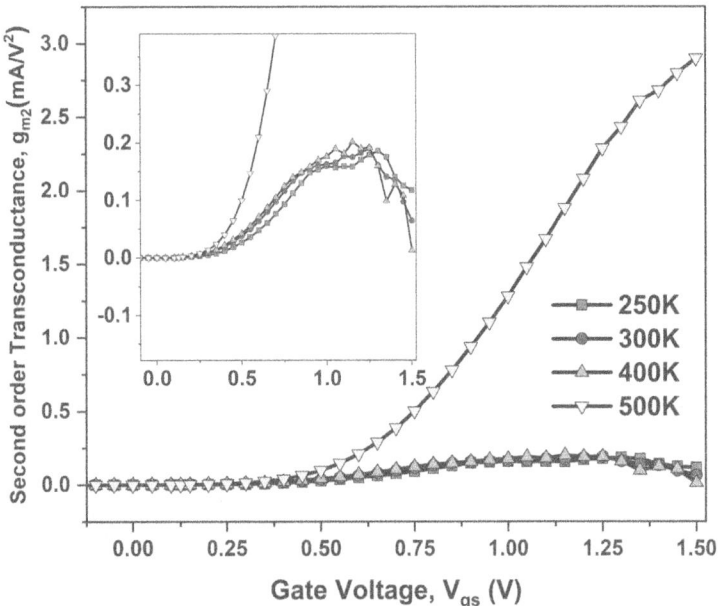

FIGURE 8.13   Second-order transconductance (g_{m2}) at different temperatures.

reaches a saturation point. At this point, gate voltage takes over as the dominating factor, and any further increase in gate voltage causes g_{m2} to fall.

The g_{m3} is the third derivative of I_{ds}. The g_{m3} value needs to be low for a linearity analysis to yield good results. From Equation (8.12), g_{m3} can be derived. Figure 8.14 illustrates how g_{m3} changes with gate voltage.

$$\text{Third-Order Transconductance, } g_{m3} = \frac{\partial^3 I_{ds}}{\partial^3 v_{gs}} \tag{8.12}$$

Figure 8.14 illustrates how g_{m3} degrades as temperature rises. The 400 K temperature has the lowest value of the g_{m3} parameter as compared to other temperature values. The 500 K temperature has an increase toward the positive values of g_{m3} and then falls down after a peak value. At 500 K, the simulated device does not have better linearity performance.

The first and second harmonic voltages are equivalent to the input V_{gs}, and represented by the voltage intercept points VIP$_2$ and VIP$_3$, respectively. For a device to be distortion-free, VIP$_2$ should have a high peak according to Equation (8.13). VIP$_2$ changes as the V_{gs} increases at different temperatures in Figure 8.15. Due to the fact that the impact of g_{m2} dominates over the effect of g_{m1}, the value of VIP$_2$ falls as temperature rises.

$$\text{Second-Order Voltage Intercept Point } \left(VIP_2 \right) = 4 \left(\frac{g_{m1}}{g_{m2}} \right) \tag{8.13}$$

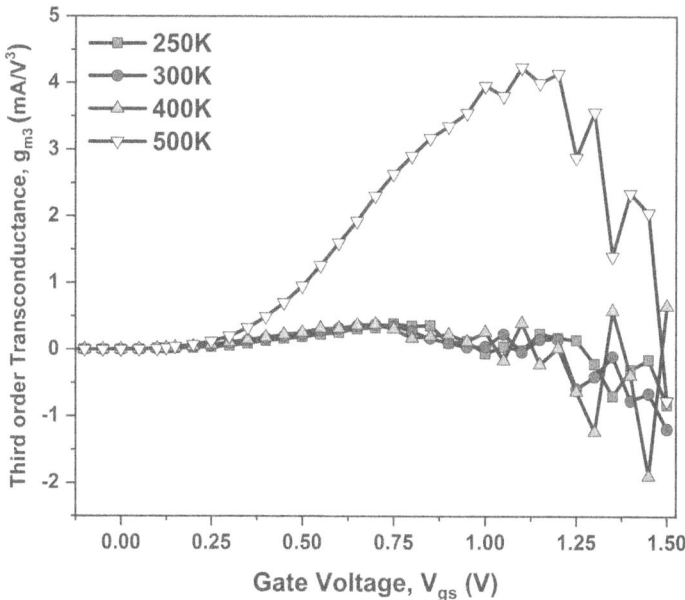

FIGURE 8.14 Third-order transconductance (g_{m3}) at different temperatures.

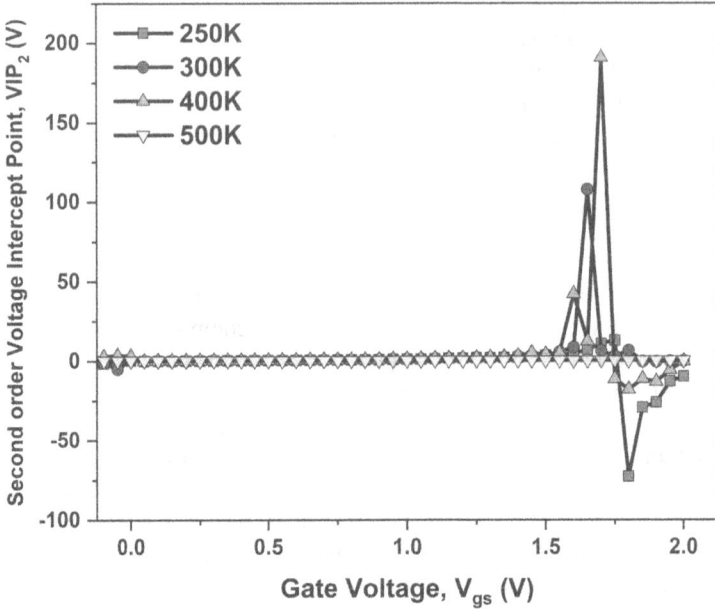

FIGURE 8.15   Second-order voltage intercept point (VIP$_2$) at different temperatures.

VIP$_3$ is a significant characteristic that affects the linearity of the device. For the device to have a good linearity, a high VIP$_3$ value is desired. VIP$_3$ is provided in Equation (8.14), which illustrates how it relates to g_{m1} and g_{m3}.

$$\text{Third-Order Voltage Intercept Point}, VIP_3 = \sqrt{24\frac{g_{m1}}{g_{m3}}} \quad (8.14)$$

Figure 8.16 illustrates a plot of VIP$_3$ versus V_{gs} at distinct temperatures, demonstrating that VIP$_3$ performs best when the temperature is at 250 K.

The third-order input intercept point (IIP$_3$) is an important component in the device's linearity examination. The maximum point of IIP$_3$ should be large for the device's linearity to be optimized. IIP$_3$ is determined by Equation (8.15).

$$\text{Third-Order Input Intercept Point, } IIP_3 = \frac{2}{3}\left(\frac{g_{m1}}{g_{m3}R_s}\right) \quad (8.15)$$

where R_s is set to 50 Ω for analog/RF operations. Third-order intermodulation distortion (IMD$_3$) is an important measure for determining the device's linearity. IMD$_3$ is mathematically denoted by Equation (8.16), demonstrating its direct relationship to VIP$_3$ and g_{m3}.

$$\text{Third-Order Intermodulation Distortion, } IMD_3 = \left(\frac{9}{2}\left(VIP_3\right)^2 g_{m3}\right)^2 R_s \quad (8.16)$$

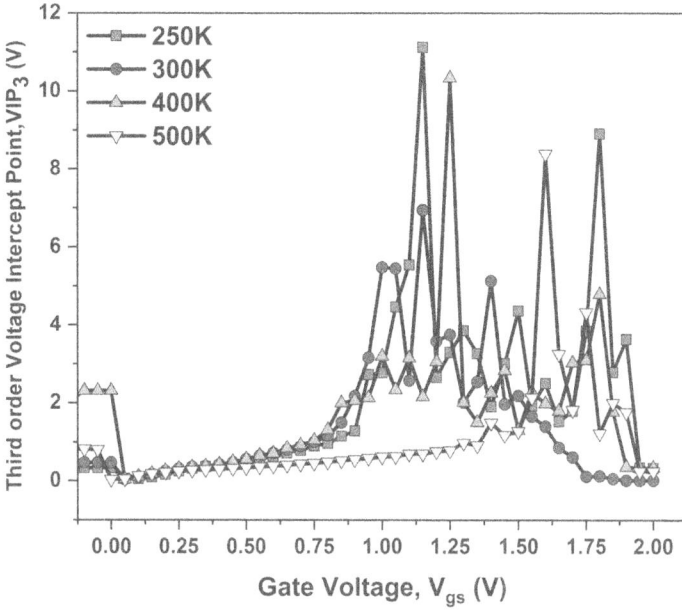

FIGURE 8.16   Third-order voltage intercept point (VIP$_3$) at different temperatures.

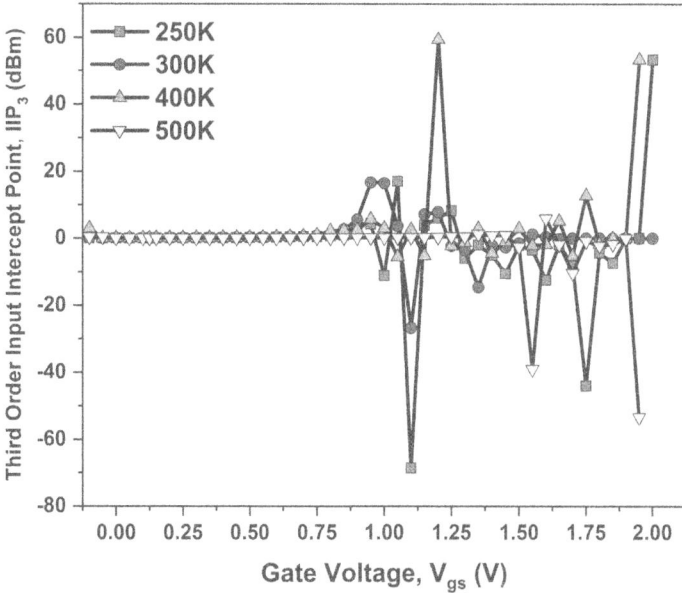

FIGURE 8.17   Third-order input intercept point (IIP$_3$) at different temperatures.

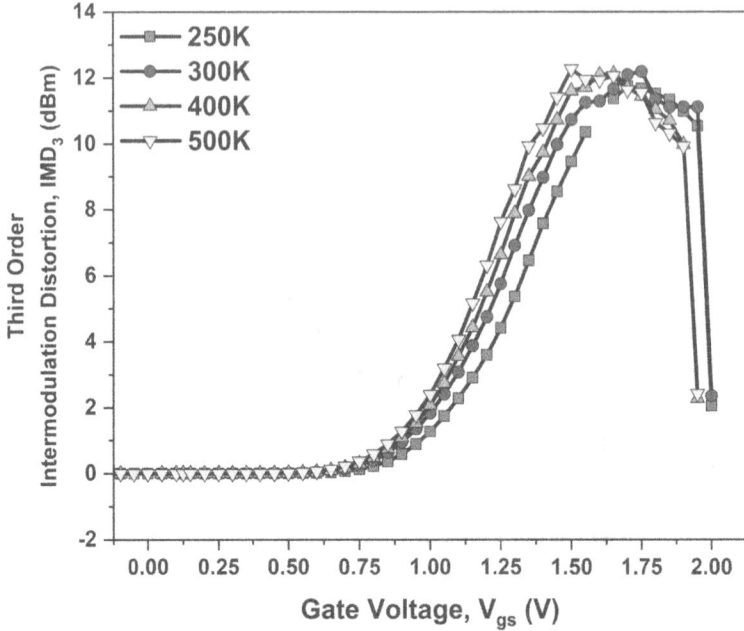

FIGURE 8.18   Third-order intermodulation distortion (IMD$_3$) at different temperatures.

The IIP$_3$ and IMD$_3$ device performance metrics are inversely connected. For greater linearity, IIP$_3$ should be higher and IMD$_3$ should be lower. The IIP$_3$ value specifies how much a signal can be amplified before IMD$_3$ develops. Figure 8.17 shows a hike in the maximum point of IIP$_3$ as the temperatures fall, indicating an improvement in linearity at low temperatures. The low IMD$_3$ for analog/RF operations denotes low distortion. Figure 8.18 depicts the IMD$_3$ graph versus V_{gs} at various temperatures. The value of IMD$_3$ rises as temperature rises, indicating a loss of device linearity at high temperatures. Furthermore, the device exhibits great stability because there is little variation in the value of IMD$_3$ at high V_{gs}. As a result, the device is more efficient in parameters of IIP$_3$ and IMD$_3$ at lower temperatures [33, 34].

8.4   CONCLUSION

In this work, a temperature-based investigation was conducted for the device's DC and analog properties, as well as a linearity analysis. According to the study, our suggested device, the Si-doped MoS$_2$-SS-DG-TFET, has high temperature sensitivity. Temperatures between 250 K and 500 K see an increase in ON-state current, while higher temperatures see an increase in OFF-state current as well, which lowers the quality of the I_{on}/I_{off} relation. Also, the OFF-state current has been investigated for a variety of models and temperatures. SRH and TAT modeling for excessive temperatures increase the current in the OFF-state. Furthermore, we looked at how temperature affects the behavior of the transfer characteristics, energy band diagram, and carrier concentration. Analog parameters are likewise temperature sensitive and produce better results at higher temperatures. The g_{m1}, g_{ds}, f_t, GBP, and TFP all reach their highest peaks at higher temperatures, whereas intrinsic gain, TGF, and τ reach their lowest peak at 500 K. A device works better if its transit time is the shortest. In comparison to other devices, the device's transit time at 500 K is quite short. We also investigated linearity measures in aspects of g_{m2}, g_{m3}, VIP$_2$, VIP$_3$, IIP$_3$, and IMD$_3$ to study the device's nonlinearity concerns. As a result, the Si-doped MoS$_2$-SS-DG-TFET has better current driving capabilities for small V_{gs} and exhibits less distortion, proving to be a viable device for high-frequency applications as well as for linearity analysis, DC analysis, and analog analysis.

REFERENCES

1. Chhowalla, M., Jena, D., & Zhang, H. (2016). Two-Dimensional Semiconductors for Transistors. Nature Reviews Materials, 1(11), 1–15.
2. Ionescu, A. M., & Riel, H. (2011). Tunnel Field-Effect Transistors as Energy-Efficient Electronic Switches. Nature, 479(7373), 329–337.
3. Mehra, R., Gagneja, A., & Kaushal, P. (2017). Comparative Analysis of Shift Registers in Different Nanometer Technologies. Sensors & Transducers, 213(6), 30.
4. Mehra, R., Kaushal, P., & Gagneja, A. (2017, March). Area and Speed Efficient Layout Design of Shift Registers using Nanometer Technology. In Proceedings of the 2nd International Conference on Advances in Sensors (pp. 58–62).
5. Walia, A., Kaushal, P., & Khanna, G. (2021). Impact of Temperature on the Performance of Tunnel Field Effect Transistor Processing in International Conference on Emerging Technologies: AI, IoT, and CPS for Science & Technology Applications.
6. Goma, P., & Rana, A. K. (2022). Analysis and Evaluation of Si (1-x) Ge (x) Pocket on Sensitivity of a Dual-Material Double-Gate Dopingless Tunnel FET Label-Free Biosensor. Micro and Nanostructures, 170, 207393.
7. Ferain, I., Colinge, C. A., & Colinge, J. P. (2011). Multigate Transistors as the Future of Classical Metal–Oxide–Semiconductor Field-Effect Transistors. Nature, 479(7373), 310–316.
8. Avci, U. E., Morris, D. H., & Young, I. A. (2015). Tunnel Field-Effect Transistors: Prospects and Challenges. IEEE Journal of the Electron Devices Society, 3(3), 88–95.
9. Kaushal, P., & Mehra, R. (2017). Energy Efficient CNTFET Based Full Adder Using Hybrid Logic. International Journal on Recent and Innovation Trends in Computing and Communication, 5(7), 98–103.
10. Kaushal, P., & Mehra, R. (2017). A Novel CNTFET Based Power and Delay Optimized Hybrid Full Adder. International Journal of Electrical, Electronics And Data Communication, 5(9), 21–27.
11. Ajayan, P., Kim, P., & Banerjee, K. (2016). van der Waals Materials. Physics Today, 69(9), 38.
12. Lin, Z., McCreary, A., Briggs, N., Subramanian, S., Zhang, K., Sun, Y., & Terrones, M. (2016). 2D Materials Advances: from Large Scale Synthesis and Controlled Heterostructures to Improved Characterization Techniques, Defects and Applications. 2D Materials, 3(4), 042001.
13. Er, D., & Ghatak, K. (2020). "Atomistic modeling by density functional theory of two-dimensional materials." Synthesis, Modeling, and Characterization of 2D Materials, and Their Heterostructures, 113–123. https://doi.org/10.1016/B978-0-12-818475-2.00006-4
14. Xie, K., Li, X., & Cao, T. (2021). Theory and Ab Initio Calculation of Optically Excited States—Recent Advances in 2D Materials. Advanced Materials, 33(22), 1904306.

15. Momeni, K., Ji, Y., Wang, Y., Paul, S., Neshani, S., Yilmaz, D. E., & Chen, L. Q. (2020). Multiscale Computational Understanding and Growth of 2D Materials: a Review. Npj Computational Materials, 6(1), 1–18.

16. Kaushal, P., Chaudhary, T., & Khanna, G. (2021). Effect of Tensile Strain on Performance Parameters of Different Structures of MoS_2 Monolayer. Silicon, 1–9.

17. Mitta, S. B., Choi, M. S., Nipane, A., Ali, F., Kim, C., Teherani, J. T., & Yoo, W. J. (2020). Electrical Characterization of 2D Materials-Based Field-Effect Transistors. 2D Materials, 8(1), 012002.

18. Kaushal, P., & Khanna, G. (2022). The Role of 2-Dimensional Materials for Electronic Devices. Materials Science in Semiconductor Processing, 143, 106546.

19. Ghosh, R. K., & Mahapatra, S. (2013). Monolayer Transition Metal Dichalcogenide Channel-Based Tunnel Transistor. IEEE Journal of the Electron Devices Society, 1(10), 175–180.

20. Ilatikhameneh, H., Tan, Y., Novakovic, B., Klimeck, G., Rahman, R., & Appenzeller, J. (2015). Tunnel Field-Effect Transistors in 2-D Transition Metal Dichalcogenide Materials. IEEE Journal on Exploratory Solid-State Computational Devices and Circuits, 1, 12–18.

21. Liu, F., Wang, J., & Guo, H. (2015). Atomistic Simulations of Device Physics in Monolayer Transition Metal Dichalcogenide Tunneling Transistors. IEEE Transactions on Electron Devices, 63(1), 311–317.

22. Khan, M. A. U., Srivastava, A., Mayberry, C., & Sharma, A. K. (2020). Analytical Current Transport Modeling of Monolayer Molybdenum Disulfide-Based Dual Gate Tunnel Field Effect Transistor. IEEE Transactions on Nanotechnology, 19, 620–627.

23. Radisavljevic, B., Radenovic, A., Brivio, J., Giacometti, V., & Kis, A. (2011). Single-Layer MoS2 Transistors. Nature Nanotechnology, 6(3), 147–150.

24. Kim, S. W., Kim, J. H., Liu, T. J. K., Choi, W. Y., & Park, B. G. (2015). Demonstration of L-Shaped Tunnel Field-Effect Transistors. IEEE Transactions on Electron Devices, 63(4), 1774–1778.

25. Kim, S. W., Choi, W. Y., Sun, M. C., Kim, H. W., & Park, B. G. (2012). Design Guideline of Si-Based L-Shaped Tunneling Field-Effect Transistors. Japanese Journal of Applied Physics, 51(6S), 06FE09.

26. Meshkin, R., Ziabari, S. A. S., & Jordehi, A. R. (2020). Representation of an Engineered Double-Step Structure SOI-TFET with Linear Doped Channel for Electrical Performance Improvement: A 2D Numerical Simulation Study. Semiconductor Science and Technology, 35(6), 065006.

27. Dutta, R., & Sarkar, S. K. (2019). Analytical Modeling and Simulation-Based Optimization of Broken Gate TFET Structure for Low Power Applications. IEEE Transactions on Electron Devices, 66(8), 3513–3520.

28. Zhang, M., Guo, Y., Zhang, J., Yao, J., & Chen, J. (2020). Simulation Study of the Double-Gate Tunnel Field-Effect Transistor with Step Channel Thickness. Nanoscale Research Letters, 15(1), 1–9.

29. Kaushal, P., & Khanna, G. (2022). MoS_2 Based Thickness Engineered Tunnel Field-Effect Transistors for RF/analog Applications. Materials Science in Semiconductor Processing, 151, 107016.

30. Der Agopian, P. G., Martino, J. A., Rooyackers, R., Vandooren, A., Simoen, E., & Claeys, C. (2013). Experimental Comparison between Trigate p-TFET and p-FinFET Analog Performance as a Function of Temperature. IEEE Transactions on Electron Devices, 60(8), 2493–2497.

31. Pandey, C. K., Dash, D., & Chaudhury, S. (2020). Improvement in analog/RF Performances of SOI TFET Using Dielectric Pocket. International Journal of Electronics, 107(11), 1844–1860.

32. Saha, R., Panda, D. K., Goswami, R., Bhowmick, B., & Baishya, S. (2021). Analysis on Effect of Lateral Straggle on Analog, High Frequency and DC Parameters in Ge-source DMDG TFET. *International* Journal of RF and Microwave Computer-Aided Engineering, 31(4), e22579.

33. Kumar, S., Singh, K. S., Nigam, K., Tikkiwal, V. A., & Chandan, B. V. (2019). Dual-Material Dual-Oxide Double-Gate TFET for Improvement in DC Characteristics, analog/RF and Linearity Performance. Applied Physics A, 125(5), 353.

34. Priyadarshani, K. N., Singh, S., & Naugarhiya, A. (2021). RF & Linearity Distortion Sensitivity Analysis of DMG-DG-Ge Pocket TFET with Hetero Dielectric. Microelectronics Journal, 108, 104973.

9 Implementation of Logic Gates Using Step-Channel TFETs

Varun Bharti, Prabhat Singh, and
Dharmendra Singh Yadav

9.1 INTRODUCTION

In the past, complementary metal–oxide–semiconductor (CMOS) technology was used to implement various digital integrated circuits (ICs), such as adders, multipliers, memory chips, and logical operations [1–3], due to its inherent advantages over n-type metal–oxide–semiconductor (NMOS) technology, including excellent dependability, robust performance, in addition to low power consumption even in standby mode. Hence, in the semiconductor industry, CMOS is therefore the optimal solution for digital circuits. In contrast, the requirement for continuing downscaling demands further modifications to integrate numerous Boolean gates on the single Si substrate [4]. That necessitated extra space, which increased the delay of the logic gates and the capacitance [5]. The logic gates in Refs. [6–10] are created with a single tunneling field-effect transistor (TFET) structure to reduce space overhead, in comparison to CMOS-based implementations. In addition, the TFET design offers inherent benefits, such as a low rate of power loss due to static resistance and a high level of energy efficiency in its operation. In comparison to metal–oxide–semiconductor field-effect transistors (MOSFETs), TFETs have a subthreshold swing that is lower than 60 mV/dec.

In spite of the difficulties of employing a TFET in digital circuits because of its low I_{ON} current and huge ambipolar current, certain digital circuits have been shown in the literature [6–8]. Step-channel tunnel field-effect transistors (SCTFETs) comprise a novel device shape that is designed to improve I_{ON} current, decrease ambipolar behavior, and enhance subthreshold swing. This device improves its performance by including a thin layer of dielectric material at the source–channel (S-C) junction. This reduced dielectric thickness at the S-C interface enhances device subthreshold swing, whereas increased dielectric thickness at the drain–channel (D-C) junction inhibits ambipolar current [9–17]. In order to achieve an even greater rise in the I_{ON} current, the oxide layer of the high-K SCTFET is constructed using high-k dielectric material. In the previously mentioned work [18–24], logic gates utilizing a single SCTFET structure were not built. In this study, all four basic gates are implemented using a single SCTFET architecture. These Boolean functions are implemented utilizing the appropriate work function (WF, or ϕ) of the SCTFET gate and the gate-to-source overlap (GSO) technique. The functionality of every logic function is represented by energy band diagrams (EBDs) [25–27] and the I-V characteristic of

 DOI: 10.1201/9781003393542-9

a SCTFET at different logic inputs; here, an n-type SCTFET is used to implement OR logic, whereas for implementation of NAND Boolean logic, a p-type SCTFET is used. In addition, the GSO approach is commonly used for the construction of AND and NOR logic gates, but SCTFET types vary. AND logic gates utilize n-type SCTFETs, whereas NOR logic gates utilize p-type SCTFETs. Using the Silvaco ATLAS simulator, the recommended devices were simulated [28].

9.2 SIMULATION PARAMETERS OF SCTFETS

Figure 9.1(a) and (b) represents n- and p-type SCTFETs, which have been used to implement different logic functions. Figure 9.2 represents the transfer characteristics of SCTFETs [36] at $V_{DD} = 1.2$ V. The transfer characteristics of simulated SCTFETs are quite similar to the curve shown in previous literature [36]. The device dimensions and specifications are assumed to be the same as those stated in Refs. [36, 37]. The gate oxide thickness (T_{OX}) of 1 nm has been used at the S-C interface, while the D-C junction has a T_{OX} of 2 nm. As a result, this device is asymmetrical, since its S-C

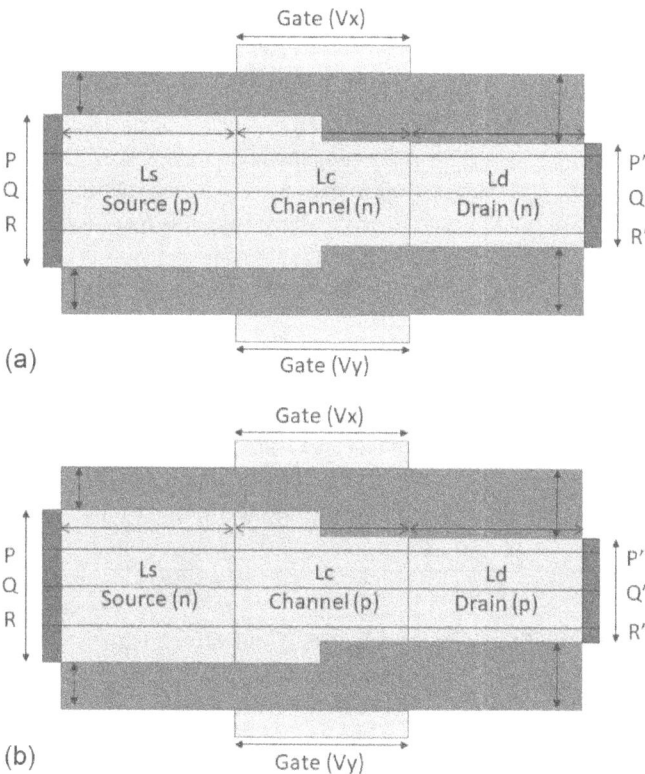

FIGURE 9.1 (a) A structural view of an n-type step-channel tunnel field-effect transistor (SCTFET). (b) A structural view of a p-type SCTFET.

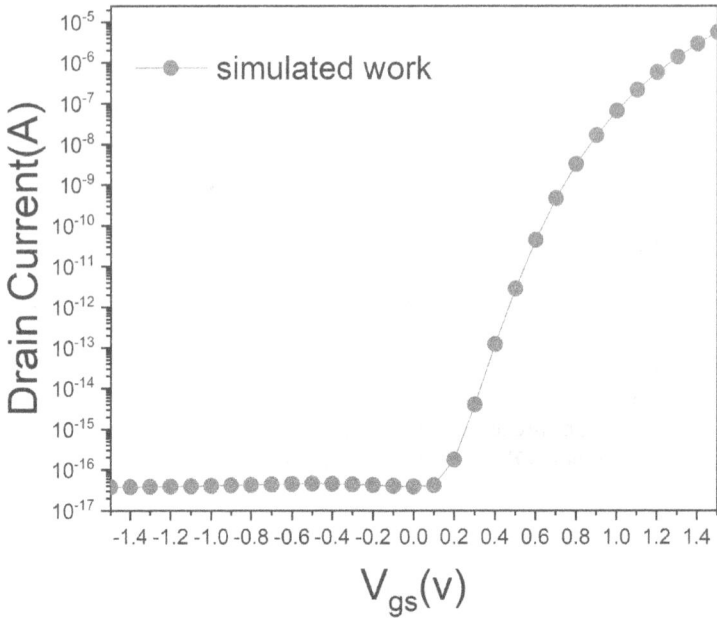

FIGURE 9.2 Transfer characteristics of a standard SCTFET [36].

interface and D-C junction have different gate oxide thicknesses. Under metal gate1 (G1) and gate2 (G2), two distinct dielectrics are employed. On the source side, the silicon body is 8 nm thick, whereas on the drain side, it is 6 nm thick. Furthermore, the gate length is 10 nm. For logic "1," the bias voltage (V_{DD}) is considered 1.2 V. ATLAS Version 5.22.1.R has been used to simulate the outcomes of SCTFET-based logic functions. To examine the tunneling phenomena, a nonlocal band-to-band tunneling (BTBT) model is incorporated, which integrates spatial energy band distribution. This BTBT model agrees well with simulation and theoretical tunneling data [15–17]. In contrast to that, the influence of the Fermi–Dirac distribution on the narrowing of the bandgap is taken into consideration. The Shockley–Read–Hall (SRH) model is utilized in order to compute the generation rate as well as the recombination rate [18–22]. Table 9.1 shows the drain current for different input combinations for

---

TABLE 9.1

Drain Currents of a Step-Channel Tunnel Field-Effect Transistor (SCTFET) for Different Combinations of Logic Inputs

Input Combinations (X and Y)	OR	NAND	AND	NOR
00	I_{OFF}	I_{ON}	I_{OFF}	I_{ON}
01	I_{ON}	I_{ON}	I_{OFF}	I_{OFF}
10	I_{ON}	I_{ON}	I_{OFF}	I_{OFF}
11	I_{ON}	I_{OFF}	I_{ON}	I_{OFF}

SCTFET-based Boolean functions. The I_{ON} current denotes a high logic "1," whereas the I_{OFF} current denotes a low logic "0."

9.3   IMPLEMENTATION OF SCTFET-BASED "OR" AND "NAND" FUNCTIONS

In this section, n- and p-type SCTFETs are used to realize the OR and NAND logic gates, as illustrated in Figure 9.1(a) and (b). Several input combinations are used over the upper and lower gates of the SCTFET structure. Because of this, the supply voltage, which is denoted by $V_{DD} = 1.2$ V, is seen as representing logic "1," whereas ground, which is denoted by 0 V, is supposed to represent logic "0." The voltage at the upper gate is denoted by V_X, whereas the voltage at the lower gate is denoted by V_Y. When the preset voltages at V_X and V_Y are utilized, I_{XY} is the symbol that is used to represent the drain current. Drain current is denoted by the symbol I_{10} when the upper gate is connected to 1.2 V ($V_X = V_{DD}$) and the lower gate is connected to 0 V ($V_Y = 0$).

This device contains two gates: the top gate (V_X) and bottom gate (V_Y). Both are biased with various input voltages to implement SCTFET-based OR logic. If both gates are biased with V_{DD}, then it is observed that a high amount of drain current flows through the SCTFET that is designated as I_{11}, and if they are biased with low logic, then low I_D flows through the device; however, if only one gate, either V_X or V_Y, is set to high logic, then a huge amount of I_D, marked I_{01} or I_{10}, flows through the SCTFET. The I_{ON}/I_{OFF} ratio must be increased for proper implementation of SCTFET as a logic function. Table 9.2 shows the drain current characteristics and I_{ON}/I_{OFF} ratio of SCTFET-based OR logic for various input configurations. The varied gate WFs for G1 and G2 are used to improve the I_{ON} and implement the OR Boolean gate, as shown in Table 9.3. In the OR Boolean function, one gate is connected to V_{DD} ("1"), then current in the device is represented as I_{ON} (high current), while the other gate is biased with ground and represents I_{OFF}. The functioning of the OR gate is exemplified in Figure 9.3 by the EBD of the SCTFET. The EBD of a SCTFET for the input combination "00" is shown in Figure 9.3(a). There is no tunneling because there is a large potential barrier observed between the S-C interface while both gates are at the low logic level. This means that there is no tunneling. Nonetheless, when one of the gates is set to high logic for combinations "01" and "10" in Figure 9.3(b) and (c), tunneling occurs, and a small number of electrons can go across the potential barrier.

TABLE 9.2
I_{ON}/I_{OFF} Ratio and Drain Current (in A/μm) for Different Logic Functions

Logic Function	I_{00}	I_{01}	I_{10}	I_{11}	I_{ON}/I_{OFF}
OR	10^{-15}	10^{-7}	10^{-7}	10^{-5}	10^{10}
NAND	10^{-5}	10^{-7}	10^{-7}	10^{-17}	10^{12}
AND	10^{-16}	10^{-15}	10^{-15}	10^{-7}	10^{9}
NOR	10^{-9}	10^{-16}	10^{-15}	10^{-16}	10^{7}

TABLE 9.3

Device Parameters for All Boolean Functions

Parameter	OR	AND	NAND	NOR
Source length (L_S) (nm)	20	20	20	20
Channel length (L_C) (nm)	10	10	10	10
Drain length (L_D) (nm)	20	20	20	20
Oxide body thickness at drain–channel junction (nm)	2	2	2	2
Oxide body thickness at source–channel junction (nm)	1	1	1	1
Drain body thickness (nm)	6	6	6	6
Source body thickness (nm)	8	8	8	8
Source doping (atom/cm³)	10^{20} (p-type)	10^{20} (p-type)	10^{20} (n-type)	10^{20} (n-type)
Channel doping (atom/cm³)	10^{17} (n-type)	10^{17} (n-type)	10^{17} (p-type)	10^{17} (p-type)
Drain doping (atom/cm³)	10^{18} (n-type)	10^{18} (n-type)	10^{18} (p-type)	10^{18} (p-type)
Gate work function (G1, G2)	4.25, 4.23	5.0, 5.0	5.20, 5.15	4.4, 4.4
Gate–source overlap length (nm)	0	4	0	4

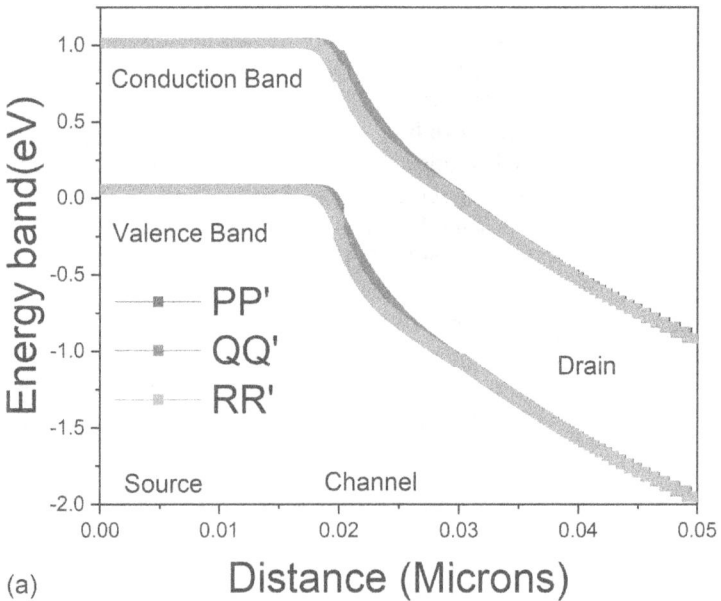

(a)

FIGURE 9.3 (a) Energy band diagram (EBD) for OR logic function for input setup "00" with PP′, QQ′, and RR′ cutlines. *(Continued)*

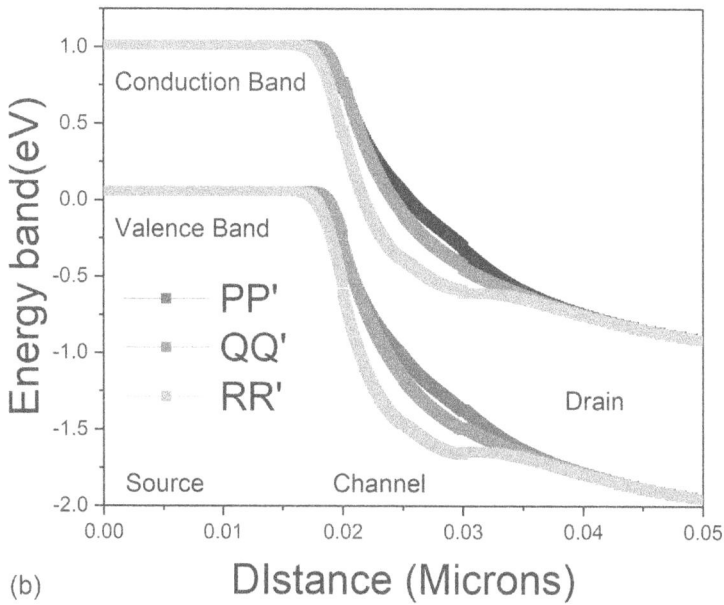

(b)

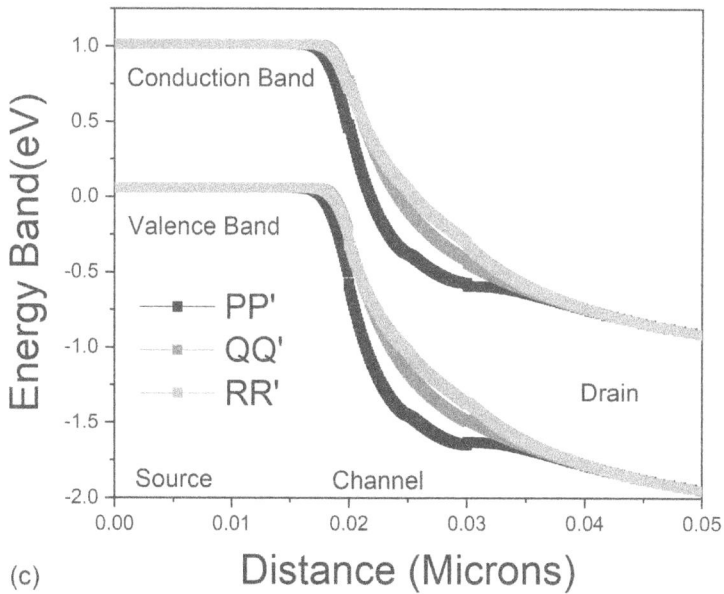

(c)

FIGURE 9.3 *(Continued)* (b) EBD for OR logic function for input setup "01" with PP', QQ', and RR' cutlines. (c) EBD for OR logic function for input setup "10" with PP', QQ', and RR' cutlines.

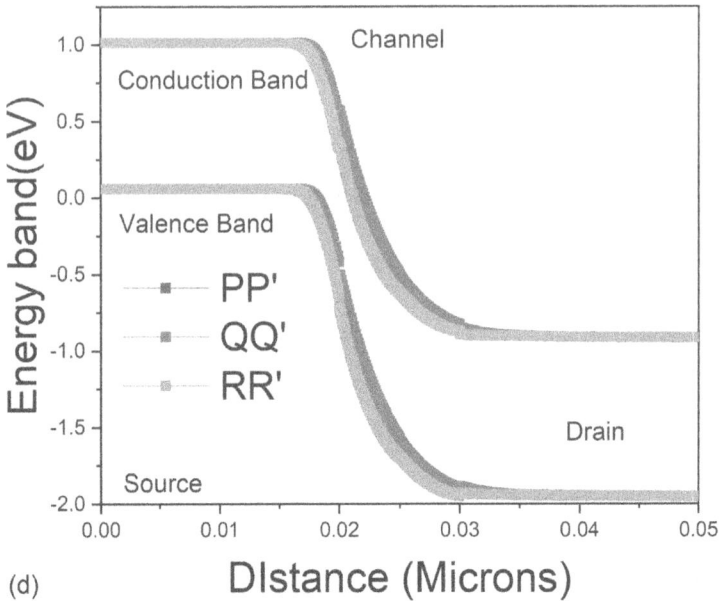

(d)

FIGURE 9.3 *(Continued)* (d) EBD for OR logic function for input setup "11" with PP', QQ', and RR' cutlines.

Figure 9.3(d) illustrates significant electron tunneling and high ON-current when both gates are connected to V_{DD} ("11") [34, 35]. Along the x-axis, the energy band diagrams for the PP', QQ', and RR' cutlines are drawn up to illustrate them, as shown in Figure 9.3. Figure 9.4 depicts the I_D-V_G curve of SCTFET-based OR logic. It is noticed that logic level "00" achieves a low current, whereas logic level "11" achieves a large current.

The NAND logic has been constructed using the p-type SCTFET seen in Figure 9.1(b). If both inputs are connected to "0" logic or one of them is connected to "0" logic, electrons tunnel from source to channel. If both inputs are at logic "1," no tunneling will occur. Table 9.3 lists the different device parameters and characteristics for the NAND Boolean function. Figure 9.5 depicts the EBD for various input configurations. When top and bottom gates are biased with logic level "0," the input combination "00" exhibits maximal tunneling, resulting in significant overlapping and band bending at the S-C interface, as seen in Figure 9.5(a). Significant tunneling is discovered for logic levels "01" and "10," but, on the other hand, there is no tunneling detected for "11"; see Figure 9.5(b)–(d). Figure 9.6 depicts the "transfer characteristics" of NAND gate implementation using a p-type SCTFET; here, the supply voltage V_{DD} is equal to 1.2 V and V_{DS} = −1.0 V, and I-V characteristics are shown. Observing the findings, we can conclude that for NAND logic, the "00" input configuration results in the highest possible I_{ON} current, whereas the "11" input configuration results in the lowest possible current. The I_{ON}/I_{OFF} ratio is in the region of 10^{12} for NAND logic, which is a tenfold improvement over previous studies [29, 30].

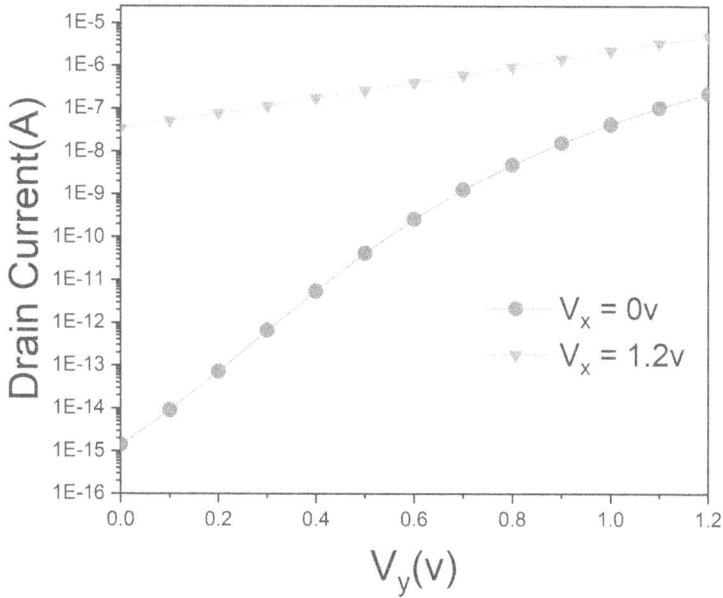

FIGURE 9.4 I_D-V_G curve of n-type SCTFET-based OR logic.

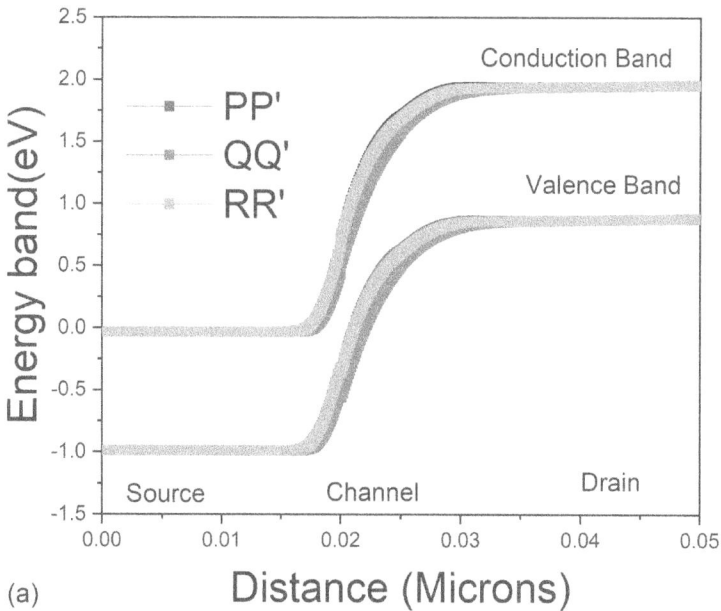

FIGURE 9.5 (a) EBD of a p-type SCTFET used for NAND for input configuration "00" with PP′, QQ′, and RR′ cutlines as shown in Figure 9.1(b) *(Continued)*

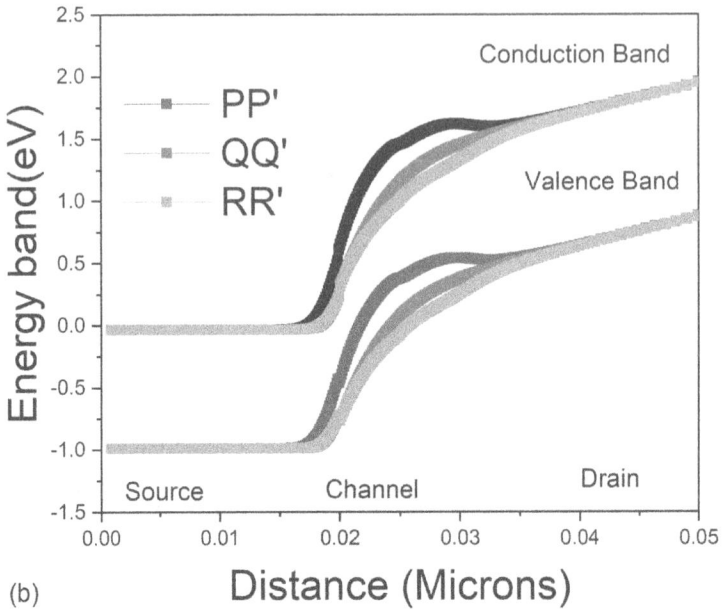

(b)

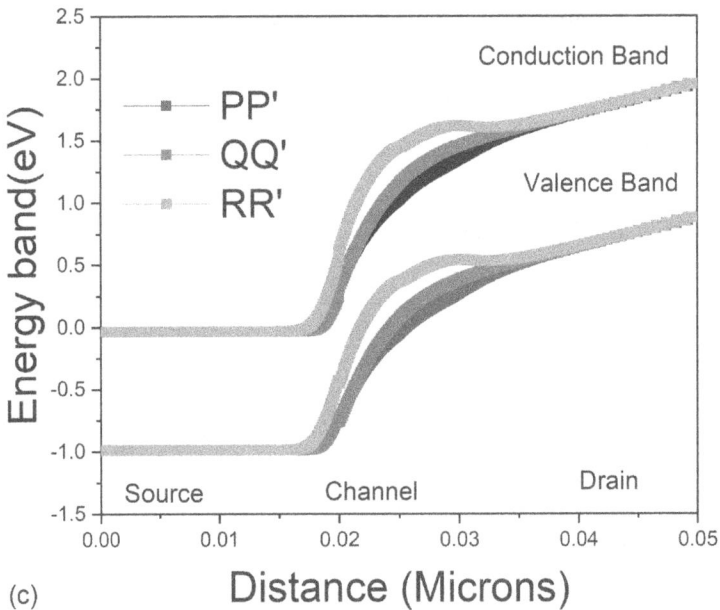

(c)

FIGURE 9.5 *(Continued)* (b) EBD of a p-type SCTFET used for NAND logic for input configuration "01" with PP′, QQ′, and RR′ cutlines, as depicted in Figure 9.1(b) (c) EBD of a p-type SCTFET used for NAND logic for input configuration "10" with PP′, QQ′, and RR′ cutlines, as depicted in Fig. 9.1(b).

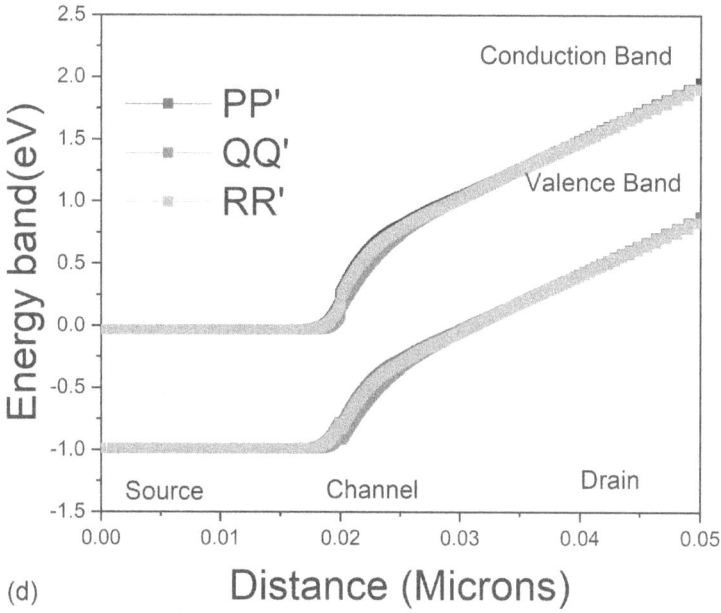

FIGURE 9.5 *(Continued)* (d) EBD of a p-type SCTFET used for NAND logic for input configuration "11" with PP′, QQ′, and RR′ cutlines, as depicted in Fig. 9.1(b).

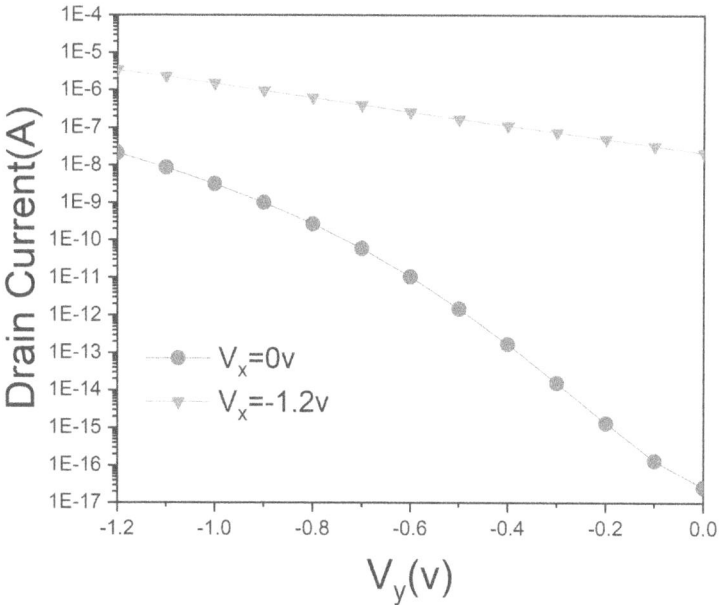

FIGURE 9.6 I_D-V_G curve of p-type SCTFET NAND logic.

9.4 IMPLEMENTATION OF "AND" AND "NOR" FUNCTIONS USING SCTFETS

As shown in Figure 9.7, AND logic is created by using a gate source overlap technique for an n-type SCTFET architecture.

The L_{OV} is assumed to be 4 nm in this case. The logic functioning of the AND gate is further proven using the EBD of the SCTFET, as shown in Figure 9.8. As the length of overlap between the gate and source is extended, the barrier height in the channel region increases; and, because of that, tunneling is not possible for input combinations "01" and "10," as seen in Figure 9.8(b) and (c), the EBD of the SCTFET across PP' and RR' cutlines. As a result, electron tunneling is not conceivable for the input configurations "01" and "10"; if both the top and bottom gates are connected to the high input logic "11," then tunneling is achievable over the cutline QQ', as illustrated in Figure 9.8(d). This represents the functionality of the SCTFET-based AND Boolean gate. "Body tunneling" refers to tunneling deep into the silicon structure. As a result, the L_{OV} allows body tunneling only when both gates relate to a high logic voltage, while limiting tunneling for input logic configurations "01," "10," and "00." If the L_{OV} is deleted, this indicates a configuration like that seen in Figure 9.1(a). Nevertheless, once $L_{OV} = 4$ nm, tunneling is completely blocked, and no further reduction in I_{ON} is seen. As a result, the L_{OV} is used as an optimal value for the I_{OFF}. Table 9.3 lists the different parameters of SCTFET-based AND logic.

Moreover, the oxide thickness has a considerable impact on AND gate functioning [37–39]. Previous research has shown that when oxide thickness increases, the effective tunneling reduces [30]. This results in an increase in the tunneling width between the valence band and the conduction band, which in turn leads to a decrease in the BTBT rate and, as a consequence, a decrease in the I_{ON}/I_{OFF} ratio [40–45]. As a result, the significant T_{OX} is ignored [32, 33] in order to achieve a suitable I_{ON}/I_{OFF} ratio. In the channel, there is no quantum confinement (QC) effect when both n-type SCTFET gates (top and bottom) are linked to logic "11 (described as I_{11}) [29]. As a result, this effect is not taken into account while implementing logic functionality.

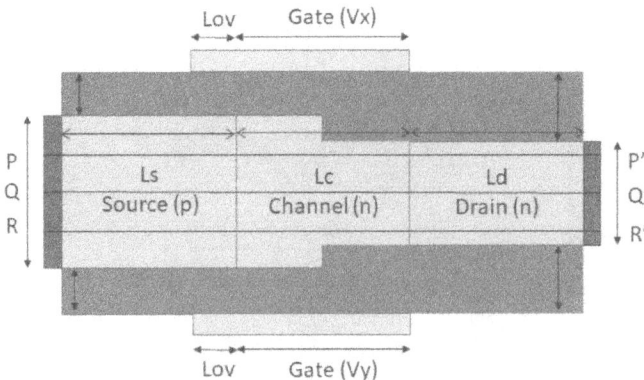

FIGURE 9.7 Cross-sectional view of a conventional SCTFET for AND gate implementation with PP', QQ', and RR' cutlines.

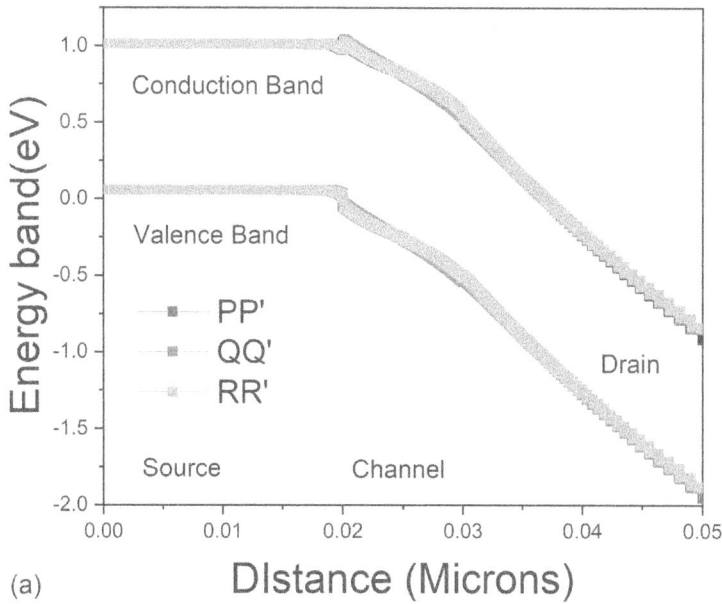

(a)

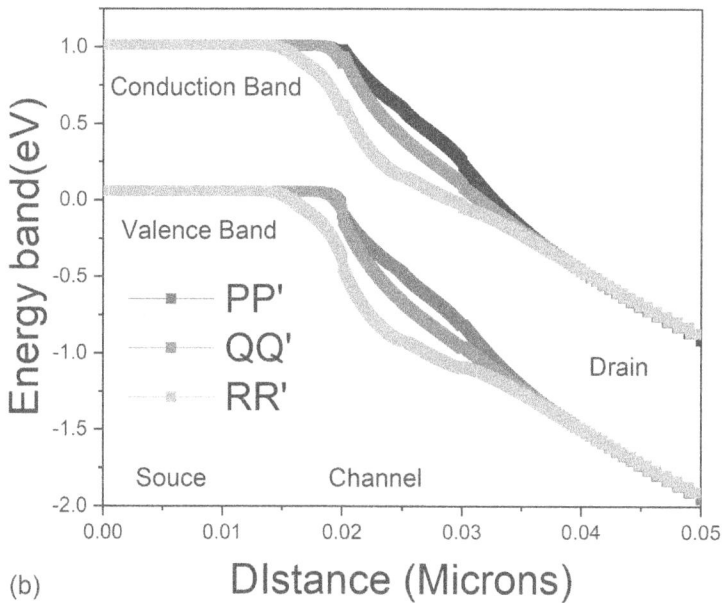

(b)

FIGURE 9.8 EBDs of an n-type SCTFET that are being used for AND logic with a variety of inputs, including (a) "00," (b) "01," along with PP', QQ', and RR' cutlines. *(Continued)*

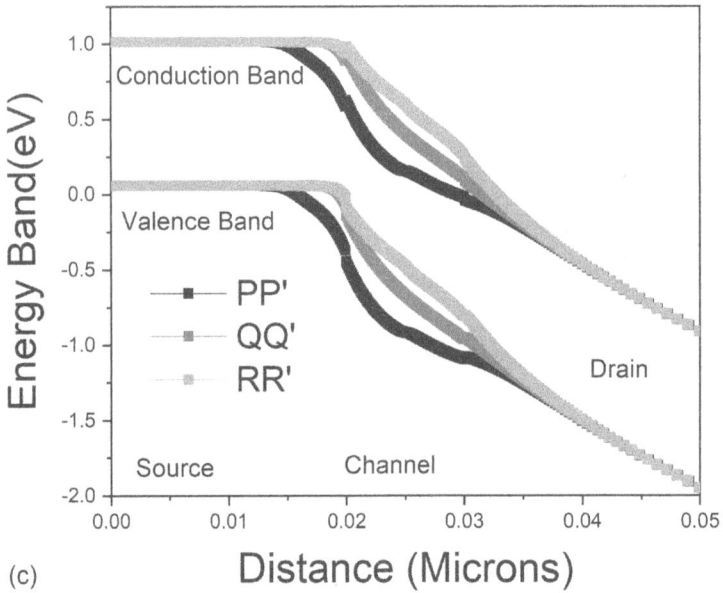

(c)

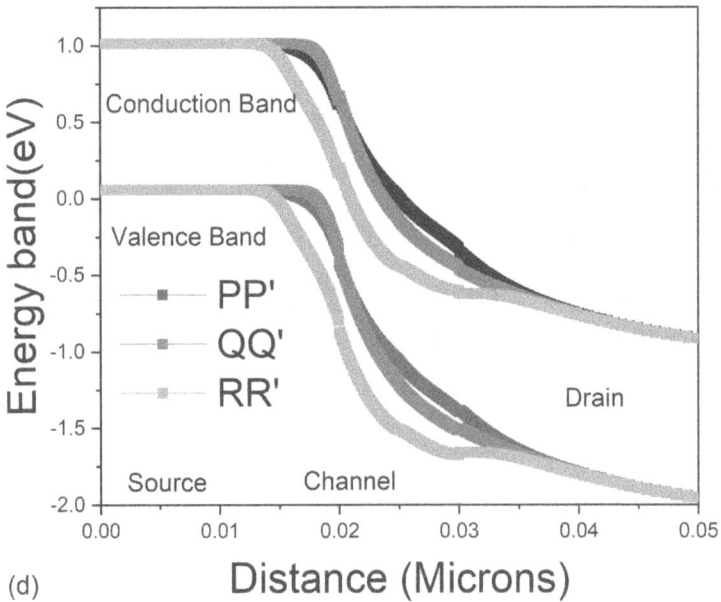

(d)

FIGURE 9.8 *(Continued)* EBDs of an n-type SCTFET that are being used for AND logic with a variety of inputs, including ("01," (c) "10," and (d) "11," along with PP', QQ', and RR' cutlines

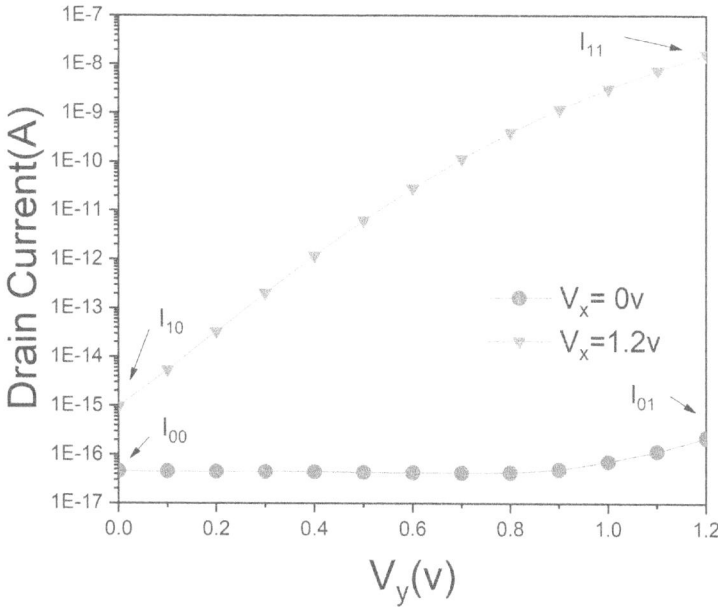

FIGURE 9.9 I_D-V_G curve of n-type SCTFET AND logic.

When $V_{DD} = 1.2$ V and $V_{DS} = -1.0$ V, the I-V characteristics of the SCTFET-based AND logic with L_{OV} are depicted in Figure 9.9. It has been seen that a large I_{ON} current has acquired exclusively for the input combination "11," with an I_{ON}/I_{OFF} ratio in the region of 10^9, while the I_{ON} detected for the remaining input combinations is very low and inconsequential, implying the AND logic operation.

To construct NOR logic, electron tunneling must occur exclusively for the "00" input combination, but for the other input combinations—"11," "01," and "10"—there should be almost no tunneling. To achieve this type of specialized tunneling, L_{OV} is incorporated in the p-type SCTFET architecture, as illustrated in Figure 9.10. Figure 9.11 illustrates the EBDs of p-type SCTFETs for various input combinations spanning cutlines PP′, QQ′, and RR′. If the given input combination is "00," tunneling at the QQ′ is conceivable, as can be seen in Figure 9.11(a). If one or both input logics are "1," there is no overlapping or band bending in the conduction and valence for the specified L_{OV}, as illustrated in Figure 9.11(b)–(d) [29, 30]. Seeing the figures, one may conclude that tunneling is not possible with given inputs. Table 9.3 shows the implementation of SCTFET-based NOR logic using different combinations of voltage supply and WF. The band diagrams at various voltage levels indicate the detection of equivalent BTBT.

Figure 9.12 depicts the "transfer characteristics" for the NOR gate implementation using a p-type SCTFET [31] with L_{OV}; here, the supply voltage V_{DD} is equal to 1.2 V and $V_{DS} = -1.0$ V, and I-V characteristics are shown. Observing the findings, we can conclude that for NOR logic, the "00" input configuration is the only one that gets the high ON-state current, while for all other input combinations the I_{ON} is insignificant and the I_{ON}/I_{OFF} current ratio is in the region of 10^7.

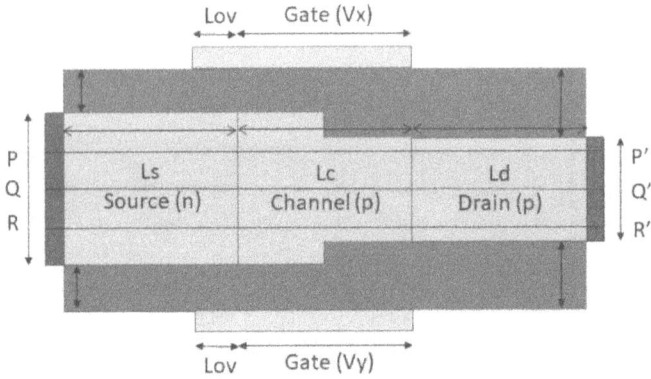

FIGURE 9.10 Cross-sectional view of a conventional SCTFET for NOR gate implementation with PP′, QQ′, and RR′ cutlines.

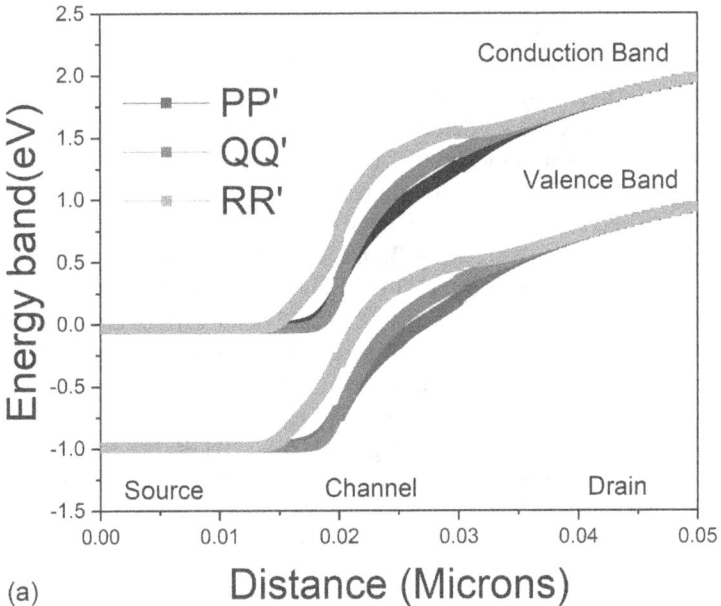

(a)

FIGURE 9.11 (a) A SCTFET's EBD for the NOR logic function for input setup "00" with PP′, QQ′, and RR′ cutlines. *(Continued)*

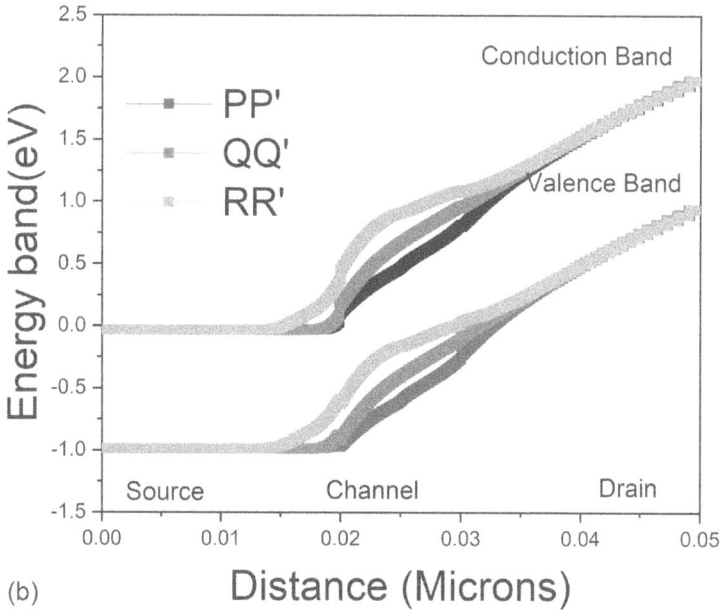

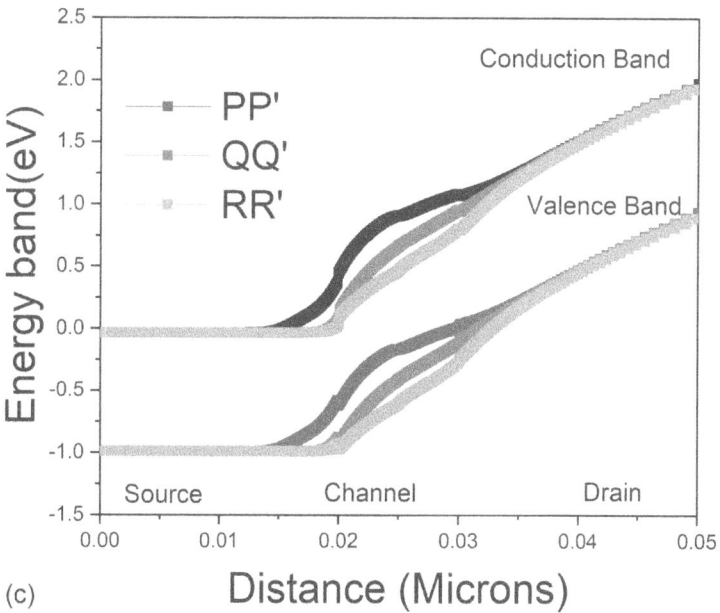

FIGURE 9.11 *(Continued)* (b) A SCTFET's EBD for the NOR logic function for input setup "01" with PP', QQ', and RR' cutlines. (c) A SCTFET's EBD for the NOR logic function for input setup "10" with PP', QQ', and RR' cutlines.

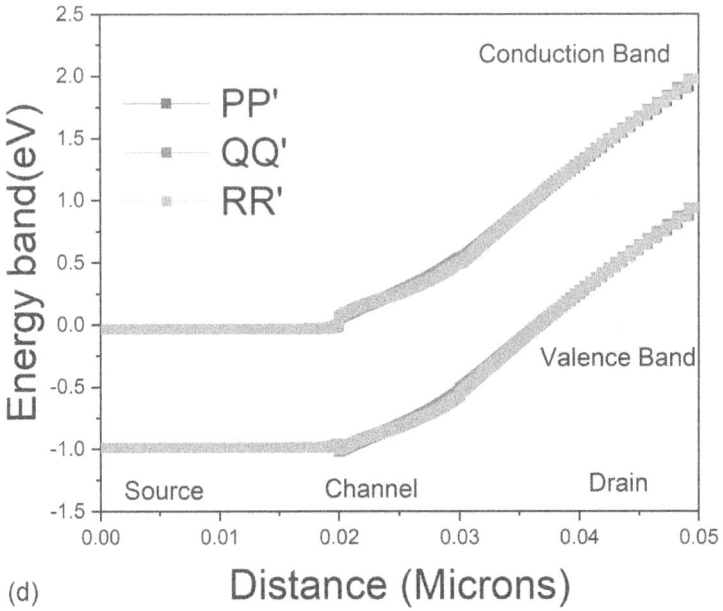

(d)

FIGURE 9.11 *(Continued)* (d) A SCTFET's EBD for the NOR logic function for input setup "11" with PP', QQ', and RR' cutlines.

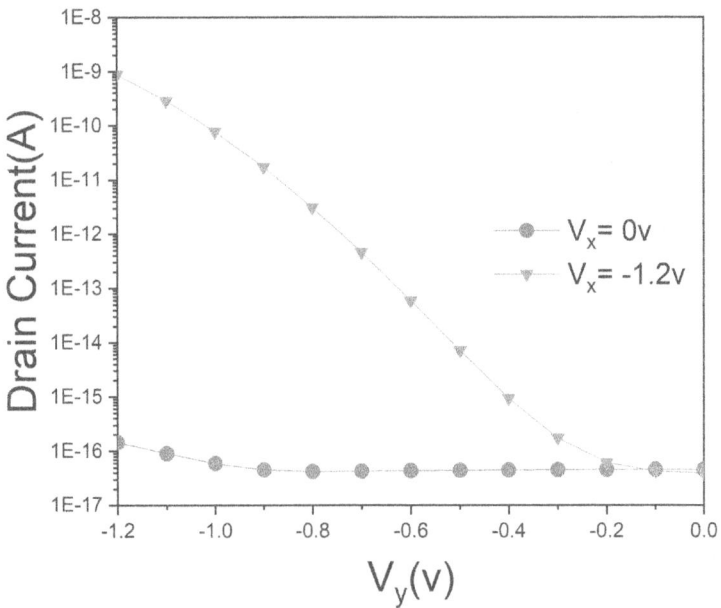

FIGURE 9.12 I_D-V_G curve of p-type SCTFET NOR logic.

9.5 CONCLUSION

We demonstrated in this chapter that Boolean functions can be achieved by utilizing a single SCTFET architecture, by altering the design and selecting the relevant parameters and device characteristics. AND and NOR logic can be implemented using the GSO technique in both p- and n-type SCTFET structures. It is essential that the p-type and n-type SCTFET structures each have their upper and lower gates separately regulated. The transfer characteristic of the SCTFET device is used in conjunction with an EBD to demonstrate the fundamental functionality of logic gates. The EBD of the SCTFET shows that the greatest tunneling is reached for the "11" input in AND logic, Hence, the expected I_{ON}/I_{OFF} current ratio is around 10^9, which is 10 times higher than the prior literature. When it comes to NAND logic, the "00" input combination is the one that achieves maximum tunneling. This, in turn, results in a large I_{ON} current and an I_{ON}/I_{OFF} current ratio in the range of 10^{12}.

As a consequence, the simulation findings that were given in this study are helpful for building high-speed and concise logic functions utilizing a single SCTFET, which permits designs that are more energy efficient.

REFERENCES

1. Verhulst AS, Soree B, Leonelli D, Vandenberghe WG, Groe-´seneken G (2010) J Appl Phys 107:024518. https://doi.org/10.1063/1.3277044
2. Narang R, Saxena M, Gupta RS, Gupta M (2013) Drain current model for a gate all around (GAA) p–n–p–n tunnel FET. Microelectron J 44:479–488. https://doi.org/10.1016/j.mejo.2013.04.002
3. Sakurai T (1990) Alpha-power law MOSFET model and its applications to CMOS inverter delay and other formulas. IEEE J Solid-State Circ 25(2):584–594. https://doi.org/10.1109/4.52187
4. Koga J, Toriumi A (1999) Three-terminal silicon surface junction tunnelling device for room temperature operation. IEEE Electron Device Lett 20(10):529–531. https://doi.org/10.1109/55.791932
5. Knoch J, Appenzeller J (2010) Modeling of high-performance ptype III–v heterojunction tunnel FETs. IEEE Electron Device Lett 31(4):305–307. https://doi.org/10.1109/55.791932
6. Boucart K, Ionescu AM (2007) Double gate tunnel FET with high k gate dielectric. IEEE Trans Electron Devices 54(7):1725–1733. https://doi.org/10.1109/TED.2007.899389
7. Kumar MJ, Janardhanan S (2013) Doping-less tunnel field effect transistor: Design and investigation. IEEE Trans Electron Devices 60(10):3285–3290. https://doi.org/10.1109/TED.2013.2276888
8. Wadhwa G, Raj B (2020) An analytical modeling of charge plasma based tunnel field effect transistor with impacts of gate underlap region. Superlattice Microst 142:106512. https://doi.org/10.1016/j.spmi.2020.106512
9. Knoch J (2009) Optimizing tunnel FET performance—impact of device structure, transistor dimensions and choice of material. In: Proceedings of the international symposium VLSI-TSA, pp 45–46. https://doi.org/10.1109/VTSA.2009.5159285
10. Choi WY, Park B-G, Lee JD, Liu T-JK (2007) Tunnelling field effect transistors (TFETs) with subthreshold swing (SS) less than 60 mV/dec. IEEE Electron Device Lett 28(8):743–745. https://doi.org/10.1109/LED.2007.901273

11. Wadhwa G, Singh J (2020) Implementation of linearly modulated work function gate electrode and Si 0.55 Ge 0.45 N+ pocket doping for performance improvement in gate stack vertical-TFET. Appl Phys A 126(11):1–11. https://doi.org/10.1007/s00339-020-04065-5

12. Ionescu AM, Riel H (2011) Tunnel field-effect transistors as energy efficient electronic switches. Nature 479(7373):329–337. https://doi.org/10.1038/nature10679

13. Saurabh S, Kumar MJ (2016) Fundamentals of tunnel field effect transistors. CRC Press, Boca Raton. https://doi.org/10.1201/9781315367354

14. Singh J, Wadhwa G (2020) Novel linear graded binary metal alloy gate electrode and middle N+ pocket Si0.5 Ge0.5 vertical TFET for high performance. Silicon 1–8. https://doi.org/10.1007/s12633-020-00654-4

15. Guo PF, Yang L-T, Yang Y, Fan L, Han G-Q, Samudra GS, Yeo Y-C (2009) Tunneling field-Effect transistor: Effect of strain and temperature on tunneling current. IEEE Electron Device Lett 30(9):981–983. https://doi.org/10.1109/LED.2009.2026296

16. Choi WY, Lee W (2010) Hetero-gate-dielectric tunneling field effect transistors. IEEE Trans Electron Devices 57(9):2317–2319. https://doi.org/10.1109/TED.2010.2052167

17. Ghosh B, Akram MW (2013) Junctionless tunnel field effect transistor. IEEE Electron Device Lett 34(5):584–586. https://doi.org/10.1109/LED.2013.2253752

18. Lu H, Seabaugh A (2014) Tunnel field-effect transistors: State-of-the ar. IEEE *J* Electron Devices Soc 2(4):44–49. https://doi.org/10.1109/JEDS.2014.2326622

19. Saurabh S, Kumar MJ (2009) Impact of strain on drain current and threshold voltage of nanoscale double gate tunnel field effect transistor. Jpn J Appl Phys 48(6):064503. https://doi.org/10.1143/JJAP.48.064503

20. Priyanka SS, Panchore M (2022) Dopingless-TFET leakyintegrated-fire (LIF) neuron for high-speed energy efficient applications. IEEE Trans Nanotechnol 21:110–117. https://doi.org/10.1109/TNANO.2022.3151241

21. Verhulst AS et al (2008) Complementary silicon-based heterostructure tunnel-FETs with high tunnel rates. IEEE Eletron Device Lett 29(12):1398–1401. https://doi.org/10.1109/LED.2008.2007599

22. Jain G, Sawhney RS, Kumar R, Wadhwa G (2021) Analytical modeling analysis and simulation study of dual material gate underlap dopingless TFET. Superlattice Microst 153:106866. https://doi.org/10.1016/j.spmi.2021.106866

23. Cheung KP (2010) On the 60 mV/dec @300 K limit for MOSFET sub-threshold swing. In: Proceedings of 2010 international symposium on VLSI technology, system and application, Hsinchu, pp 72–73. https://doi.org/10.1109/VTSA.2010.5488941

24. Lahgere A, Panchore M, Singh J (2016) Dopingless ferroelectric tunnel FET architecture for the improvement of performance of dopingless n-channel tunnel FETs. Superlattice Microst 96:16–25. https://doi.org/10.1016/j.spmi.2016.05.004

25. Zhao QT et al (2015) Strained Si and SiGe nanowire tunnel FETs for logic and analog applications. IEEE J Electron Devices Soc 3(3):103–114. https://doi.org/10.1109/JEDS.2015.2400371

26. Betti Beneventi G, Gnani E, Gnudi A, Reggiani S, Baccarani G (2015) Optimization of a pocketed dual-metal-gate TFET by means of TCAD simulations accounting for quantization-induced bandgap widening. IEEE Trans Electron Devices 62(1):44–51. https://doi.org/10.1109/TED.2014.2371071

27. Kumar S, Nigam K, Chaturvedi S et al (2022) Performance improvement of double-gate TFET using metal strip technique. Silicon 14:1759–1766. https://doi.org/10.1007/s12633-021-00982-z

28. Atlas User's Manual Silvaco (2016) Int. Softw., Silvaco Inc., Santa Clara.

29. Banerjee S, Garg S, Saurabh S (2018) Realizing logic functions using single double-gate tunnel FETs: A simulation study. IEEE Electron Device Lett 39(5):773–776. https://doi.org/10.1109/LED.2018.2819205

30. Garg S, Saurabh S (2020) Implementation of boolean functions using tunnel field-effect transistors. J Exploratory Solid-State Comput Devices Circ 6(2). https://doi.org/10.1109/JXCDC.2020.3038073

31. Kumar S et al (2018) 2-D analytical drain current model of double-gate heterojunction TFETs with a SiO2/HfO2 stacked gate-oxide structure. IEEE Trans Electron Devices 65(1):331–338. https://doi.org/10.1109/TED.2017.2773560

32. Kamal N, Panchore M, Singh J (2018) 3-D simulation of junction- and doping-free field-effect transistor under heavy ion irradiation. IEEE Trans Device Mater Reliab 18(2):173–179. https://doi.org/10.1109/TDMR.2018.2811493

33. Kumar MJ, Nadda K (2012) Bipolar charge-plasma transistor: A novel three terminal device. IEEE Trans Electron Devices 59(4):962–967. https://doi.org/10.1109/TED.2012.2184763

34. Vishnoi R, Kumar MJ (2014) A pseudo-2-D-analytical model of dual material gate all-around nanowire tunnelling FET. IEEE Trans Electron Devices 61(7):2264–2270. https://doi.org/10.1109/TED.2014.2321977

35. Prabhat V, Dutta AK (2016) Analytical surface potential and drain current models of dual-metal-gate double-gate tunnel-FETs. IEEE Trans Electron Devices 63(5):2190–2196. https://doi.org/10.1109/TED.2016.2541181

36. Yadav DS, Kamal M (2022) Performance analysis of hetero gate oxide with work function engineering based SCTFET with impact of ITCs. Silicon 14, 11429–11441. https://doi.org/10.1007/s12633-022-01792-7

37. Ambekar V, Panchore M (2022) Realization of boolean functions using heterojunction tunnel FETs. Silicon 14, 6467–6475. https://doi.org/10.1007/s12633-022-01888-0

38. Saurabh S, Kumar MJ, Fundamentals of Tunnel Field-Effect Transistors. Boca Raton, FL, USA: CRC Press, Oct. 2016, doi: 10.1201/9781315367354.

39. Roat P, Singh P, Yadav DS, "Linearity and Analog Performance Analysis of Double Gate Tunnel FET With Source Pocket," 2022 10th International Conference on Emerging Trends in Engineering and Technology - Signal and Information Processing (ICETET-SIP-22), Nagpur, India, 2022, pp. 1–6, doi: 10.1109/ICETET-SIP-2254415.2022.9791609.

40. Singh P, Parmar N, Yadav DS, "Assessment of ITCs on Dual Metal Strip Based Charge plasma TFET: Analog to Linearity analysis," 2022 IEEE Delhi Section Conference (DELCON), New Delhi, India, 2022, pp. 1–6, doi: 10.1109/DELCON54057.2022.9753244.

41. Sharma R, Yadav DS, "Interface Trap Charges and their impact on Linearity and RF Performance metrics for a Heterodielectric Dual Metal Gate GaSb-Si Nanowire TFET," 2021 International Conference on Communication, Control and Information Sciences (ICCISc), Idukki, India, 2021, pp. 1–6, doi: 10.1109/ICCISc52257.2021.9485006.

42. Singh P, Samajdar DP, Yadav DS, "Doping and Dopingless Tunnel Field Effect Transistor," 2021 6th International Conference for Convergence in Technology (I2CT), Maharashtra, India, 2021, pp. 1–7, doi: 10.1109/I2CT51068.2021.9418076.

43. Kumar S, Yadav DS, Saraswat S, Parmar N, Sharma R, Kumar A, "A Novel Step-Channel TFET for Better Subthreshold Swing and Improved Analog/RF Characteristics," 2020 IEEE International Students' Conference on Electrical,Electronics and Computer Science (SCEECS), Bhopal, India, 2020, pp. 1–6, doi: 10.1109/SCEECS48394.2020.104.

44. Sharma R, Yadav DS, Kumar S, Parmar N, Saraswat S, Kumar A, "Novel Perspective Approach to Improve Performance of Nanowire TFET," 2020 IEEE International Students' Conference on Electrical,Electronics and Computer Science (SCEECS), Bhopal, India, 2020, pp. 1–6, doi: 10.1109/SCEECS48394.2020.113.

45. Yadav DS, Sharma D, Sharma DG, Bajpai S, "High Frequency Analysis of GaAsP/InSb Hetero-Junction Double Gate Tunnel Field Effect Transistor," 2018 3rd International Conference for Convergence in Technology (I2CT), Pune, India, 2018, pp. 1–6, doi: 10.1109/I2CT.2018.8529594.

10 CMOS-Based SRAM with Odd-Numbered Transistor Configurations
An Extensive Study

Dharmendra Singh Yadav, Prabhat Singh, and Vibhash Choudhary

10.1 INTRODUCTION

Researchers working with electronic devices face the challenge of making devices with greater performance but lower power consumption. Immense usage of electronic products is causing demand for faster speed, less power consumption, and greater stability [1, 2]. Accessing data takes a lot of time, hindering faster operating speeds, so a circuit is made that will be close to the processor to supply required data [3]. One faster memory device is named "cache memory," which is made up of static random-access memory (SRAM) [4]. It is static in nature, so no refreshing mechanism is needed. SRAM is a type of volatile memory, which means that it holds data until the supply voltage is high, i.e. logic 1. SRAM is used in devices that need processors, like mobiles, laptops, and computers [5, 6]. Dynamic random-access memory (DRAM) also stores data, but it is refreshed periodically to retain the data that needs a refreshing circuit. Due to this reason, DRAM works slower. SRAM became a better choice for speed of operation and power dissipation. A basic variant of SRAM is made up of complementary metal–oxide–semiconductor (CMOS)-based six-transistor (6T) cells; this became the most frequently used commercial topology for cache memory design [7, 8]. Due to increases in the operating speed of devices and high-speed clocks, however, we need memory that responds quicker [9, 10]. There is always a gap between processor speeds and memory speed. To make memory faster, we explore different topologies of SRAM. In this chapter, we explore 7T SRAM, 9T SRAM, 11T SRAM, and 13T SRAM topologies. To achieve greater speed, we sacrifice power because power dissipation and leakages rise with an increasing number of transistors. Therefore, we need to make trade-offs with speed and power dissipation [11, 12]. Stability is also one of the major parameters; due to a decrease in technology nodes, we reduce supply voltages, and from that stability also decreases. To improve stability, we use the cell ratio (CR) and pullup ratio (PR), and by varying them we improve the stability of the circuit.

As the CR and PR rise, the capacitance of the device increases, making it slower and dissipating more energy [13, 14]. As a result, we must design a device that has low power consumption and is reliable.

DOI: 10.1201/9781003393542-10

10.2 VARIOUS TOPOLOGIES OF SRAM AND ITS OPERATION

10.2.1 OPERATION OF 7T SRAM

Resembling positive feedback, a set of back-to-back connected inverters is present in the 7T SRAM (as shown in Figure 10.1), holding bit values "0" or "1." Two pass transistors P_{Tran} and an extra transistor M7 are used to enable the route for reading and writing necessary data to the SRAM cell. We use a distinct pre-charge circuit to charge the circuit to maximum voltage before performing read or write actions. The write driver of the device puts data into the SRAM. The data in the cell is extracted based on the difference in voltage levels in the bit lines (BLs) and bit line bar (BLB) or invert logic of BL, which is sensed by the sensing amplifier S_{amp} circuit [15, 16]. Different operating modes of SRAM are as follows:

1. *HOLD mode*: In this condition, the pair of access transistors A_{Tran} are switched off by whatever data was present before; this data will be kept in an inverter cell that has positive feedback connected to it [17].
2. *Read mode*: To retrieve data from storage, first activate high value on the word lines (WLs) and then pre-charge the BL to high levels. Any of the BL or BLB bit lines can be easily discharged from Q and QB via the bottom component. A sense amplifier is used to calculate the deviation among the signals at each component and displays the result as an output [17].
3. *Write mode*: Bit lines must be pre-charged to high value before reading, and high value must be enabled by WLs, for the SRAM cell to be able to hold the necessary data. There will be a single bottom node between Q and QB, allowing for a direct route to discharge from any of the BL or BLB bit lines. The information for the cell is written by a circuit called the "write controller" [17, 18].

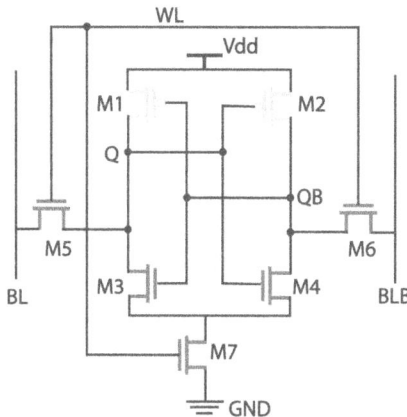

FIGURE 10.1 Seven-transistor (7T) static random-access memory (SRAM) cell for analysis.

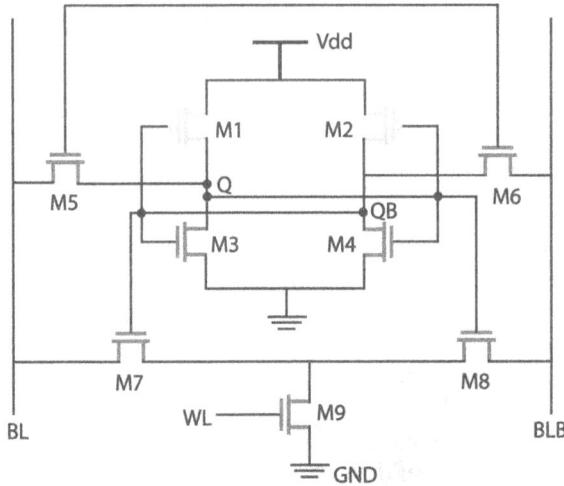

FIGURE 10.2   9T SRAM cell for analysis.

10.2.2   9T SRAM AND ITS OPERATION

Also resembling positive feedback, a set of back-to-back connected inverters is present in the 9T SRAM (as shown in Figure 10.2), which holds bit values "0" or "1." Two P_{Trans} are operated to facilitate the path for read/write required data to the SRAM cell, and M7 and M8 are added extra to the 7T SRAM circuit. Prior to read/write operations, maximum voltage is given to the circuit with the ease of a discrete pre-charge circuit. To write the data into the SRAM, use a circuit named "write driver." To extract the data present in the cell based on the difference in voltage levels in bit lines BL and BLB, use the S_{amp} circuit, which helps to calculate the voltage levels [19, 20].

10.2.3   11T SRAM AND ITS OPERATION

The 11T SRAM contains a pair of inverters (coupled as back-to-back), similar to how it is shown in Figure 10.3. This configuration resembles positive feedback and stores bit values of either "0" or "1." Two pass transistors are utilized in order to enable the path that is necessary to receive and transfer the required data to the SRAM cell. With the assistance of a distinct pre-charge circuit, the circuit tries to achieve its maximum voltage before beginning any reading or writing activities. By utilizing a write controller, the data will be written into the SRAM memory. We are able to retrieve the data that is stored in the cell by analyzing the difference in voltage levels that are found in the BL and BLB, which is something that is detected by the S_{amp} circuit [21–25].

10.2.4   13T SRAM AND ITS OPERATION

The 13T SRAM contains a pair of inverters that are connected back to back, just like it is shown in Figure 10.4. This configuration is similar to positive feedback and

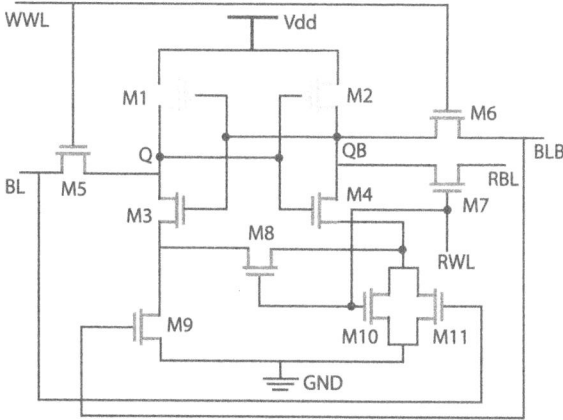

FIGURE 10.3 11T SRAM cell for analysis.

stores bit values of either "0" or "1." In order to enable the path that is necessary to receive and write the essential data to the SRAM cell, two pass transistors are utilized. The distinct pre-charge circuit is utilized in succession to bring the circuit's voltage up to its maximum level before carrying out any read or write activities. The information will be penned into the SRAM with the assistance of the write controller circuit. The differentiation in voltage levels between BL and BLB is used to determine which data is stored in the cell. This is gleaned from the S_{amp} circuit, which is a segment of the device. Read and write procedures both make use of signals such as pre-charge and Write Enable (WE) in order to detect the output through the use of a

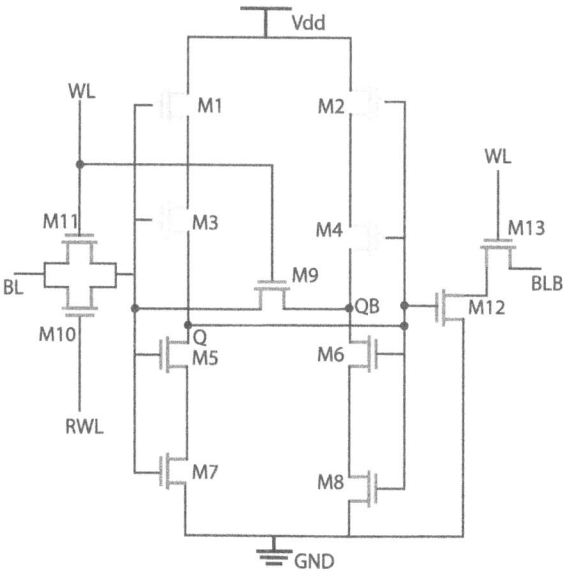

FIGURE 10.4 13T SRAM cell for analysis.

S_{amp} circuit. The CR and PR are determined by Equations (10.1) and (10.2), and their values change in accordance with those equations. Operations in the read, write, and hold phases are very comparable to those of 7T SRAM [26, 27].

$$CR = \frac{\frac{W}{L} \ of \ Pulldown \ Network}{\frac{W}{L} \ of \ Access \ Transistor} \qquad (10.1)$$

$$PR = \frac{\frac{W}{L} \ of \ Pullup \ Network}{\frac{W}{L} \ of \ Access \ Transistor} \qquad (10.2)$$

10.3 READ DELAY CALCULATION AND ITS COMPARISON

Because reading data from SRAM cells competes with higher clock speed, the signals linked to access transistors and read enable must be maintained at high values in read mode. Read delay is an essential parameter for determining circuit speed; it is determined as the time interval among the WL signal when it affects 50% of its maximum value and 50% of the signal present at output [28]. We calculated the read delay measure for SRAM odd-numbered transistors by providing discrete voltage values ranging from 1 to 1.8 V with 0.1 V increments. Figure 10.5 depicts the result of a comparison of odd numbers of transistors, after adding extra transistors to the circuit, which increases the strength of the read current value and gives the advantage of a lower read delay with an increase in transistor numbers and supply voltage [29]. Figures 10.6(a) and 10.7(a) show read delay variations from the 7T SRAM cell for the increase in PR and CR values as per Equations (10.1) and (10.2). From Figures 10.6(a)–(d) and 10.7(a)–(d), we can see noteworthy improvements with respect to voltage variations compared to 1.8 V with respect to 1 V, 7T enhanced by 1.61× with the PR and 1.5× with the CR, 9T improved by 1.4× with the PR and 1.28× with

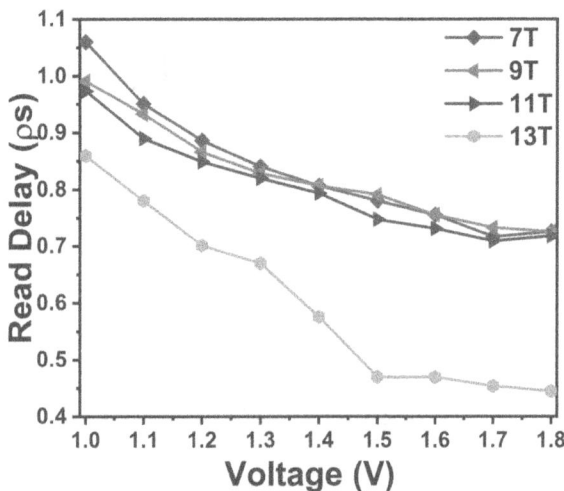

FIGURE 10.5 SRAM read delay comparison for odd numbers of transistors.

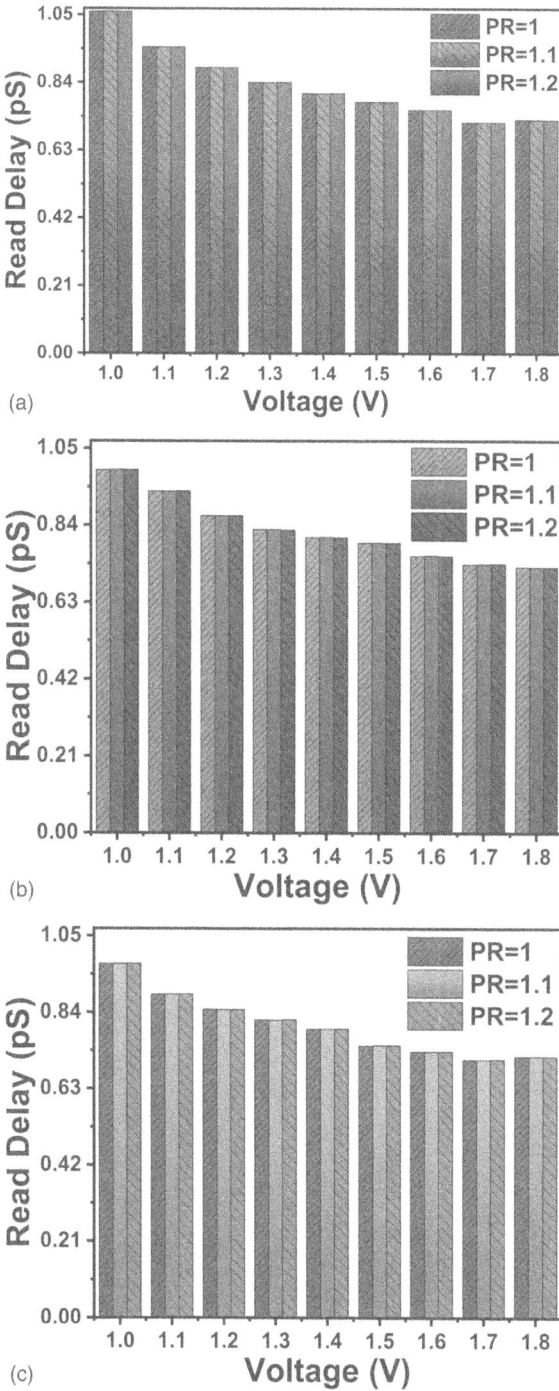

FIGURE 10.6 Read delay variation with the pullup ratio (PR) for (a) a 7T SRAM cell, (b) a 9T SRAM cell, (c) an 11T SRAM cell. *(Continued)*

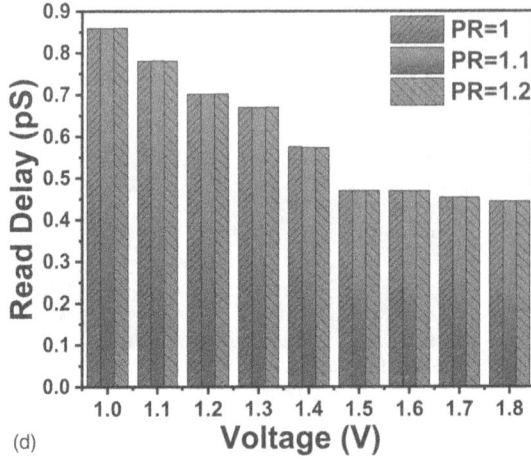

(d)

FIGURE 10.6 *(Continued)* Read delay variation with the pullup ratio (PR) for (d) a 13T SRAM cell.

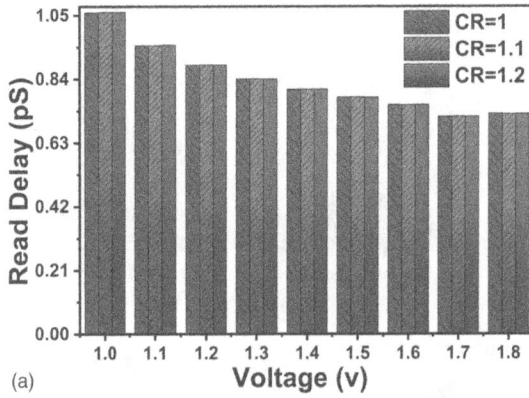

(a)

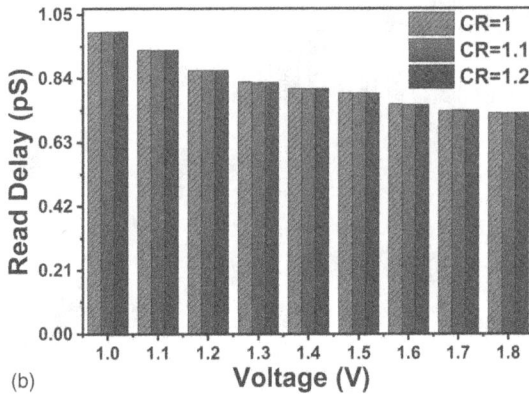

(b)

FIGURE 10.7 Analysis of read delay variations with the cell ratio (CR) for (a) a 7T SRAM cell, (b) a 9T SRAM cell, *(Continued)*

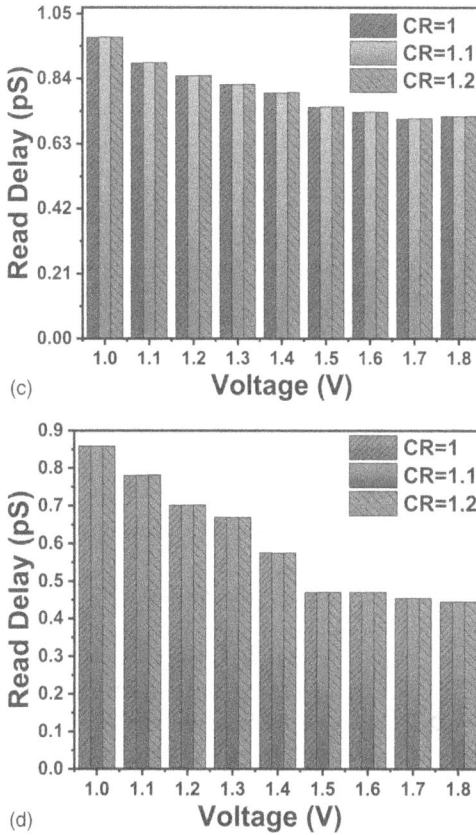

FIGURE 10.7 *(Continued)* Analysis of read delay variations with the cell ratio (CR) for (c) an 11T SRAM cell, and (d) a 13T SRAM cell.

the CR, 11T improved by 1.36× with the PR and 1.30× with the CR, and finally 13T increased by 1.9× with the PR and 2.12× with the CR. This shows that the delay for the read operation is improved as we add more transistors and supply voltage.

10.4 WRITE DELAY (WD)

The write operation should be speedy, and it has to keep up with current high-speed clocks; in write mode operation, the signals related to access transistors and read enable must be kept at high value. WD is computed as the time space relating the WL signal when it reaches 50% of the maximum value of the signal to the 50% of signal Q [30]. We extracted the WD parameter for even-numbered transistors and odd-numbered transistors of SRAM by giving discrete voltage values from 1 to 1.8 V in increments of 0.1 V [31–34]. Figure 10.8 depicts the outcome of a comparison of odd numbers of transistors, after adding extra transistors to the circuit, which raises the intensity of the write current value and provides the benefit of reduced read latency with an increase in transistor numbers and supply voltage. The addition of further transistors

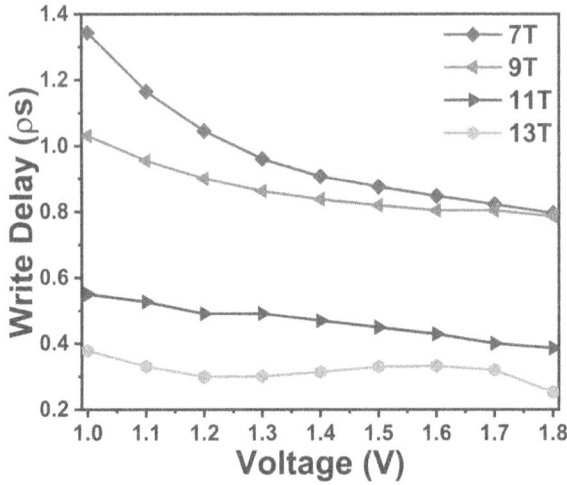

FIGURE 10.8 SRAM write delay comparison for odd numbers of transistors.

will create ease in writing the data into the SRAM cell, which increases the write current and outcomes with less WD.

Figures 10.9(a) and 10.10(a) show WD variations for 7T SRAM cells with increases in PR and CR values, as per Equations (10.1) and (10.2). Thus, it shows that with respect to WD, 7T increased 1.5× with the CR and PR, 11T was enhanced by 1.62× with the PR and 1.28× with the CR, 11T improved 2.9× with the PR and 2.36× with the CR, and 13T improved by 4.21× with the PR and 3.42× with the CR.

Figures 10.9(a)–(d) and 10.10(a)–(d) show significant improvements with respect to voltage variations compared to 1.8 V with 1 V: 7T improved by 1.88× with the PR and 1.58× with the CR, 9T improved by 1.23× with the PR and 1.52× with the CR, 11T increased by 1.44× with the PR and 1.3× with the CR, and finally 13T

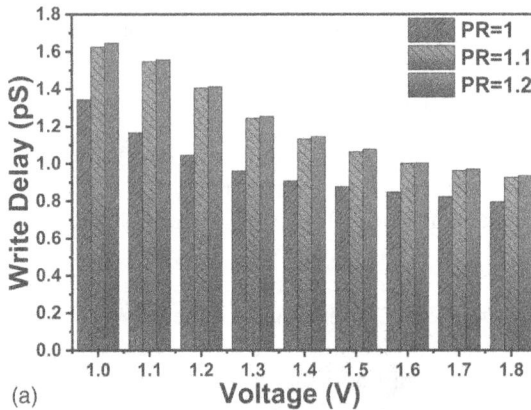

FIGURE 10.9 Write delay (WD) variation with the PR for (a) a 7T SRAM cell, *(Continued)*

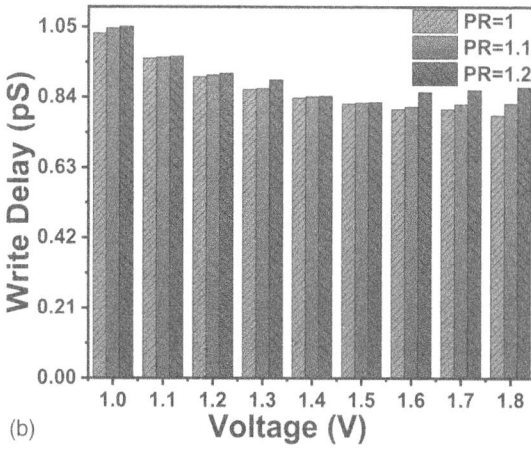

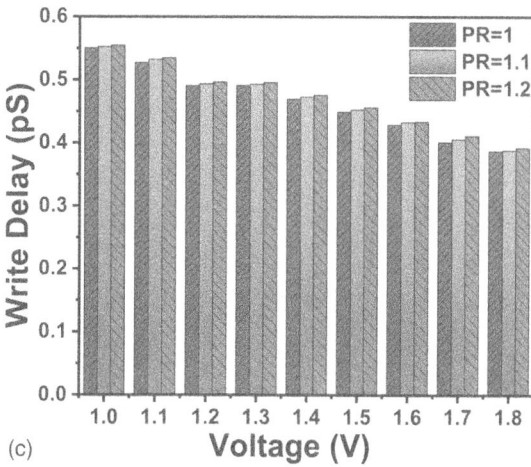

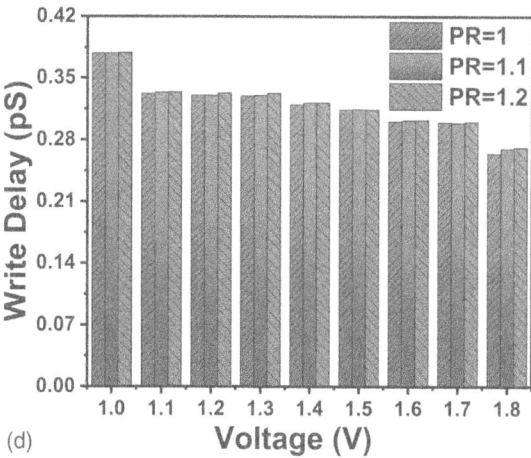

FIGURE 10.9 *(Continued)* Write delay (WD) variation with the PR for (b) a 9T SRAM cell, (c) an 11T SRAM cell, and (d) a 13T SRAM cell.

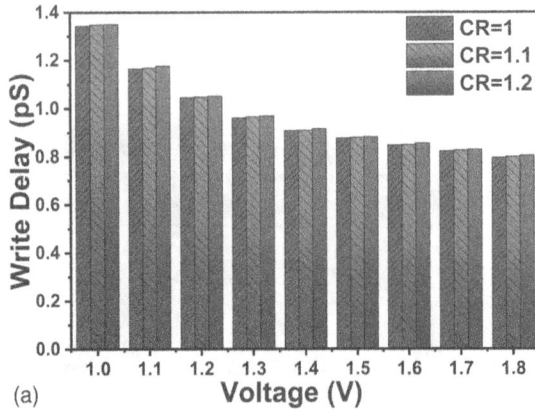

(a)

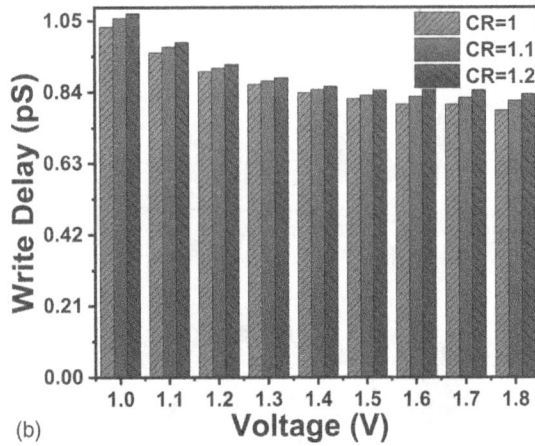

(b)

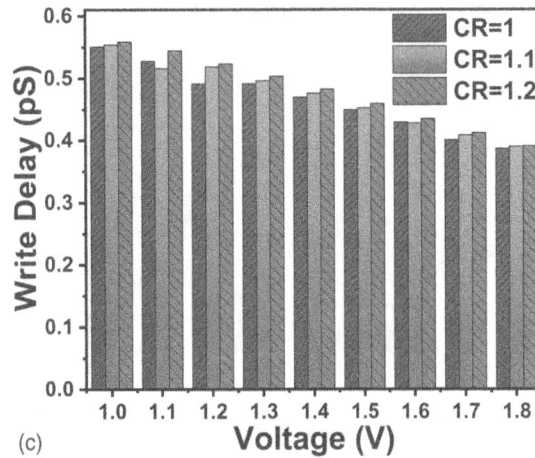

(c)

FIGURE 10.10   Write delay variation with the CR for (a) 7T, (b) 9T, (c) 11T *(Continued)*

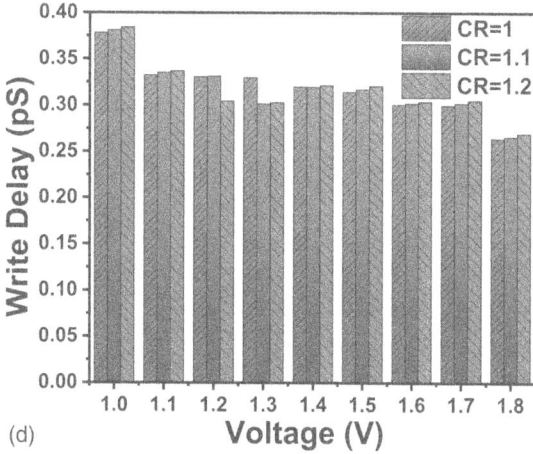

FIGURE 10.10 *(Continued)* Write delay variation with the CR for (d) 13T SRAM cells.

performance increased by 1.52× with the PR and 1.6× with the CR. This shows that WD is improved as we increase the number of transistors, the PR and CR, and the supply voltage.

10.5   AVERAGE POWER DISSIPATION (APD)

APD is the measure of circuit power exhausted during the entire operation of a SRAM cell. It's an important parameter because, to produce a reliable device, we need to reduce the power leakage [35, 36]. Figure 10.11 shows the results of an

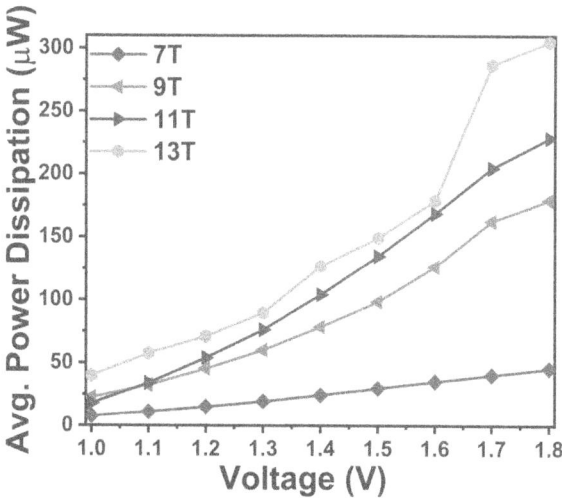

FIGURE 10.11   SRAM average power dissipation (APD) comparison for odd numbers of transistors.

assessment of odd transistor numbers; the circuit with more transistors is consuming high power, so a rise in APD is observed. So, the APD of a 13T SRAM cell is higher than that of an 11T SRAM cell. Also, the APD is relative to the number of transistors and supply voltage [37–39]. Therefore, an increase in either of the values will increase the power dissipation value. Figures 10.12(a) and 10.13(a) show the variation of APD with increases in PR and CR values. Thus, an increase in either the CR or PR will also result in a rise in APD. For portable devices, power dissipation is a crucial parameter, so it's better to opt for low-power techniques for reduction of power dissipation. Thus, it shows that the APD in comparison with 7T got 1.1× with the CR and PR, 11T was enhanced by 4× with the PR and 1.38× with the CR, 11T improved 5× with the PR and 1.92× with the CR, and the APD of 13T increased by 6× with the PR and 2.53× with the CR. Significant improvements can be seen with respect to voltage variations compared to 1.8 V with respect to 1 V: 7T enhanced by 4.8× with the PR and 2× with the CR, 9T performance improved by 7.8× with the PR and 8× with the CR, 11T performance improved by 10× with the PR and 9.2× with the CR, and finally performance of 13T SRAM boosted by 7.5× with the PR and 7.4× with the CR. This is shown in all of the APD graphs of 7T to 13T SRAM in Figures 10.12(a)–(d) and 10.13(a)–(d) for CR and PR variations. This shows that average power consumption rises as we increase the total of transistors, supply voltage, CR, and PR.

10.6 ANALYSIS OF THE STATIC NOISE MARGIN (SNM)

To find the stability of SRAM, the best technique is SNM. Here, we will estimate the noise voltage value to check the deviation in circuit performance. For the inverter circuits, voltage transfer curves (VTCs) are extracted and combined to form the shape of a butterfly and draw a square that fits in it to find the length of one side of the square, which denotes the SNM voltage [40]. Along with this is the need to find

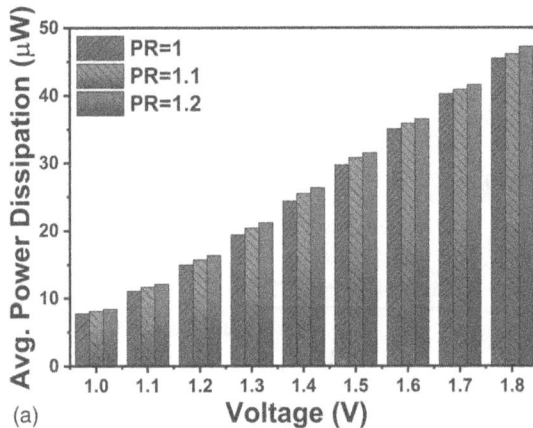

(a)

FIGURE 10.12 APD variation with the PR for (a) a 7T cell *(Continued)*

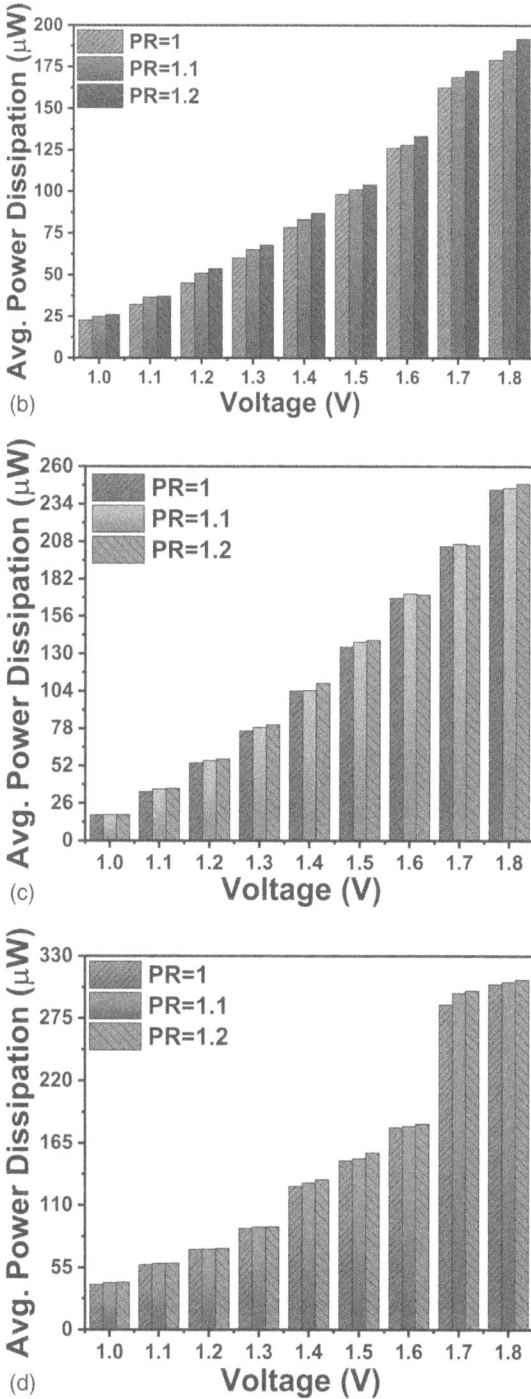

(b)

(c)

(d)

FIGURE 10.12 *(Continued)* APD variation with the PR for (b) a 9T cell, (c) an 11T cell, and (d) a 13T SRAM cell.

(a)

(b)

(c)

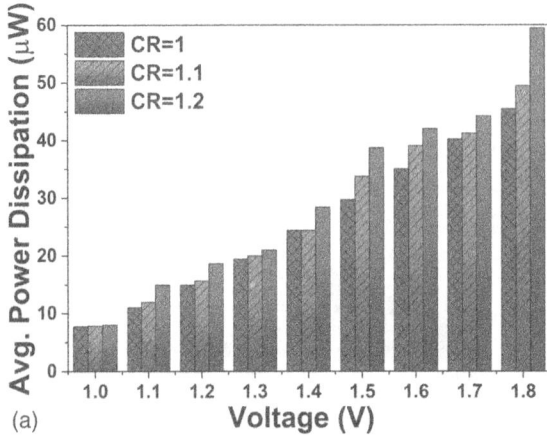

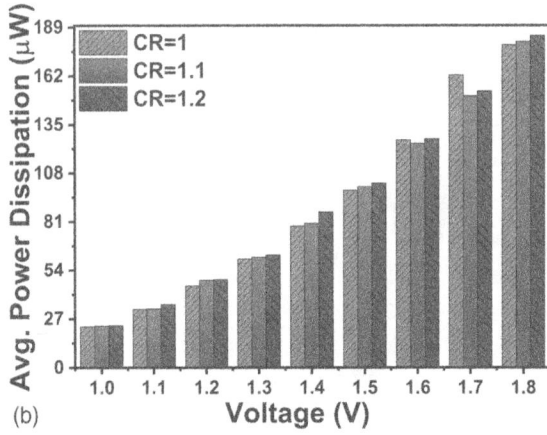

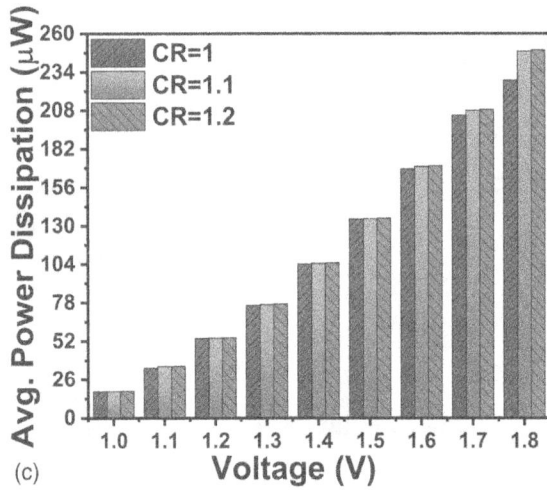

FIGURE 10.13 Investigation of (a) 7T, (b) 9T, (c) 11T *(Continued)*

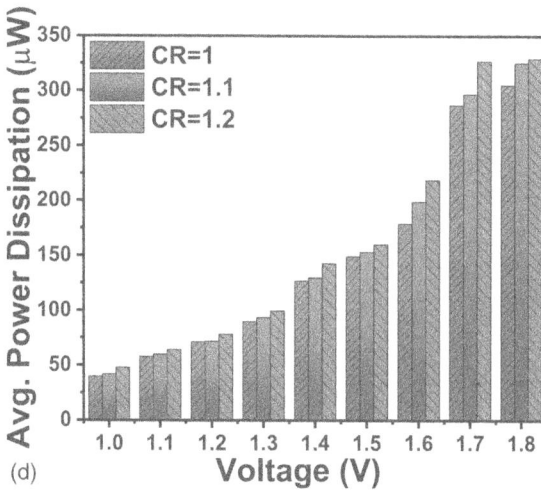

(d)

FIGURE 10.13 *(Continued)* Investigation of (d) 13T SRAM cells' APD with deviation in the CR.

stability for the three modes of operation: (a) hold SNM (SNMH), (b) read SNM (SNMR), and (c) write SNM (SNMW).

10.6.1 HOLD STATIC NOISE MARGIN (SNMH)

An important measuring parameter for stability is the noise voltage added while the processing of the circuit is in HOLD mode, so it flips the state of the data stored in the SRAM cell; this is the static noise margin in hold mode [41, 42]. In the HOLD mode of operation, we turn off pass transistors (T5, T6); only transistors related to cross-coupled inverters are in ON-condition (active state). When we pick one inverter cell and feed input signal Q, then we get output signal QB, which is the VTC waveform [43]. Similarly, by feeding signal QB as input and getting signal Q as output, we get one more VTC waveform; after that, we must merge both waveforms in the origin software (data plotting tool) so that we get a butterfly-shaped waveform and draw a square that fits inside of our butterfly diagram [44]. Hence, the SNM voltage will be the length of one side of the square. Figure 10.14 shows the SNMH for 7T, 9T, 11T, and 13T (odd-transistor-configuration CMOS-based SRAM); all configurations show the same hold margin as per the calculation only for the inverter, irrespective of topology. The optimized SNMH value obtained is approximately 480 mV.

10.6.2 READ STATIC NOISE MARGIN (SNMR)

A crucial measuring factor for stability is the noise voltage added while the processing of the circuit is in HOLD mode, so it flips the state of the data stored in the SRAM cell; this is the SNM in read mode [45, 46]. In the read mode of operation, we turn on pass transistors (T5, T6), and make bit lines BL and BLB high with the help of a pre-charge circuit. When we pick one inverter cell and feed input signal Q, then we

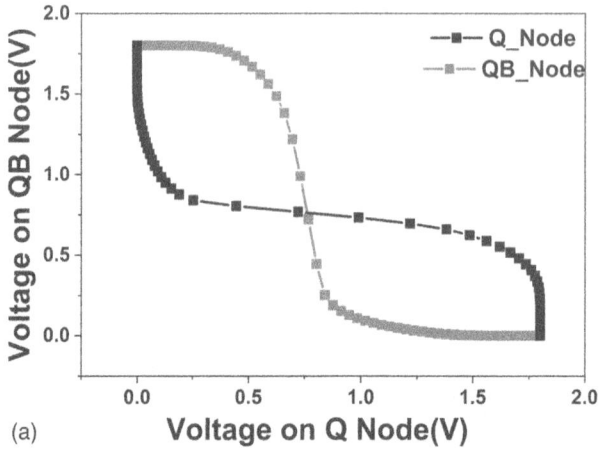

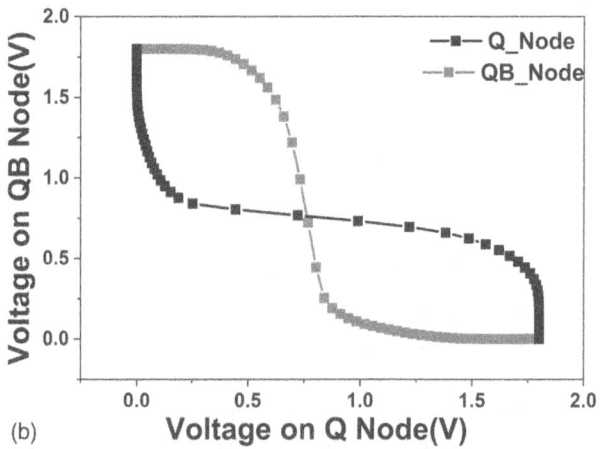

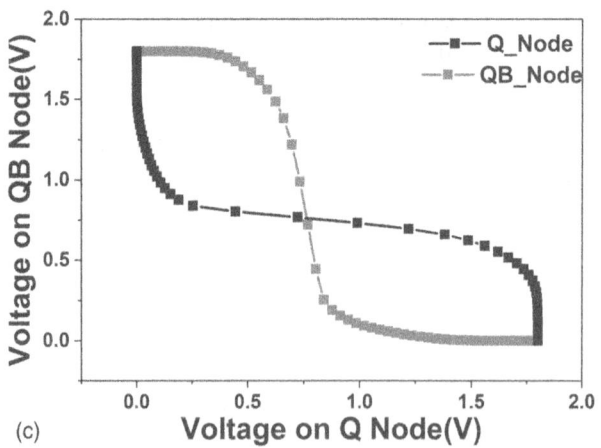

FIGURE 10.14   Analysis of the (a) 7T, (b) 9T, (c) 11T *(Continued)*

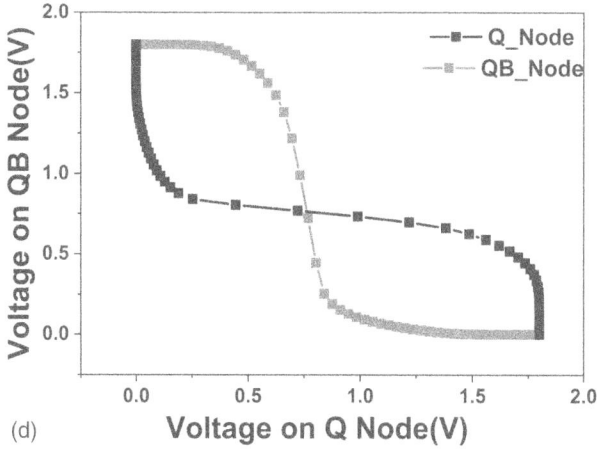

FIGURE 10.14 *(Continued)* Analysis of the (d) 13T SRAM cells' hold static noise margin (SNMH).

get the output signal QB, which is the VTC waveform [47]. Similarly, by maintaining signal QB as input and signal Q as output, one more VTC waveform is obtained; at last, we must merge both waveforms in the origin software to get a butterfly-shaped waveform and draw a square that fits inside of our butterfly diagram. Hence, the SNM voltage will be the length of one side of the square [47, 48]. Figure 10.15 shows a comparative analysis of 7T, 9T, 11T, and 13T for the SNMR. From that, we can observe that, as the number of transistors increases in the SRAM cell, the SNMR will improve accordingly. Furthermore, Figure 10.16(a) depicts the SNMR comparison for a 7T SRAM cell, which shows 235 mV for 1 CR and improved to 252 mV

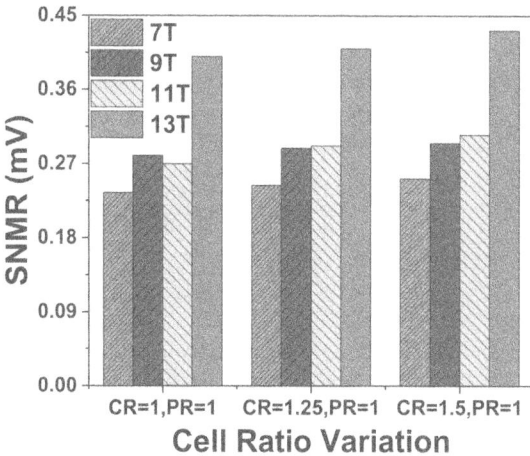

FIGURE 10.15   SRAM read delay comparison for odd numbers of transistors.

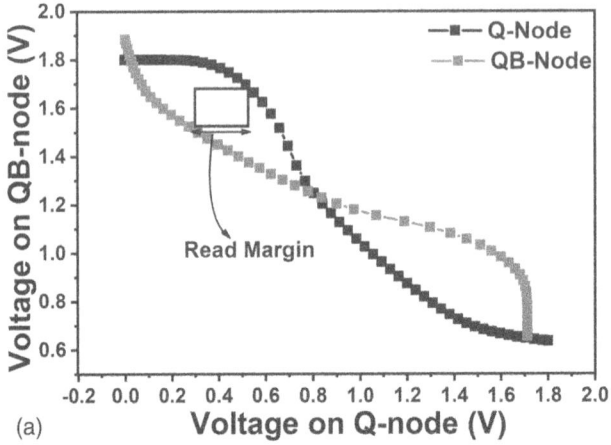

(a)

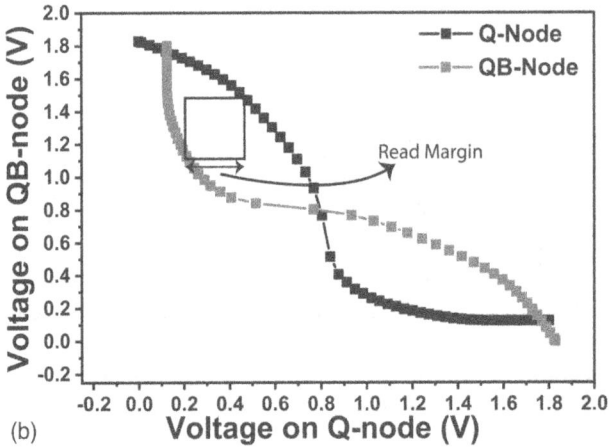

(b)

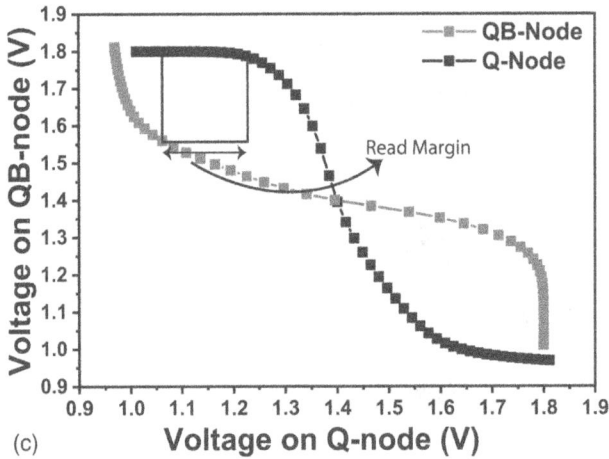

(c)

FIGURE 10.16 Observation of the static noise margin (SNM) for the read mode operation of (a) 7T, (b) 9T, (c) 11T *(Continued)*

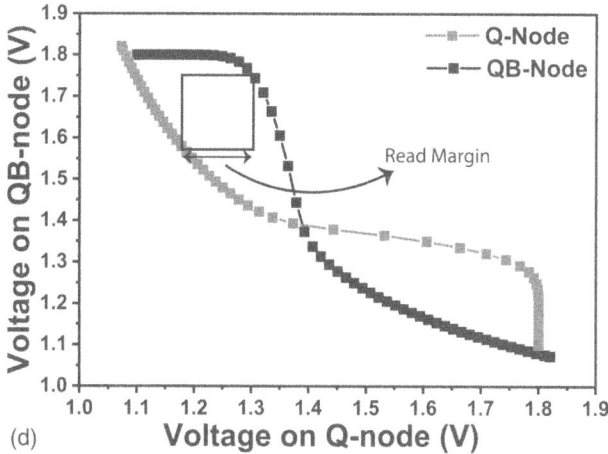

(d)

FIGURE 10.16 *(Continued)* Observation of the static noise margin (SNM) for the read mode operation of (d) 13T SRAM cells.

for 1.5 CR. Similarly, 9T shows 280 mV for 1 CR and increased to 295 mV for 1.5 CR, 11T shows deviation from 270 to 305 mV when the CR varies from 1 to 1.5, and 13T shows an increment of 32 mV for a 0.5 V deviation in the CR (as depicted in Figure 10.16(b)–(d). The read noise margin is related to the CR directly; by increasing the CR, an improvement can be seen in the SNMR. Along with this, an increasing number of transistors in the cell also improves the values of the SNMR, and, through them, the stability of the configuration increases significantly.

10.6.3 WRITE STATIC NOISE MARGIN (SNMW)

An important measuring parameter for stability is the noise voltage added while the processing of the circuit is in HOLD mode, so it flips the state of the data stored in the SRAM cell; this is the SNM in write mode [35]. The write mode of operation turns on pass transistors (T5, T6), and the BL and BLB are at opposite logic at a time so this makes us calculate separate VTC curves [49]. For situations where the BLB is "0," connect corresponding pass transistors by merging with bit line BLB to ground and by applying signal Q, and find out the QB that is the output signal waveform. By considering the state where the BL is high and the gate terminal of the pass transistor is also high, give the input signal Q to get QB, then merge these two signals in the software and make a square that fits inside of our butterfly figure [50]. Hence, SNM voltage will be the length or width of the square. Figure 10.17 shows a comparison of the SNMW for odd-numbered transistor SRAM configurations. In Figure 10.18(a), the SNMW of odd numbers of transistors like 7T shows 790 mV for 1 CR and improved to 838 mV for 1.5 CR. Similarly, 9T shows 820 mV for 1 CR and improved to 852 mV for 1.5 CR (Figure 10.18(b)), 11T shows improvement from 860 to 880 mV for 1 to 1.5 CR (Figure 10.18(c)), and 13T shows 890 mV for 1 CR and improved to 896 mV for 1.5 CR (Figure 10.18(d)).

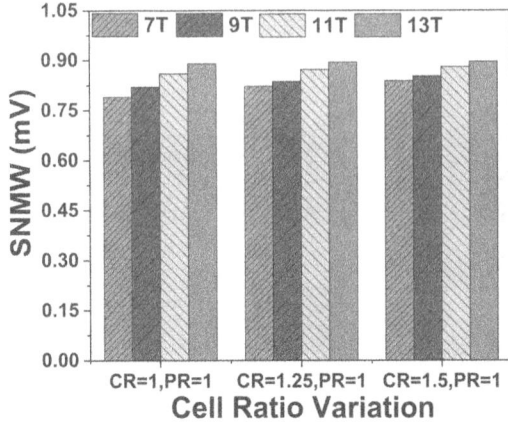

FIGURE 10.17   SRAM write delay comparison for 7T, 9T, 11T, and 13T SRAM cells.

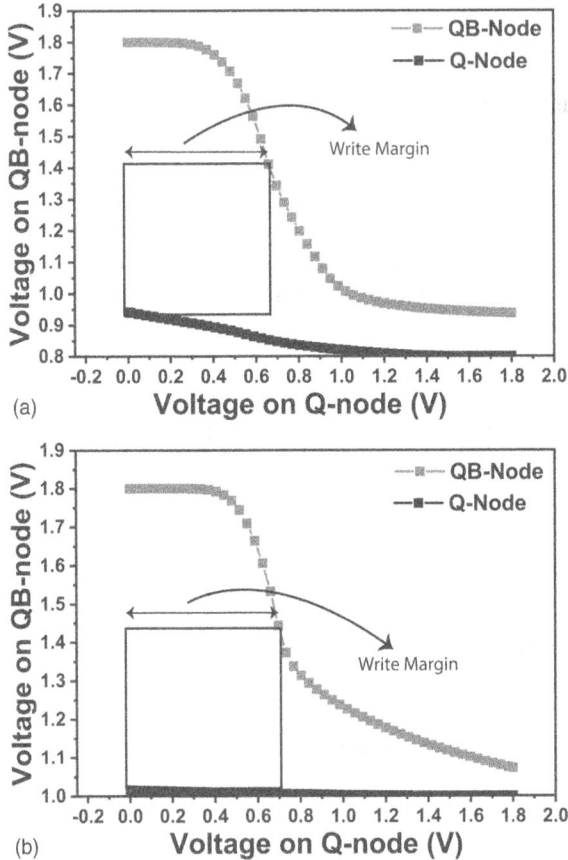

FIGURE 10.18   Data plot for the (a) 7T SRAM, (b) 9T SRAM, *(Continued)*

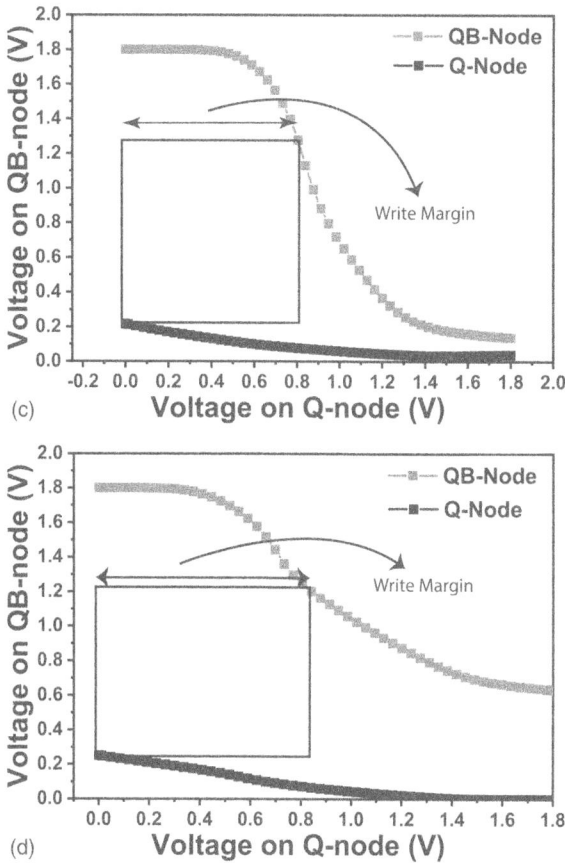

FIGURE 10.18 *(Continued)* Data plot for the (c) 11T SRAM, and (d) 13T SRAM SNMW.

10.7 CONCLUSION

The designed 7T, 9T, 11T, and 13T SRAM topologies employed the Cadence Virtuoso tool and used a technology node (180 nm) to examine different parameters from which we need to opt the best SRAM cell configuration. The read delay is decreasing with increments in voltage because higher voltage drives more read current, and 13T got better in delay compared to 7T as we add the number of transistors in each topology that also contributes to more current. In the same way that WD also reduces with a rise in voltage, a significant reduction can be seen in 13T; as with read delay, write delay also reduces with an increase in voltage due to intensification in the write current. WD rises with modifications in the CR and PR due to an increase in capacitance nature. 7T, 9T, 11T, and 13T show better in WD. APD rises as we add more transistors with a fixed supply voltage. There is a significant rise in power dissipation with a rise in voltage, the CR, and the PR. This shows that there is a substitution nature with delay and power dissipation. Stability is analyzed by the SNM for the hold, read, and write modes, and then the topologies are evaluated by changing the values of the CR and PR. From these observations, we conclude that as CR rises, the read margin also improves. 13T shows better results and, in the same way that PR rises the write margin with better write noise margin. The CR and PR deviations contribute to increased stability with considerable negative changes in delay and power.

REFERENCES

1. Sharma, K., & Mehta, M, & Tyagi, S. "Design and Performance Analysis of 6T SRAM on 130 nm Technology." 2019 International Conference on Signal Processing and Communication (ICSC), pp. 363–368, 2019. IEEE.
2. Singh, V., Singh, S. K., & Kapoor, R. "Static Noise Margin Analysis of 6T SRAM." 2020 IEEE International Conference for Innovation in Technology (INOCON), Bangluru, India, pp. 6–9, 2020, doi:10.1109/INOCON50539.2020.9298431.
3. Yao, M., Cabanas-Holmen, M. F., & E. H. Cannon, "Direct Measurement Structure of SRAM SNM," pp. 273–276. https://apps.dtic.mil/sti/pdfs/AD1075360.pdf.
4. Rajput, A. S., Pattanaik, M., & Tiwari, R. "Estimation of Static Noise Margin by Butterfly Method using Curve-Fitting Technique." Journal of Active and Passive Electronic Devices, vol. 13, no. 1, pp. 1–9, 2018.
5. Mittal, D., & Tomar, V. K. "Performance Evaluation of 6T, 7T, 8T, and 9T SRAM cell Topologies at 90 nm Technology Node." 2020 llth International Conference on Computing, Communication and Networking Technologies, Kharagpur, India, pp. 4–8, 2020, doi:10.1109/ICCCNT49239.2020.9225554.
6. Viswash, B V and Chinmaye, R. "A Comparative Study of SRAM Cells." International Journal of Innovative Research in Electrical, Electronics, Instrumentation and Control Engineering, vol. 3, no. 11, pp. 92–94, 2015, doi: 10.17148/ijireeice.2015.31119.
7. Rukkumani, V., Saravanakumar, M., & Srinivasan, K. "Design and analysis of SRAM cellsfor power reduction using low power techniques." IEEE Region 10 Conference (TENCON), pp. 3058–3062, 2017, doi: 10.1109/TENCON.2016.7848609.
8. Shalini, C., & Rajendar, S. (2017). "CSI-SRAM: Design of CMOS Schmitt trigger inverter based SRAM cell for low power applications." 2017 International Conference on Energy, Communication, Data Analytics and Soft Computing (ICECDS). doi:10.1109/icecds.2017.8389823
9. Kiran, P. N. V., & Saxena, N. (2015). "Design and analysis of different types SRAM cell topologies." 2015 2nd International Conference on Electronics and Communication Systems (ICECS). doi:10.1109/ecs.2015.7124742
10. Sil, A., Ghosh, S., Gogineni, N., & Bayoumi, M. (2008). "A novel high write speed, low power, read-SNM-free 6T SRAM cell." 2008 51st Midwest Symposium on Circuits and Systems. doi:10.1109/mwscas.2008.4616913
11. Rollini, R., Sampson, J., & Sivakumar, P. (2017). "Comparison on 6T, 5T and 4T SRAM cell using 22nm technology." 2017 IEEE International Conference on Electrical, Instrumentation and Communication Engineering (ICEICE). doi:10.1109/iceice.2017.8191924
12. Akashe, S., Tiwari, N. K., & Sharma, R. (2012). "Simulation and stability analysis of 6T and 9T SRAM cell in 45 nm era." 2012 2nd International Conference on Power, Control and Embedded Systems. doi:10.1109/icpces.2012.6508061
13. Suneja, D., Chaturvedi, N., & Gurunarayanan, S. (2017). "A comparative analysis of read/write assist techniques on performance & margin in 6T SRAM cell design." 2017 International Conference on Computer, Communications and Electronics (Comptelix).

14. Giterman, R., & Fish, A. (2014). "Towards a black-box methodology for SRAM stability analysis." 2014 IEEE 28th Convention of Electrical & Electronics Engineers in Israel (IEEEI). doi:10.1109/eeei.2014.7005777

15. Sachdeva, A., & Tomar, V. K. (2020). "Statistical Stability Characterization of Schmitt Trigger Based 10-T SRAM Cell Design." 2020 7th International Conference on Signal Processing and Integrated Networks (SPIN).

16. Kolhal, R., & Agarwal, V. (2019). "A Power and Static Noise Margin Analysis of different SRAM cells at 180nm Technology." 2019 3rd International Conference on Electronics, Communication and Aerospace Technology (ICECA). doi:10.1109/iceca.2019.8821868

17. Krishnan, A. R., & Shekar, G. (2020). "Static Noise Margin Analysis of Various SRAM Array." 2020 International Conference on Communication and Signal Processing (ICCSP).

18. Reddy, B. N. K., Sarangam, K., Veeraiah, T., & Cheruku, R. (2019). "SRAM cell with better read and write stability with Minimum area." TENCON 2019 - 2019 IEEE Region 10 Conference (TENCON). doi:10.1109/tencon.2019.8929593

19. Manan, A. (2018). "Efficient 16 nm SRAM Design for FPGA's." 2018 5th International Conference on Signal Processing and Integrated Networks (SPIN). doi:10.1109/spin.2018.8474069

20. Singh, C., Grover, A., & Grover, N. (2013). "Implementation of a Modified Model-SRAM Using Tanner EDA." 2013 Fifth International Conference on Computational Intelligence, Modelling and Simulation. doi:10.1109/cimsim.2013.69

21. Madan, R., Gupta, R., Nirwan, B. S., & Grover, A. (2015). "Comparative analysis of SRAM cells in sub-threshold region in 65nm." 2015 International Conference on Advances in Computer Engineering and Applications. doi:10.1109/icacea.2015.7164763

22. Ho, W.-G., Zheng, Z., Chong, K.-S., & Gwee, B.-H. (2018). "A Comparative Analysis of 65nm CMOS SRAM and Commercial SRAMs in Security Vulnerability Evaluation." 2018 IEEE 23rd International Conference on Digital Signal Processing (DSP). doi:10.1109/icdsp.2018.8631874

23. Shaik, S., & Jonnala, P. (2013). "Performance evaluation of different SRAM topologies using 180, 90 and 45 nm technology." 2013 International Conference on Renewable Energy and Sustainable Energy (ICRESE). doi:10.1109/icrese.2013.6927819

24. Keshavapura, S., Jain, S., & Pattnaik, M. (2012). "A new assist technique to enhance the read and write margins of low voltage SRAM cell." International Symposium on Electronic System Design.

25. Bellerimath, P., & Banakar, R. M. "Implementation of 16x16 SRAM memory array using 180nm technology." IJCET - International Journal of Current Engineering and Technology, pp. 2277–4106, 2013.

26. Madhavi, B. K., & Reddy, T. V. (2017). "Design strategy & analysis of Subthreshold SRAM in power & delay for wearable applications." 2017 2nd International Conference on Communication and Electronics Systems (ICCES). doi:10.1109/cesys.2017.8321209

27. Kapre, R., Shakeri, K., Puchner, H., Tandigan, J., Nigam, T., Jang, K., & Whately, M. (2007). "SRAM Variability and Supply Voltage Scaling Challenges." 2007 IEEE International Reliability Physics Symposium Proceedings. 45th Annual. doi:10.1109/relphy.2007.369863.

28. Bhaskar, A. (2017). "Design and analysis of low power SRAM cells." 2017 Innovations in Power and Advanced Computing Technologies (i-PACT), Vellore, India (2017.4.21–2017.4.22), pp. 1–5. doi:10.1109/IPACT.2017.8244888.

29. Praveen, K. N., & Shivaleelavathi, B. G., "SRAM Memory Layout Design in 180nm Technology," IJERT - International Journal of Engineering Research and Technology, vol. 4, no. 8, 2015.

30. Patel, P., Zafar, S., & Soni, H. "Performance of Various Sense Amplifier Topologies in sub100nm Planar MOSFET Technology." IJETTCS - International Journal of Emerging Trends & Technologies in Computer Science, vol. 3, no. 2, 2014.

31. Singh, S., & Lakhmani, V. "Read and Write Stability of 6T SRAM." IJARECE - International Journal of Advanced Research in Electronics and Communication Engineering, vol. 3, no. 5, 2014.

32. Mann, R. W., Nalam, S., Wang, J., & Calhoun, B. H. "Limits of Bias Based Assist Methods in Nano-Scale 6T SRAM." 11th Int'l Symposium on Quality Electronic Design, 2010, pp. 1–6.

33. Sharma, R., & Kumar, G. "Reliability Aware Negative Bit-Line Voltage Write Assist Scheme for SRAM." International Journal of Advanced Research in Computer Science and Software Engineering, vol. 3, no. 8, 2013.

34. Rahman, N., & Singh, B. P. "Static-Noise-Margin Analysis of Conventional 6T SRAM Cell at 45nm Technology." International Journal of Computer Applications, vol. 66, no. 20, 2013.

35. Chung, Y., & Song, S. H. "Implementation of Low-Voltage Static RAM with Enhanced Data Stability and Circuit Speed." Microelectronics Journal, vol. 40, no. 6, pp. 944–951, 2009.

36. Yoo, S.-M., Han, J. M., Hag, E., Yoon, S. S., Jeong, S.-J., Kim, B. C., Lee, J.-H., Jang, T.-S., Kim, H.-D., Park, C. J., Seo, D. H., Choi, C. S., Cho, S.-I., & Hwang, C. G. "A 256 M DRAM with simplified register control for low power self refresh and rapid burn-in." Symp. VLSI Circuits Dig. Tech. Papers, 1994, pp. 85–86.

37. Birla, S., Singh, R. K., & Pattnaik, M. "Static Noise Margin Analysis of Various SRAM Topologies." IACSIT - International Journal of Engineering and Technology, vol. 3, No. 3, 2011, pp. 304–309.

38. Choudhary, D., & Bhatnagar, V., "Low Voltage and Low Power in SRAM Read and Write Assist Techniques," IJSTE - International Journal of Science Technology & Engineering, vol. 3, no. 02, 2016.

39. Mann, R. W., Wang, J., Nalam, S., Khanna, S., Braceras, G., Pilo, H., & Calhoun, B. H., "Impact of Circuit Assist Methods on Margin and Performance in 6T SRAM," Solid State Electron., vol. 54, no. 11, pp.1398–1407, Nov. 2010.

40. Calhoun, B. H., & Chandrakasan, P. "A 256-kb 65-nm Sub-threshold SRAM Design for Ultra-Low-Voltage Operation." IEEE Journal of Solid-State Circuits, vol. 42, no. 3, pp. 680–688, 2007.

41. Grossar, E., Stucchi, M., & Maex, K. "Read Stability and Write-Ability Analysis of SRAM Cells for Nanometer Technologies." Solid-State Circuits, IEEE Journal, vol. 41, no. 11, pp. 2577–2588, 2006.

42. Liu, Z., & Kursun, V. "Characterization of a Novel Nine-Transistor SRAM Cell." IEEE Transactions on Very Large Scale Integration (VLSI) Systems, vol. 16, no. 4, pp. 488–492, 2008.

43. Lin, S., Kim, Y.-B., & Lombardi, F., "A Low Leakage 9T SRAM Cell for Ultra-Low Power Operation." ACM Great Lakes Symposium on VLSI 2008, May 2008, pp. 123–126.

44. Chandrakasan, A., Bowhill, W. J., & Fox, F. "Design of High-Performance Microprocessor Circuits." IEEE Press, 2000.

45. Thomas, O. "Impact of CMOS Technology Scaling on SRAM Standby Leakage Reduction techniques." ICICDT, May 2006.

46. De, V. et al. "Techniques for Leakage Power Reduction," in Design of High-Performance Microprocessor Circuit, Circuits, A. Chandrakasan, W. J. Bowhill, and F. Fox, Eds. Piscataway, NJ: IEEE, 2001, pp. 285–308.

47. Calhoun, Benton H., & Chandrakasan, Anantha P. "Static Noise Margin Variation for Sub-threshold SRAM in 65 nm CMOS." IEEE Journal of Solid-State Circuits, vol. 41, no. 7, pp. 1673–1679, 2006

48. Singh, P., & Yadav, D. S. "Impact of work function variation for enhanced electrostatic control with suppressed ambipolar behavior for dual gate L-TFET." Current Applied Physics, vol. 44, pp. 90–101, 2022.
49. Singh, P., & Yadav, D. S. "Performance analysis of ITCs on analog/RF, linearity and reliability performance metrics of tunnel FET with ultra-thin source region." Applied Physics A, vol. 128, no. 7, 612, 2022.
50. Pavlov, A., & Sachdev, M. CMOS SRAM Circuit Design AND Parametric Test in Nano-Scaled Technologies. Springer, 2008

11 Gate-All-Around Nanosheet FET Device Simulation Methodology Using a Sentaurus TCAD

Anushka Singh and Archana Pandey

11.1 INTRODUCTION TO THE SYNOPSYS SENTAURUS TCAD SUITE AND DEVICE SIMULATION

Technology computer-aided design (TCAD) is any design software that is used for developing and optimizing semiconductor process technologies and devices using computer simulations [1]. Synopsys TCAD provides a varied range of products that include industry-leading devices and process simulation tools, along with an extremely powerful graphical user interface (GUI)-based/enabled simulation atmosphere to manage simulation tasks and analyses of simulation results. Moreover, Synopsys TCAD has tools for interconnecting modeling and extraction, and for providing critical parasitic information to enable optimum chip performance [1–3].

In three-dimensional (3D) technology modeling, Synopsys TCAD is the industry lead and successfully provides a vast integrated simulation platform for design co-optimization. TCAD device simulation and process simulation tools back a wide range of applications, like fin field-effect transistors (FinFETs), gate-all-around nanosheet field-effect transistors (GAA-NSFETs), gate-all-around nanowire field-effect transistors (GAA-NWFETs), complementary metal–oxide–semiconductor field-effect transistors (CMOS-FETs), analog/radiofrequency (RF) devices, and so on. In addition, TCAD provides proven production and modeling solutions for memory, logic, and power applications. Moreover, it offers intelligent technology modeling that encourages further cost and time savings in path-finding, development, and production ramping of semiconductor technologies [3, 4]. Time consumption due to expensive device fabrication processes is significantly reduced due to access to a TCAD simulation at the design stage for analysis of physical as well as electric properties [4].

In summary, Sentaurus is an entourage of TCAD tools that perform several simulation operations, check for device structure reliability, and eventually fabricate a semiconductor device. Physical models are used by Sentaurus simulators to deputize fabrication procedures and operations of the device, thus eventually leading to better optimization and exploration of next-generation semiconductor devices. Hence, Sentaurus's TCAD tools function coherently and thus are easy to combine in simulation process flows in two-dimensional (2D) and 3D design. Thus, the TCAD software supports an array of semiconductor technologies, and hence covers a wide

DOI: 10.1201/9781003393542-11

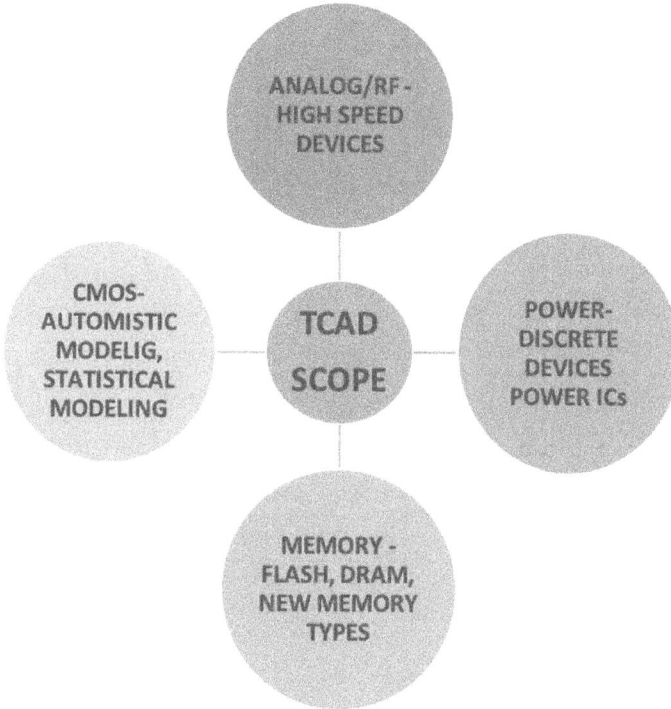

FIGURE 11.1 Synopsys Sentaurus TCAD (technology computer-aided design) suite: Application scope.

range of semiconductor utilization areas [2–4]. The myriad scopes of TCAD are illustrated in Figure 11.1.

The intention and focus of the chapter are to build a solid knowledge base regarding Synopsys Sentaurus TCAD to eventually enable the design and simulation of 3D devices like GAA-NSFETs.

11.1.1 Simulation in Synopsys Sentaurus TCAD and its Benefits

Sentaurus TCAD holds immense value in technology development and optimization and provides tools for device performance simulation. As feedback to exterior thermal, electric, or ocular boundary situations inflicted over the structure, the tools simulate the electric attributes of semiconductor devices. Device design steps are done with the use of tools such as the Sentaurus Structure Device Editor (SDE). Hence, SDE as a tool enables simulation of both 2D and 3D devices [2–3].

The semiconductor manufacturing industry faces immense challenges in developing process technologies with limited time and cost restrictions. The most significant factor that impacts consistent development cost and time is the amount of wafers required to bring the development of any new process to completion. Hence, TCAD enables a significant reduction in wastage of engineering wafers, time, and cost by

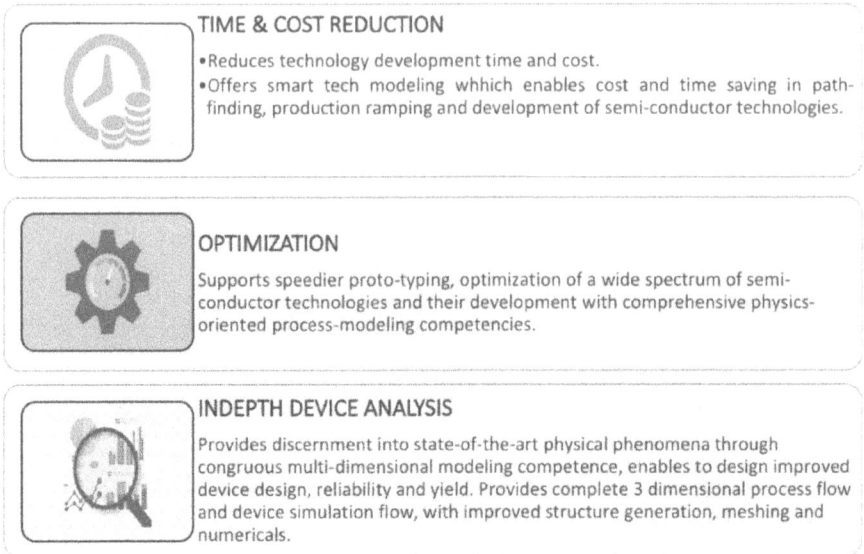

TIME & COST REDUCTION
- Reduces technology development time and cost.
- Offers smart tech modeling whhich enables cost and time saving in path-finding, production ramping and development of semi-conductor technologies.

OPTIMIZATION

Supports speedier proto-typing, optimization of a wide spectrum of semi-conductor technologies and their development with comprehensive physics-oriented process-modeling competencies.

INDEPTH DEVICE ANALYSIS

Provides discernment into state-of-the-art physical phenomena through congruous multi-dimensional modeling competence, enables to design improved device design, reliability and yield. Provides complete 3 dimensional process flow and device simulation flow, with improved structure generation, meshing and numericals.

FIGURE 11.2 Benefits accrued by using the simulation tools of the Synopsys Sentaurus TCAD Suite [3].

providing alternative tools to simulate the process flows, device operations, and so on before the actual production of a wafer. Moreover, these simulations furnish engineering experts with significant cognizance over the performance of semiconductor devices under varied environments, thereby leading to the development of new device conceptions [4]. All of the benefits accrued by using TCAD software have been summarized in Figure 11.2.

11.1.2 TYPICAL FLOW OF MODELING A SEMICONDUCTOR DEVICE

11.1.2.1 Process Emulation

Process steps are emulation results, in which device structure is obtained by performing and following certain procedures that are similar to the actual process flow. First-order devices are analyzed using the process emulation methodology. Process simulations are done after optimization of new device design architecture by using device simulation methodology in order to explore process non-idealities and target process specifications [2].

11.1.2.2 Compact Modeling

Compact modeling methodology is precisely related to only TCAD. As soon as the physics of a device gets verified by TCAD, device electrical characteristics are then "synthesized" through several analytical functions that are either physically based or simply behavioral. Compact modeling results are required for providing "device model cards" to circuit engineers for circuit simulations [2]. The typical flow model of a semiconductor device is given in the form of a flowchart in Figure 11.3.

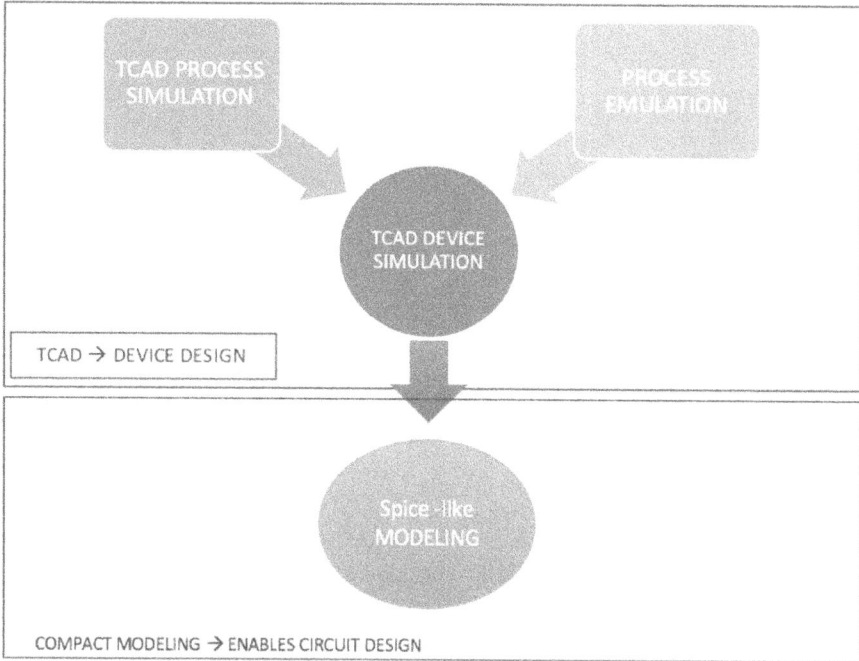

FIGURE 11.3 Typical flow of modeling a semiconductor device.

11.2 THE SENTAURUS TCAD SUITE

Synopsys Sentaurus TCAD encapsulates several steps for device simulation, and for each step there are different tools available for simulation [1–4]. Both the steps and tools are given in Figures 11.4 and 11.5, respectively.

11.3 SENTAURUS PROCESS SIMULATION

- Sentaurus Process simulates the device fabrication steps that are included in silicon-based process technology in 2D and 3D devices. Sentaurus Process simulation provides a comprehensive framework for simulation of a wide array of technologies, varying from nanoscale CMOS to power 2D and 3D device designs.
- Sentaurus Process enables easy simulation of process modules for users. Hence, it further eases the integration of the modules with the complete frontend of line process flows. It includes an advanced cumulation of various models, such as implantation, diffusion, oxidation, and mechanical, along with vigorous mesh generation as well as structure-editing potentiality. Furthermore, it also encapsulates significant process modules like a high dielectric (k) metal gate, strained silicon, and ultra-shallow junction formation, to name a few.

SDE •Getting the device geometry and doping concentrations (from process emulation)

SNMESH •Generating a grid (mesh) for numerical computation

SDEVICE •Solve for Poisson equations, Current continuity and Transport equations on the defined mesh for some given boundary conditions

INSPECT; SVISUAL •Visualizing the results (both electrical results and internal quantities)

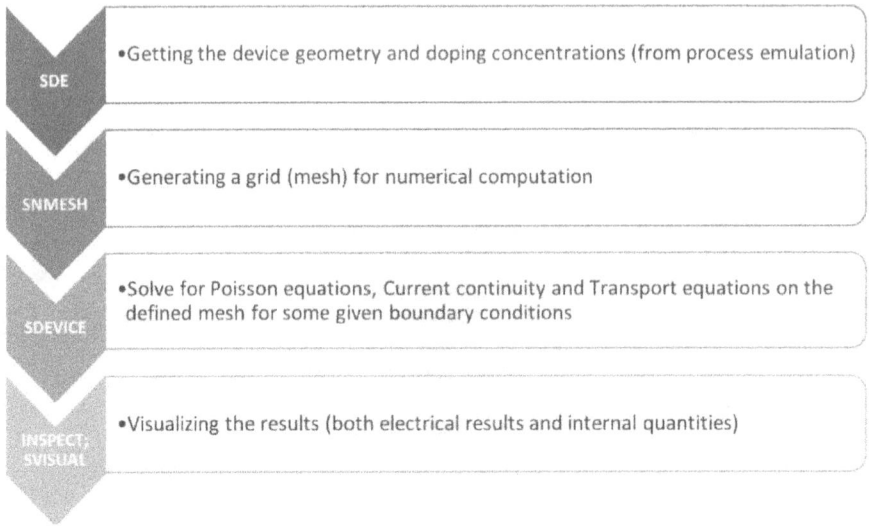

FIGURE 11.4 Going through the device simulation steps.

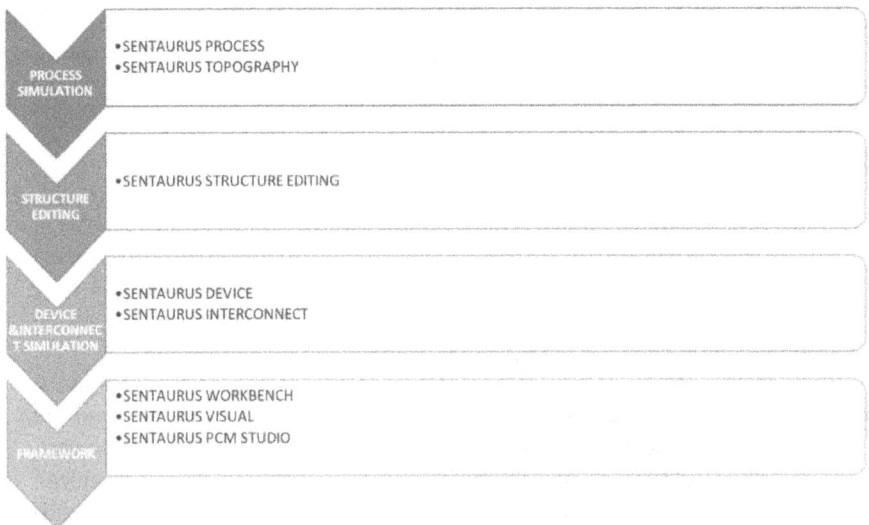

PROCESS SIMULATION
•SENTAURUS PROCESS
•SENTAURUS TOPOGRAPHY

STRUCTURE EDITING
•SENTAURUS STRUCTURE EDITING

DEVICE &INTERCONNECT SIMULATION
•SENTAURUS DEVICE
•SENTAURUS INTERCONNECT

FRAMEWORK
•SENTAURUS WORKBENCH
•SENTAURUS VISUAL
•SENTAURUS PCM STUDIO

FIGURE 11.5 Sentaurus TCAD Suite: Tools for various simulations.

- An array of highly developed implantation and diffusion models are present in the Sentaurus process simulation. A wide energy range varying from sub-kiloelectronvolts to various mega-electronvolts is covered by analytic implant charts. Moreover, Monte Carlo implantation models, being highly methodical and precise, handle conditions better than other analytical models.
- The overall stress history while processing is simulated, and then the eventual results as stress fields are easily exported to Sentaurus Device. This enables a thorough evaluation of these stress effects on the electrical performance of the device. Hence, due to its vast versatility, various models such as clustering, oxidation, diffusion, and silicidation models are easily implementable [3].

11.4   STRUCTURE DEVICE EDITOR (SDE)

- Sentaurus Structure Editor is a device editor that enables 2D and 3D device structure editing and designing using geometric operations. It is backed by the ACIS® geometry kernel, which is in several computer-aided design (CAD) application tools.
- The GUI of SDE consists of a command-line window at the bottom half of the window. In this window, the script of command line generated as per the GUI operations is generated and displayed. Scripts can be easily and directly entered at the command scheme window as well. Similarly, a GUI user can easily define the meshing strategies and doping profiles interactively.
- The meshing tools are called from the Structure Editor GUI, and hence, the meshing profiles as well as doping profiles can be seen simultaneously in the SDE window.
- Users can rerun the journal script file and reconstruct the device structure, since the journal file records all the operations.
- 2D as well as 3D devices can be created using 2D and 3D shapes provided in the GUI. Some of the available shapes are circles, spheres, polygons, rectangles, cuboids, and so on [2].

11.5   SENTAURUS DEVICE SEMICONDUCTOR DEVICE SIMULATOR

Figure 11.6 illustrates the varied simulation and optimization support provided, and Figure 11.7 illustrates physical models and their simulation utility.

11.6   SENTAURUS VISUAL: TCAD VISUALIZATION

- Sentaurus Visual yields an advanced interactive 1D, 2D, and 3D visualization and data examination platform. It substructures TCL scripting, thereby further enabling the post-processing of output data, and generates fresh curves and other obtained parameters [3].

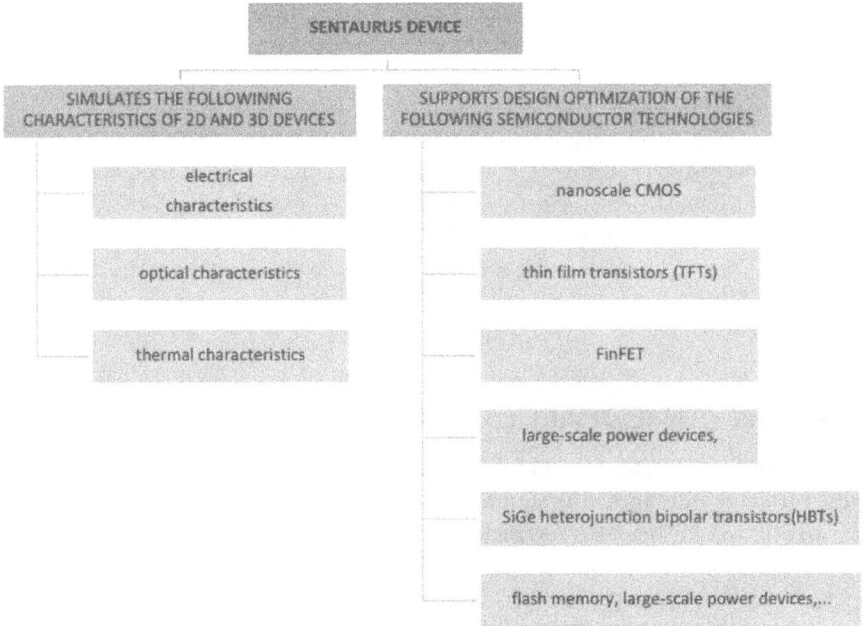

FIGURE 11.6   Sentaurus Device: Simulation and design optimization support.

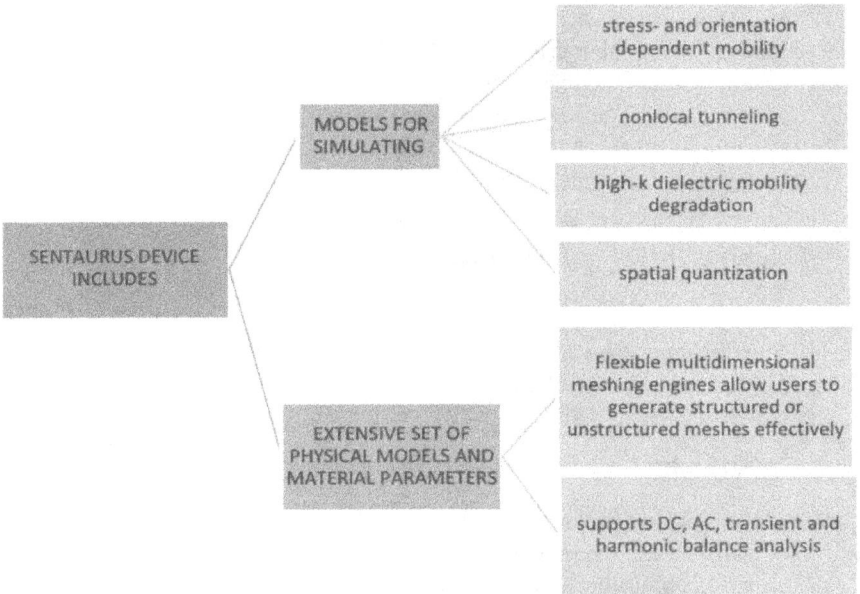

FIGURE 11.7   Sentaurus Device: Physical models and their simulation utility.

11.7 EVOLUTION OF NOVEL DEVICE ARCHITECTURE: GATE-ALL-AROUND NANOSHEET FETs

11.7.1 TECHONOLOGICAL PROGRESSION OF SEMICONDUCTOR TRANSISTORS

Digital informative datasets are coded in binary language and thereby transformed into electrical signals by using a semiconductor device called a "transistor." A transistor consists of a channel region, through which electric current flows between the source region and drain region by way of a gate region, which manages the electric current flowing through the channel [4–6]. The gate region works as a switch along with generating binary system data through amplification of electrical signals. Due to this functioning process of a transistor, it has become an essential basic building block of a semiconductor chip [7–9].

Chip sizes have been consistently shrunk in the industry to achieve the maximum number of semiconductor chips possible on the given size of the silicon wafer/substrate. Moreover, to enhance the adaptability of chips to incorporate new and complex functions, the basic building block (i.e., the transistor) is being made smaller, and its power consumption is being brought to minimum with state-of-the-art technology advancements. Hence, low power consumption will provide a better and longer battery period, and at the same time a reduction in heat generated will also be achieved simultaneously [8–9]. As can be seen in Figure 11.8, the industry has been gradually shifting to new and better transistors.

Since the electricity consumption of a semiconductor chip depends upon the voltage it is operated on, the focus now is to develop chips that function on low voltage power. Thus, the progression of transistors over time is equivalent with the creation of transistors devices, which are smaller in size, work on low-voltage power, and work at a faster pace [10].

The metal–oxide–semiconductor (MOS) is the most significant and widely used transistor across the semiconductor industry. The major components of a MOS transistor are given in Figure 11.9.

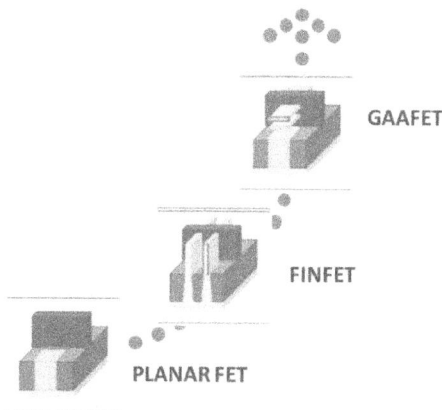

GAAFET

FINFET

PLANAR FET

FIGURE 11.8 Progressive advancement of transistors with time.

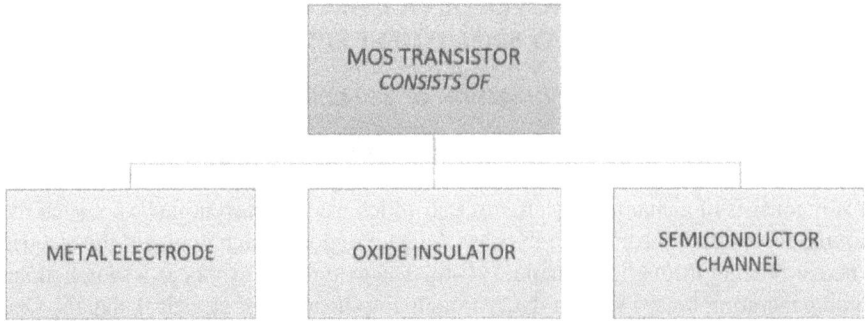

FIGURE 11.9   Key components of metal–oxide–semiconductor (MOS) transistors.

The first MOS transistor was a planar device in which the gate and channel had contact only on one planar surface. But, with the gradual decrease in transistor size, the source and drain distance also declined, resulting in a difficulty for the device to function as a switch. Thus, due to increasing limitations arising due to SCEs and limiting voltage scaling down, planar transistors could be scaled down to only 20 nm nodes [11].

Fully depleted (FD) transistors were developed as next-generation transistors to overcome the issue of SCEs. In FD transistors, the gates' ability to adjust the channel is enhanced by using a thin Si (silicon) channel, and it thereby avoids SCEs. The structure of the FD transistor is an evolution from the basic planar transistor. The structure is thinner and more rugged than the standard transistor and has a thin standing rectangular channel, which has a channel presence on three sides. Due to its channel's thin standing shape, it resembles a fish's fin and thereby got its name as a "fin transistor." Today, the industry manufactures fin transistors in a variety of sizes, beginning at more 14 nm [11–13].

As mentioned, in a planar transistor, the gate and channel have only a singular plane contact. On the other hand, in a fin transistor, the channel makes a three-sided contact with the gate due to its 3D structure. Thus, due to the increase in contact regions with the gate, the overall working of the semiconductor device improves greatly. The progression of semiconductor devices with time and technological advancement is shown in Figure 11.10. This improvement in device design also led to a decrease in operating voltage, which eventually resolved the SCE issues [12].

However, due to increasing industry demands and a drive to improve the device performance, the fin transistor has been facing a performance limitation after years of development and several process transformations. Today, the industry is focused on reducing the operating voltage of devices and thus focusing on developing low-power, low-voltage devices. However, in FinFETs, only the three sides of the fin are in contact with the gate, thereby creating a limiting boundary as transistors are being scaled down and thereby becoming more compact [13].

Hence, in order to overcome the limiting factors of existing transistors, the industry is now focused on developing a new structure, the GAA transistors (performance

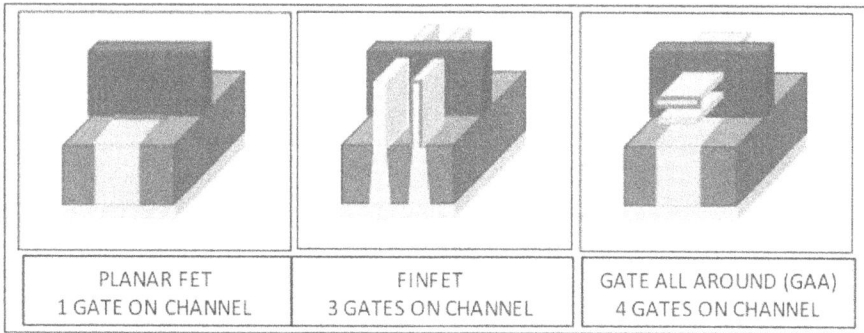

| PLANAR FET
1 GATE ON CHANNEL | FINFET
3 GATES ON CHANNEL | GATE ALL AROUND (GAA)
4 GATES ON CHANNEL |

FIGURE 11.10   The progression of semiconductor devices with time and technological advancement.

improvement is shown in Figure 11.11). The GAA-FET is designed in such a manner that it increases the gates' ability to control the channel functions, since all four faces of the channel are covered by the gate. Thus, the gate encloses the channel from all four faces, which further helps in addressing the SCE issues that further lead to reduced operating voltage [14, 15].

The GAA-FET usually consists of a long and ultrathin nanowire (GAA-NWFET). But, in order for more current to flow through it, the channel needs to be wider. Hence, the small diameter of the nanowire creates an issue to achieve higher current flow [14, 15]. A figurative comparison between a GAA-NWFET and a GAA-NSFET is shown in Figure 11.12. Hence, to overcome this problem, the GAA-NSFET has been developed, which is an optimized version of the GAA-NWFET [23]. In NSFETs, the area of contact between the channel and gate increases as we align the channel structure in a 2D cuboid (nanosheets). By incorporating stacked nanosheets, we achieve two benefits: uncomplicated device integration and an increased current flow [15].

Thus, today's GAA-NSFET is an improved and better device structure as it not just mitigates SCEs due to its GAA structure, but also provides better and improved performance due to the increased channel area. Hence, its application as a biosensor is being worked upon relentlessly across academia [16–20].

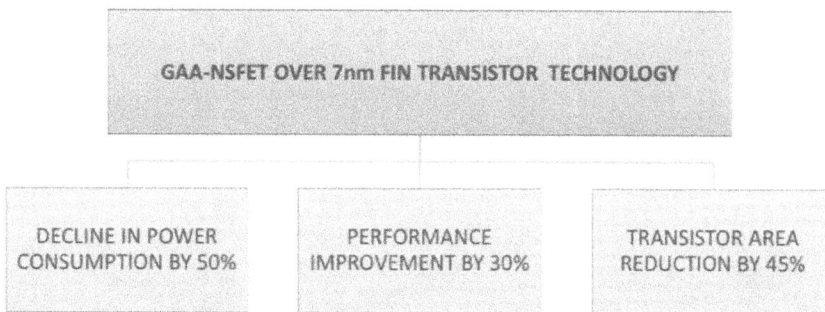

GAA-NSFET OVER 7nm FIN TRANSISTOR TECHNOLOGY

| DECLINE IN POWER
CONSUMPTION BY 50% | PERFORMANCE
IMPROVEMENT BY 30% | TRANSISTOR AREA
REDUCTION BY 45% |

FIGURE 11.11   Performance improvement of a GAA-NSFET over 7 nm FinFET transistors.

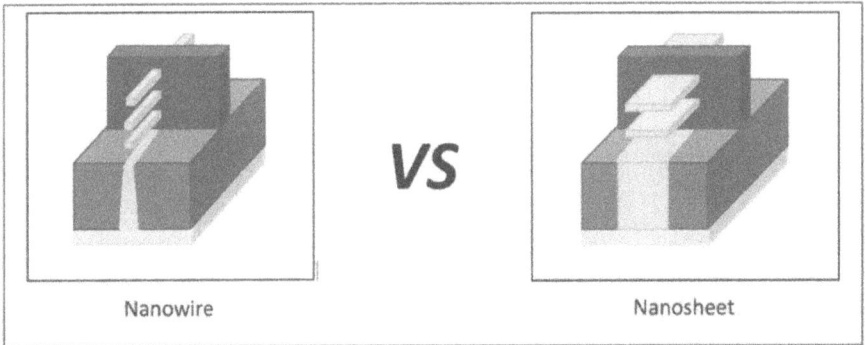

FIGURE 11.12   Structural comparison between a nanowire and nanosheet FET.

11.8 GATE-ALL-AROUND NANOSHEET FET (GAA-NSFET)

11.8.1 Introduction to a Vertically Stacked GAA-NSFET

With the fast-paced advancement of transistor/MOSFET technology, semiconductor devices are continuously scaled down to attain improved performance along with greater density [21, 22], while nonplanar FETs like FinFETs are immune to the SCEs and thus are carried through manufacturing in the improved technology node. Recently, GAA-NSFETs have been introduced as a strong contender to replace FinFETs and thus are being implemented in the industry. Recently, vertically stacked GAA-NSFETs have been an outstanding replacement to FinFETs in 10 nm technology. The GAA-NSFET shows improved gate controllability, process compatibility, and current drivability; advantages of GAA-NSFETs are shown in Figure 11.13 [22]. In GAA-NSFETs, horizontal sheets are stacked vertically one above the other, which enables maximization of the I_D (drive current) for a specific track upon a wafer [24, 25]. The performance of multi-gate devices is upgraded by implementing various geometric alterations such as process techniques at different technology nodes [26]. CAD technology is used for studying GAA-NSFETs. The fabrication of nanoscale devices has become immensely complex; hence, their performance predictions for given technology nodes are performed through TCAD [27–35].

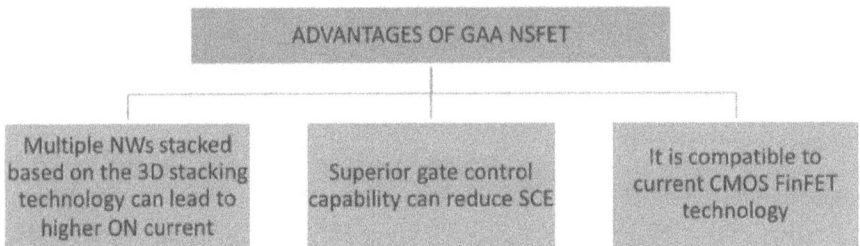

FIGURE 11.13   Advantages of GAA-NSFETs over other conventional transistors [28].

TABLE 11.1

Key Parameters of GAA-NSFETs and Their Values

Key Parameter	Value Applied (nm)
Length: Source region	25
Length: Gate region	20
Length: Drain region	25
Length: Spacer region	25
Width: Channel region	30
Thickness: Oxide region	4.4
Thickness between nanosheets	14

The following sections have step-by-step guidelines for designing a 3D GAA-NSFET device structure and its simulation.

11.9 PROPOSED GAA-NSFET

Table 11.1 shows the values of major key components of the proposed GAA-NSFET. Figure 11.14 shows the 3D structure of the device design, whereas Figure 11.15 shows a cross section of the device.

11.9.1 STEPS INVOLVED IN DESIGNING THE GAA-NSFET USING SYNOPSYS SENTAURUS TCAD, 2018 VERSION

- Open terminal → enter SDE command.
- In SDE using the GUI interface, the user can create a 3D device structure using the calculated 3D coordinates based on the set parameters of the device structure.
- In GUI, for better labeling of the regions formed and user control, select the exact coordinates and turn off auto region naming. This will enable

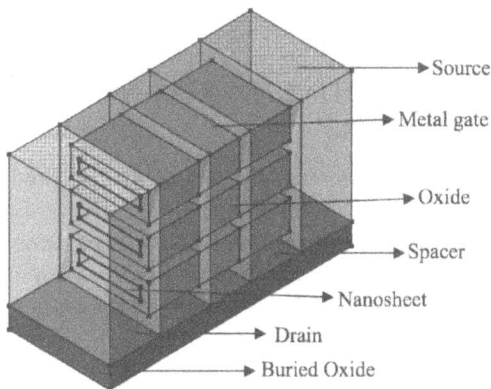

FIGURE 11.14 3D structure of the proposed GAA-NSFET.

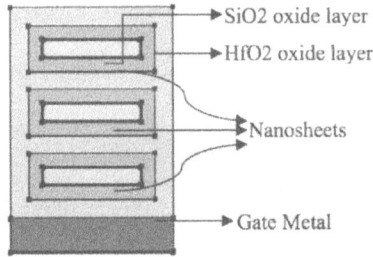

FIGURE 11.15   2D cross section of the GAA-NSFET.

the user to enter the exact calculated coordinates and then name the shape accordingly.

- Following the above step, the user can design the entire 3D structure. There is no hard-and-fast rule regarding which part of the device should be designed first in SDE.
- As the device structure creation continues, the script file gets generated simultaneously in the command scheme window.
- After creating the gate-all-around, vertically stacked three-nanosheet FET, the source, drain, and channel regions are doped.
- First, the constant doping profiles are set for the source, drain, and channel regions of the device. This is followed by Gaussian doping, which will be discussed later in this chapter.
- After doping, meshing of the device is supposed to be done to assess the electrical performance of the device more intrinsically.
- Eventually, once the meshing is successful, a .tdr file is generated that opens the device structure in SVisual.
- In SVisual, the doping profiles of the device structure can be assessed along with the meshing grids.
- The user can use various GUI tools in SVisual to assess the doping with further ease and clarity. For example, we used precision-cut z and y tools to assess the doping profile in the nanosheets.

11.10   SCRIPT FILE GUIDELINE FOR GAA-NSFET DEVICE STRUCTURE

```
;##############################################################
;                            EXAMPLE
;      Gate-All-Around Nanosheet Field-Effect Transistor (GAA-NSFET)
;                    Ns = 3; Lg = 20 nm; Tox = 4.4 nm
;##############################################################

(sde:clear)
```

11.10.1 Defining Parameters of the Device Structure

;**************************defined parameters**************************

- While designing any proposed device structure, we define the constant parameters whose values are independent and are mostly the key parameters.
- Defining these parameters enables the user to more easily transition to other values and obtain the required results accordingly.
- In the proposed GAA-NSFET, we define parameters like the length, height, and width of buried oxide (BOX); the fixed parameters of the source, drain, and the length of nanosheets; and their thickness, pitch, and so on.

```
(define Lbox 120)
(define Wbox 50)
(define Hbox 10)
```

- Define the length and width of source. Similarly, the user has to define the length and width of the drain and spacer, and the metal gate length.

```
(define Xs 25)
(define Ys 50)
```

- Define the number of nanosheets in the device structure and the thickness and width of each nanosheet.
- Similarly, define the pitch (i.e., the distance between each nanosheet).

```
(define nsheet 3)
(define Tns 5)
```

- Define the oxide thickness. If the user is using multiple oxide materials, then it is advised to define their respective thicknesses as well.
- In this example, HfO_2 and SiO_2 have been used as dielectric oxide layers.

```
(define Tox 4.4)
```

- Users must take note that, in Sentaurus TCAD SDE, the numeric data is considered to be of microunits.
- But, since we are designing a nanosized device structure, and all parameters are in nanometer (nm) unit size, the values are set to the required unit by dividing the defined parameters by 1e3.
- Below is an example of how the values of various parameters of the BOX can be set to the required unit size.
- Similarly, the values of other defined parameters have to be set to nanometer unit size.

```
(set! Xbox (/ Xbox 1e3))
(set! Ybox (/ Ybox 1e3))
(set! Zbox (/ Zbox 1e3))
```

;**************************derived parameters**************************

- After defining the constant parameters of the device structure, the other parameters whose values are dependent on the defined parameters are defined through mathematical operations.
- For example, the height of the total nanosheet stack (Hns) will be a multiplication output of the thickness of each sheet (Tns) and the number of sheets (nsheet).

```
(define Hns (* nsheet Tns))
```

- Similarly, the user can define several other parameters as per the design requirements. In this device design, some of the parameters defined include parameters to calculate the total height of the device, which is a variable parameter since it would depend upon the number of nanosheets, their thickness, their pitch, the oxide thickness, and so on.
- Defining the parameters greatly simplifies the process of structure designing. Hence, in the proposed GAA-NSFET, we defined maximum parameters such as minimum and maximum z-coordinate points of the respective nanosheets, their minimum and maximum y-coordinate points, and so on.
- For example, the following command line was used to define the minimum and maximum y-coordinate points of the nanosheets to define its width (channel width).

```
(define Ynsmin (- (/ Ys 2) (/ Yns 2)))
(define Ynsmax (+ (/ Ys 2) (/ Yns 2)))
```

- The above command line defines the following mathematical equation:

```
Ynsmin = (Ys/2) - (Yns/2)
Ynsmax = (Ys/2) + (Yns/2)
```

  where Ys = width of source, and Yns = width of the channel (nanosheet).
- The user can similarly further define parameters, such as the height of spacers, maximum length coordinate of spacers, metal gate, and so on.

11.10.2   CREATING THE DEVICE STRUCTURE

- Table 11.2 encapsulates the materials used for key regions of the device structure.
- After defining all the constant and derived parameters and setting their respective values to nanometer unit, the next step to follow is to create the structure of the device.
- Since we are dealing with a 3D semiconductor device in this chapter (i.e., a GAA-NSFET), we use a cuboid to create the various regions of the proposed device.
- The following command line is used to create the BOX region.

TABLE 11.2

Material Definitions of Key Regions of the Device Structure

Device Region	Material
Source	Silicon (Si)
Drain	Silicon (Si)
Nanosheets	Silicon (Si)
Buried oxide	Silicon dioxide (SiO_2)
Dielectric material	Hafnium oxide (HfO_2)
Spacer	Silicon nitride (Si_3N_4)

```
**************************Box********************************

(sdegeo:create-cuboid
        (position   0    0   0   )
        (position   Xbox  Ybox  Zbox  )
 "SiO2"   "box"
)co
```

- In the command line, first the x, y, and z coordinates of the two diagonal points of the cuboid are defined.
- Thereafter, the material of the region is defined, followed by the name of the region.
- After the command line is entered, the following box region is created on the SDE window, as shown in Figure 11.16.
- The same command line of "create-cuboid" is used to create other regions, such as the source, drain, spacers, and other regions of the device structure.

11.10.3 SOURCE

- But, before creating the cuboid for the source region, the following command line is used such that the old region replaces any common region created with the new region.

```
(sdegeo:set-default-boolean "ABA")
"ABA"
```

FIGURE 11.16 Buried oxide (BOX) region as created in the SDE window (z-x-axis view).

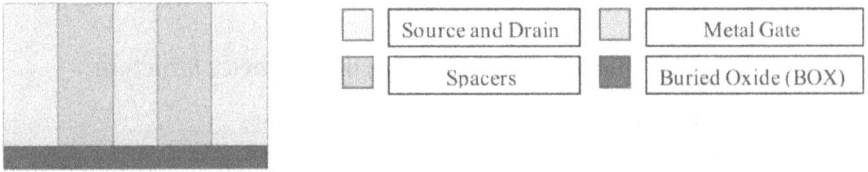

FIGURE 11.17 Source, drain, spacer, and metal gate regions obtained on the SDE window (z-x-axis view).

- Thereafter, the same "create-cuboid" command is used to create the source region, and, as per the placement of each region on the device structure, their extreme diagonal x-y-z coordinates are defined in the position brackets and followed by the material and region name.

```
(sdegeo:create-cuboid
        (position  0   0    Z₁  )
        (position   X₁  Y₁    H₁)
   "Silicon"   "source"
)
```

- The two command lines above are used to create the drain, spacer, and metal gate regions.
- After these command lines are run, the following regions of the device are created on the SDE window, as shown in Figure 11.17.

11.10.3.1 Diaelectric Material

- Now, to create the dielectric material layer, we first create the silicon dioxide layer over which the hafnium oxide layer will be created.
- To create a layer of oxide material over each nanosheet, first the user needs to have the extreme diagonal coordinates of each oxide cuboid to be created.
- The cuboid to be created to define the dielectric region will replace the overlapping region of the initially created regions.
- Hence, we call the following command line of "new replaces old" to successfully create the dielectric regions of HfO_2 and SiO_2 for each nanosheet.
- For each dielectric region, separate cuboids are created and are named accordingly.

```
***************************HfO₂---di2***************************

;new replaces old
(sdegeo:set-default-boolean "ABA")
"BAB"
---------------------for nanosheet1--------------------
(sdegeo:create-cuboid
        (position   X2   Y2    Z2 )
        (position  X3   Y3    Z3)

        "HfO₂"   "di2_ns1"
)
```

- Similarly, "create-cuboid" command lines are generated to create the HfO_2 region around nanosheets 2 and 3.
- Next, the SiO_2 region has to be created within the HfO_2 cuboid region. Thus, again we call the "new replaces old" command line, after which three separate cuboids are created for each nanosheet.

*************************** SiO_2-Di1****************************

```
new replaces old
(sdegeo:set-default-boolean "ABA")
"BAB"
;------------------------for nanosheet1---------------------

(sdegeo:create-cuboid
        (position    X4     Y4     Z4)
        (position    X5     Y5     Z5)
  "SiO2"    "Di1_ns1"
)
```

- Similarly, "create-cuboid" command lines are generated to create the SiO_2 region around nanosheets 2 and 3.

11.10.3.2   Nanosheets

- The next step involves the creation of nanosheets. In this chapter, a GAA-NSFET with three vertically stacked nanosheets is being designed; hence, three separate cuboids with their respective coordinate points are to be created.
- Similar as before, the nanosheet region has to be created within the cuboid of SiO_2, and the other overlapping region space of the two spacers is already created.
- Hence, we again call the command line "new replaces old" first, and then begin with the creation of three separate cuboids to form the nanosheet regions.
- The following is the example code for creating the nanosheet. The same can be used to define the remaining nanosheets using their respective coordinate positions and naming.
- On following the correct steps and prompting the correct command lines, we achieve the progressive design of the GAA-NSFET, as shown in Figure 11.18.

*****************************nanosheet****************************

```
(sdegeo:set-default-boolean "ABA")
"BAB"
(sdegeo:create-cuboid
        (position    X6     Y6     Z6)
        (position     X7    Y7     Z7)
  "Silicon"    "ns1"
)
```

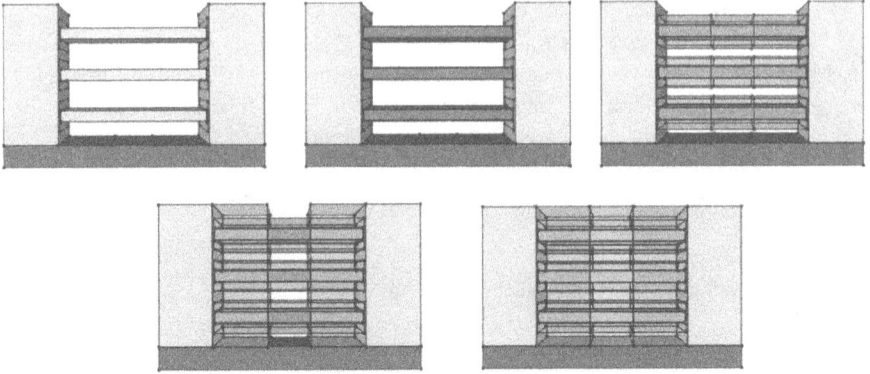

FIGURE 11.18   Progressive stages of formation of various regions of the GAA-NSFET.

11.10.4   CREATION OF DEVICE CONTACTS

- In packaging, the silicon chip is put inside a plastic or metal case that contains the contacts needed for the resulting chip to interface with external components.
- Hence, after creating the device structure, the contacts are created over the source, drain, and gate, as shown in Figure 11.18.
- The following command lines are used to create the contacts.
- For example, to create the gate contact, the contact name is defined as *G*, and color is set in a hexadecimal system, after which the contact face position coordinates are mentioned in the command line, as shown in the example below.

```
****************************gate contact****************************

(sdegeo:set-current-contact-set "G")
(sdegeo:define-contact-set "G" 4 (color:rgb 1 0 0 ) "##")
(sdegeo:set-contact-faces
        (find-face-id
        (position  X    Y    Z   )
        )
)
```

- Similarly, the same command line is used to create source and drain contacts, as shown in Figure 11.19.

11.10.5   MESHING

- Meshing enables the division of various regions of the semiconductor device into several small cells. These cells help provide better results to the governing equations, which enables a better assessment of the physical behavior.
- Hence, a good-quality, well-defined mesh ensures better accuracy and convergence in faster simulations.

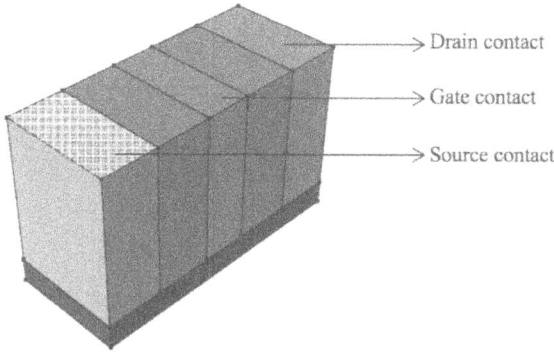

FIGURE 11.19 The GAA-NSFET device structure after creating contact faces.

- Thus, all the regions of the device structure are meshed. The cell size of meshing in various regions may vary, as per the requirement of the device performance analysis.
- For example, the following command lines were used to mesh the spacers of GAA-NSFETs.

*********************************Spacer*******************************

```
(sdedr:define-refinement-size "spacer_mesh" 0.010 0.010
0.020 0.005 0.005 0.010)
(sdedr:define-refinement-material "spacer"  "spacer_mesh"
"Si3N4"  )
```

- In the above command line, first the refinement size of the mesh is defined. The first three numbers denote the maximum mesh length in the x, y, and z coordinates, respectively.
- The latter set of numbers denotes the minimum size of the mesh length x, y, and z coordinates, respectively.
- Eventually, the mesh region is mentioned, which is "spacer" in the above command line. Then a name is given to the meshed region ("spacer_mesh"), and eventually the material of the mesh region is mentioned, as the "refinement material" command line has done.
- The same command line is used to mesh the other regions of the device structure. The following command lines can be referred to as further examples.
- *Note*: The value of the mesh lengths is to be provided in micrometer units.

***************************** metal gate*******************************

```
(sdedr:define-refinement-size "GATE_Mesh" 0.010 0.005
0.020 0.005 0.003 0.010)
(sdedr:define-refinement-material "GATE" "GATE_Mesh"
"Metal"  )
```

- Similarly, the meshing can also be done based upon the region by using the following command lines:

```
**********************************box****************************

(sdedr:define-refinement-size "box" 0.020 0.010 0.004
0.010 0.005 0.002 )
(sdedr:define-refinement-placement "box_mesh" "box" (list
"region" "box" ) )
```

- In the above command line, we define the refinement size and then the refinement placement with respect to the region.
- The name of the refinement placement in the above command line is "box_ mesh," and the region being meshed is "box."
- Similarly, other regions such as the nanosheets, drain, source, and dielectric oxide materials are to be meshed.
- Figure 11.20 shows the various meshed regions of the GAA-NSFET device structure.

11.10.6 DOPING

- In this chapter, the source and drain of the GAA-NSFET are doped with arsenic, while the channel (i.e., the nanosheets) is doped with boron.
- First, the constant-doping profile is done; the command files used are mentioned below.

11.10.6.1 Constant Doping

```
**********************************source****************************
```

- For the constant-doping profile, we name the constant profile definition, then mention the doping material and its concentration.

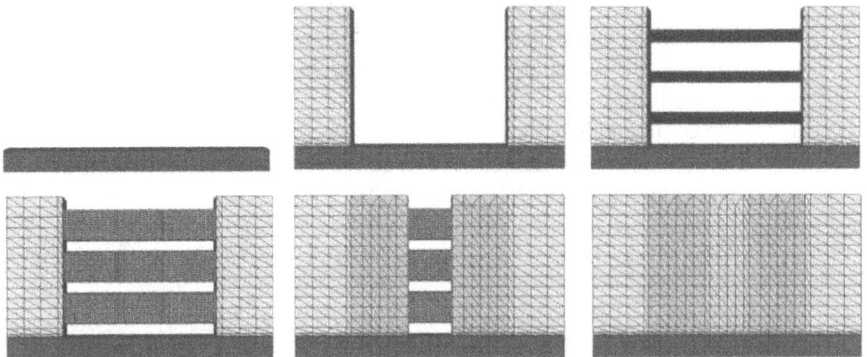

FIGURE 11.20 Meshing stages of all regions of the GAA-NSFET device structure.

```
(sdedr:define-constant-profile
            "ConstantProfileDefinition_source"
            "ArsenicActiveConcentration"    2e20
)
```

- Then, the constant profile placement region is named and the constant profile definition is mentioned, followed by the region name, which is being
  doped.

```
(sdedr:define-constant-profile-region
            "ConstantProfilePlacement_source"

            "ConstantProfileDefinition_source"   "source"
)
```

- The same command lines are used for the constant-doping profile of the
  drain region.

```
*******************************drain******************************

(sdedr:define-constant-profile
        "ConstantProfileDefinition_drain"
        "ArsenicActiveConcentration"    2e+20
)
(sdedr:define-constant-profile-region
        "ConstantProfilePlacement_drain"
        "ConstantProfileDefinition_drain"   "drain"
)
```

- The constant-doping profile of the nanosheets is done only for the region of
  nanosheets within the gate region.
- Thus, to dope a particular section of the nanosheets, Refeval windows are
  created.
- The following is an example of a set of command lines used for the constant-doping profile of nanosheet 1.

```
****************************nanosheet 1****************************

(sdedr:define-refeval-window
            "ns1_gate "     "Cuboid"
            ( position   0.05   0.010   0.020  )
            ( position   0.070  0.040   0.025  )
)

(sdedr:define-constant-profile
        "ConstantProfileDefinition_ns1"
        "BoronActiveConcentration"     1e17
)
```

```
(sdedr:define-constant-profile-placement
      "ConstantProfilePlacement_ns1"
      "ConstantProfileDefinition_ns1"    "ns1_gate "
)
```

- First, to create the window, its name and shape are mentioned, followed by the positions of the diagonal coordinates of the window.
- After creation of the window, the constant profile is defined, along with the doping material and its concentration.
- Then the constant profile placement region is named and the constant profile definition is mentioned, followed by the region name that is being doped.
- The same procedure is followed to create windows in other nanosheets and their respective constant-doping profile.

11.10.6.2 Gaussian Doping

- To do Gaussian doping, rectangular windows are created at each nanosheet and source junction as well as each nanosheet and drain junction. Thus, six separate windows were created for the given GAA-NSFET in this chapter, as shown in Figure 11.21.
- The following is the set of command lines used for Gaussian doping in nanosheet 1 and the source intersection region.

*******************************ns1_S*****************************

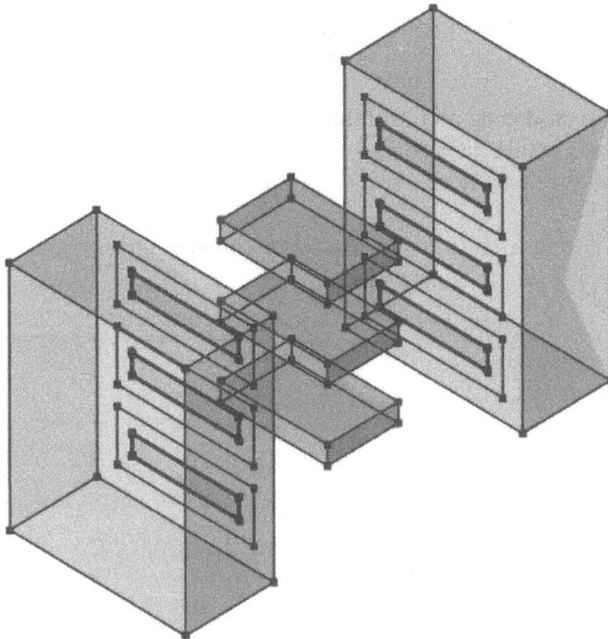

FIGURE 11.21 Windows created for doping in the GAA-NSFET device structure.

- First, the window is created wherein the name, shape, and extreme diagonal coordinates of the required window are mentioned.

```
(sdedr:define-refeval-window
          "NS1S"    "Rectangle"
          ( position   0.025   0.010   0.020  )
          ( position   0.025   0.040   0.025  )
)
```

- Following the creation of the window, the Gaussian profile, doping material, its concentration, position of peak value, peak value concentration, concentration at depth, and the depth along with the Gaussian factor are mentioned in the command line.
- Note that the peak position is the start point of the Gaussian doping and is generally considered "0," whereas "Depth" is the distance up to which the Gaussian doping has to be done.

```
(sdedr:define-gaussian-profile
"AnalyticalProfileDefinition_ns1source"
"ArsenicActiveConcentration" "PeakPos" 0 "PeakVal" 2e20
"ValueAtDepth" 1e17 "Depth" 0.025 "Gauss" "Factor" 0.8)
```

- Finally, the analytical profile placement is done, which includes the name of the placement, analytical profile placement name, and name of the window generated.

```
(sdedr:define-analytical-profile-placement
"AnalyticalProfilePlacement_ns1source"
"AnalyticalProfileDefinition_ns1source" "NS1S" "Positive"
"NoReplace" "Eval")
```

- The same set of command lines is used for Gaussian doping of the remaining five intersection junctions of nanosheets. Figures 11.22–11.24 show the doping concentrations obtained individually of boron in a section of nanosheets, arsenic doping in source and drain, and Gaussian doping in nanosheets, respectively.
- Figure 11.24 shows the total overall doping concentration of the device.

11.10.7 END OF SCRIPT FILE – GAANSFET

- Once doping is done, the mesh file of the device structure is generated by using the following command line:

```
(sde:build-mesh "mesh" "-F tdr" "name_of_your_scriptfile")
```

- After successful meshing of the device, a .tdr file is generated that can be opened in SVisual.
- In SVisual, the user can easily assess the meshing and doping concentration of the device structure.

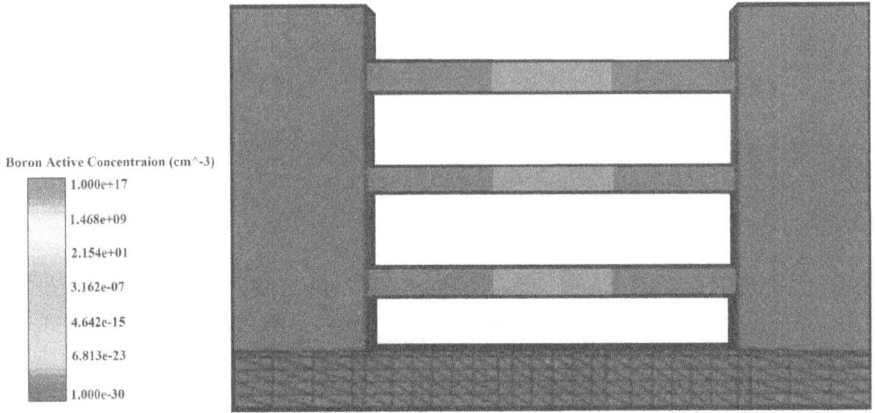

Boron Active Concentraion (cm^-3)

1.000e+17
1.468e+09
2.154e+01
3.162e-07
4.642e-15
6.813e-23
1.000e-30

FIGURE 11.22   Boron active concentration.

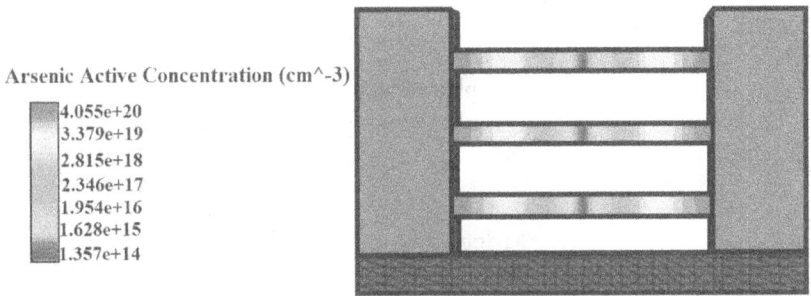

Arsenic Active Concentration (cm^-3)

4.055e+20
3.379e+19
2.815e+18
2.346e+17
1.954e+16
1.628e+15
1.357e+14

FIGURE 11.23   Arsenic active concentration doping profile.

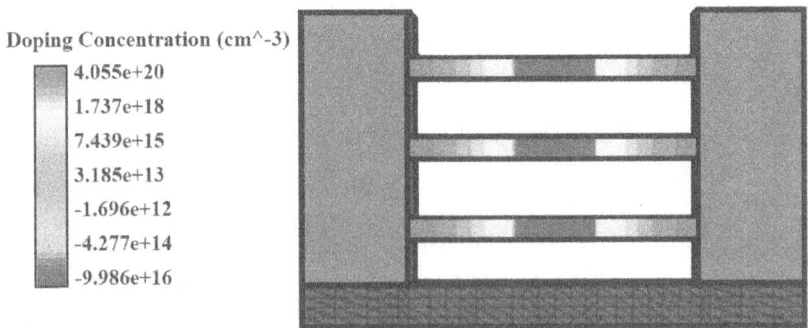

Doping Concentration (cm^-3)

4.055e+20
1.737e+18
7.439e+15
3.185e+13
-1.696e+12
-4.277e+14
-9.986e+16

FIGURE 11.24   Overall doping concentration in a GAA-NSFET.

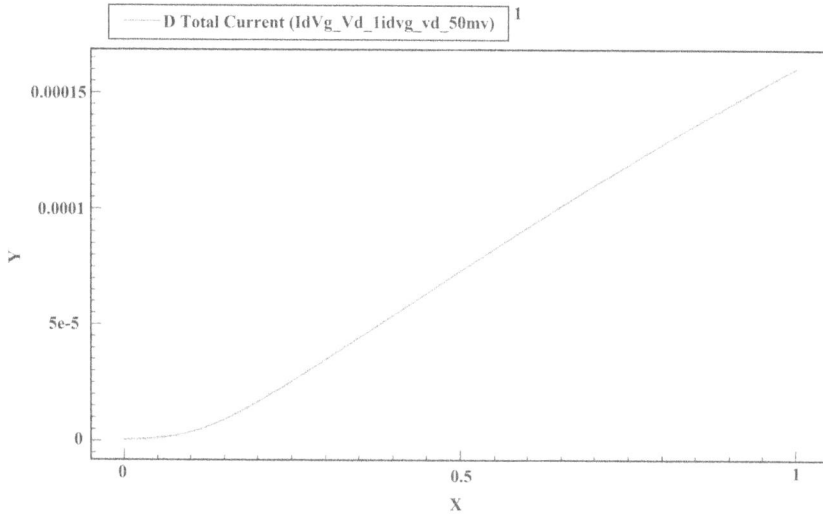

FIGURE 11.25 I_d-V_g curve of the GAA-NSFET device simulation with $Vd = 1$ V.

11.10.7.1 I-V Characteristics of the Proposed GAA-NSFET

- To obtain the I-V simulation curve of the device structure, the following three files are required:

  `nsfet.tdr, nsfet.cmd, and nsfet.par.`

- The I_d-V_g simulation result of the proposed GAA-NSFET is shown in Figure 11.25.

11.11 CONCLUSION

In this chapter, Section 11.1 comprises the scope and introduction to the Synopsys Sentaurus TCAD suite. Section 11.2 provides an in-depth understanding of various Sentaurus TCAD simulation platforms, such as Process Simulation, Structure Editing Device, and interconnect simulation along with framework Sentaurus Visual (SVisual). Section 11.3 briefly discusses the requirements of device simulation and its benefits, along with the technological progression of semiconductor transistors over time. Section 9.4 introduces vertically stacked GAA-NSFETs. The device structure elucidation along with a simulation setup comprising device models have also been included. The simulation of a GAA-NSFET device with optimum geometrical parameters and analysis of the I-V characteristics of the device have also been reported in Section 11.4.

ACKNOWLEDGMENTS

The device was designed using Synopsys TCAD, version 2018, licensed in the VLSI Lab, Department of Electronics and Communication Engineering, Jaypee Institute of Information Technology, Noida, India. We thank the institute and the faculty in charge for granting us the opportunity and permission to bring our work to fruitful completion.

REFERENCES

1. Sentaurus TCAD Industry-Standard Process and Device Simulators Datasheet, Synopsys, Mountain View, CA, USA, 2012.
2. Y. C. Wu and Y. R. Jhan, "Introduction of Synopsys Sentaurus TCAD Simulation", 3D TCAD Simulation for CMOS Nanoeletronic Devices, 2018, pp. 1–17. https://doi.org/10.1007/978-981-10-3066-6_1.
3. V. Sentaurus Device User Guide, Synopsys, Mountain View, CA, USA, 2013.
4. Betti Beneventi, Giovanni. Technology Computer Aided Design (TCAD) Laboratory Lecture 3, Overview of Synopsys Sentaurus TCAD, 2015. Available: https://www.researchgate.net/publication/340898874_Technology_Computer_Aided_Design_TCAD_Laboratory_Lecture_3_Overview_of_Synopsys_Sentaurus_TCAD.
5. A. Pandey, "Recent trends in Novel semiconductor devices", Silicon, 2022, vol. 14, pp. 1–12.
6. R. H. Yan, A. Ourmazd, and K. F. Lee, "Scaling the Si MOSFET: From bulk to SOI to bulk", IEEE Trans. on Electron Devices, 1992, pp. 1704–1710. https://doi.org/10.1109/VTSA.2011.5872206.
7. K. J. Kuhn, "CMOS scaling for the 22 nm node and beyond: Device physics and technology", International Symposium on VLSI Technology, Systems and Applications, 2011, pp. 1–2.
8. K. Roy, S. Mukhopadhyay, and H. Mahmoodi-Meimand, "Leakage current mechanisms and leakage reduction techniques in deep-submicrometer CMOS circuits", Proc IEEE, 2003, pp. 305–327.
9. D. J. Frank et al., "Device scaling limits of Si MOSFETs and their application dependencies", Proc IEEE, 2001, pp. 259–287. https://doi.org/10.1109/ASQED.2009.5206287.
10. H. W. Cheng, C. H. Hwang, and Y. Li, "Propagation delay dependence on channel fins and geometry aspect ratio of 16-nm multi-gate MOSFET inverter", 1st Asia Symposium on Quality Electronic Design, 2009, pp. 122–125.
11. S. Thompson, P. Packan, and M. Bohr, "MOS scaling: Transistor challenges for the 21st Century", Intel Technol. J., 1998, pp. 1–19.
12. D. F. Sudha and D. Sudha, "FinFET- one scale up CMOS: Resolving scaling issues", IEEE 3rd International Conference on Computing for Sustainable Global Development (INDIACom), 2016, pp. 1183–1187.
13. A. Razavieh, P. Zeitzoff, and E. J. Nowak, "Challenges and limitations of CMOS scaling for FinFET and beyond architectures", IEEE Trans. Nanotechnol., 2019, pp. 999–1004.
14. N. Loubet et al., "Stacked nanosheet gate-all-around transistor to enable scaling beyond FinFET", 2017 Symposium on VLSI Technology, Kyoto, 2017, pp. T230–T231.
15. D. Jang et al., "Device exploration of NanoSheet Transistors for Sub-7-nm Technology Node", IEEE Trans. Electron Devices, 2017, vol. 64, no. 6, pp. 2707–2713.
16. C. Li et al., "A vertically stacked nanosheet gate-all-around FET for biosensing application", IEEE Access, 2021, vol. 9, pp. 63602–63610.
17. K. Martens,"BioFET technology : Aggressively scaled pMOS FinFET as biosensor", IEDM Tech. Dig., Jun. 2019, pp. 6–18.
18. A. P. F. Turner, "Biosensors: Sense and sensibility", Chem. Soc. Rev., 2013, vol. 42, no. 8, pp. 3184–3196.
19. K. Choi, J.-Y. Kim, J.-H. Ahn, J.-M. Choi, M. Im, and Y.-K. Choi, "Inte- gration of field effect transistor-based biosensors with a digital microfluidic device for a lab-on-a-chip application", Lab Chip, 2012, vol. 12, no. 8, pp. 1533–1539.
20. E. Buitrago et al., "Vertically stacked Si nanostructures for biosensing applications", Microelectronic Eng., 2012, vol. 97, pp. 345–348.

21. R. Q. J. P. Xlooruq et al., "Performance trade-offs in Fin- FET and GAA device architectures for 7 nm-node and beyond", IEEE SOI-3D-Subthreshold Microelectronics Technology Unified Conference, 2015.
22. Y. Liu, T. Matsukawa, K. Endo, M. Masahara, S.-I. O'uchi, K. Ishii, H. Yamauchi, J. Tsukada, Y. Ishikawa, and E. Suzuki, "Cointegration of high-performance tied-gate three-terminal FinFETs and variable threshold-voltage independent-gate four-terminal FinFETs with asymmetric gate-oxide thicknesses", IEEE Electron Device Lett., 2007, vol. 28, no. 6, pp. 517–519.
23. K. Nayak et al., "CMOS logic device and circuit performance of Si gate all around nanowire MOSFET", IEEE Trans. Electron Devices, 2014, vol. 61, no. 9, pp. 3066–3074.
24. D. Yakimets et al., "Vertical GAAFETs for the ultimate CMOS scaling", IEEE Trans. Electron Devices, 2015, vol. 62, no. 6, pp. 1433–1439.
25. D. Jang, D. Yakimets, G. Eneman, P. Schuddinck, M. G. Bardon, P. Raghavan, A. Spessot, D. Verkest, and A. Mocuta, "Device exploration of nanosheet transistors for sub-7-nm technology node", IEEE Trans. Electron Devices, 2017, vol. 64, no. 6, pp. 2707–2713.
26. H. Mertens et al., "Gate-all-around MOSFETs based on vertically stacked horizontal Si nanowires in a replacement metal gate process on bulk Si substrates", IEEE Symp. VLSI Circuits Dig. Tech., 2016, pp. 1–2. https://doi.org/10.1109/IEDM.2006.346840.
27. N. Singh et al., "Ultra-narrow silicon nanowire gate-all- around CMOS devices: Impact of diameter, channel-orientation and low temperature on device performance", IEEE International Electron Devices, 2006, Meeting, pp. 1–4
28. M. De Marchi et al., "Top–down fabrication of gate-all-around vertically stacked silicon nanowire FETs with controllable polarity", IEEE Trans. Nanotechnol., 2014, vol. 13, no. 6, pp. 1029–1038.
29. B. C. Paul et al., "Impact of process variation on nanowire and nanotube device performance", IEEE Trans. Electron Devices, 2007, vol. 54, no. 9, pp. 269–270. https://doi.org/10.1109/IEDM. 2008.4796805.
30. C. Dupré et al., "15 nm-diameter 3D stacked nanowires with independent gates operation: φFET", IEEE International Electron Devices Meeting, 2008, p. 44287
31. E. Mohapatra et al., "Performance analysis of sub-10 nm vertically stacked gate-all-around FETs", IEEE VLSI Device, Circuit and System Conference (VLSI-DCS), 2020, pp. 331–334.
32. F. I. Sakib et al., "Exploration of negative capacitance in gate-all-around Si nanosheet transistors", IEEE Trans. Electron Devices, 2020, vol. 67, no. 11, pp. 5236–5242.
33. H. Kim et al., "Optimization of stacked Nanoplate FET for 3-nm node", IEEE Trans. Electron Devices, 2020, vol. 67, no.4, pp. 1537–1541.
34. D. Ryu et al., "Investigation of sidewall High-k interfacial layer effect in gate-all-around structure", IEEE Trans. Electron Devices, 2020, vol. 67, no. 4, pp. 1859–1863.
35. M.-J. Tsai et al., "Fabrication and characterization of stacked poly-Si nanosheet with gate-all-around and multi-gate junctionless field effect transistors", IEEE J. Electron Devices Soc., 2019, vol. 7, pp. 1133–1139.

12 Device Simulation Process on TCAD

Abhay Pratap Singh, R.K. Baghel, and Sukeshni Tirkey

12.1 INTRODUCTION

Technology computer-aided design (TCAD) is a powerful tool that is used to design and simulate semiconductor devices and integrated circuits (ICs). It is a simulation-based software package that integrates physics-based models and numerical algorithms to solve complex semiconductor device problems. With the help of TCAD, engineers are able to design, simulate, and check the behavior of a device and analyze its behavior with process variation parameters. Circuit simulations with electrical and electronics parameters can be explored. This is done by solving a set of nonlinear equations that are able to model the physical phenomena of the designed device [1–3]. TCAD can also be used for device designing and optimization; therefore, TCAD enhances device performance. TCAD helps engineers to simulate and analyze the behavior of semiconductor devices. It can be used to accurately predict the behavior of semiconductor devices like transistors, diodes, photodetectors, and solar cells. It is easy for engineers to understand the significance of these devices in order to develop cost-effective, reliable, and high-performance designs. The simulation of devices also enables engineers to identify and troubleshoot design issues before committing to expensive fabrication processes. By utilizing TCAD device simulation, engineers can ensure that their designs meet the required specifications and performance goals [4]. Figure 12.1 elaborates the workflow of the TCAD device-designing and simulation process.

12.1.1 Physics in Device Simulations

A TCAD simulation of a semiconductor device involves the use of physics-based principles and models to accurately predict the behavior of the device. This includes the use of quantum mechanical models of the behavior of charges present in the device, as well as the use of classical physics and thermodynamics to model the thermal and electrical properties of the device [5, 6]. Physics-based models are also used in the simulation of device aging, reliability, and system-level interactions. Engineers are able to design devices more accurately by understanding the physics behind device simulation.

DOI: 10.1201/9781003393542-12

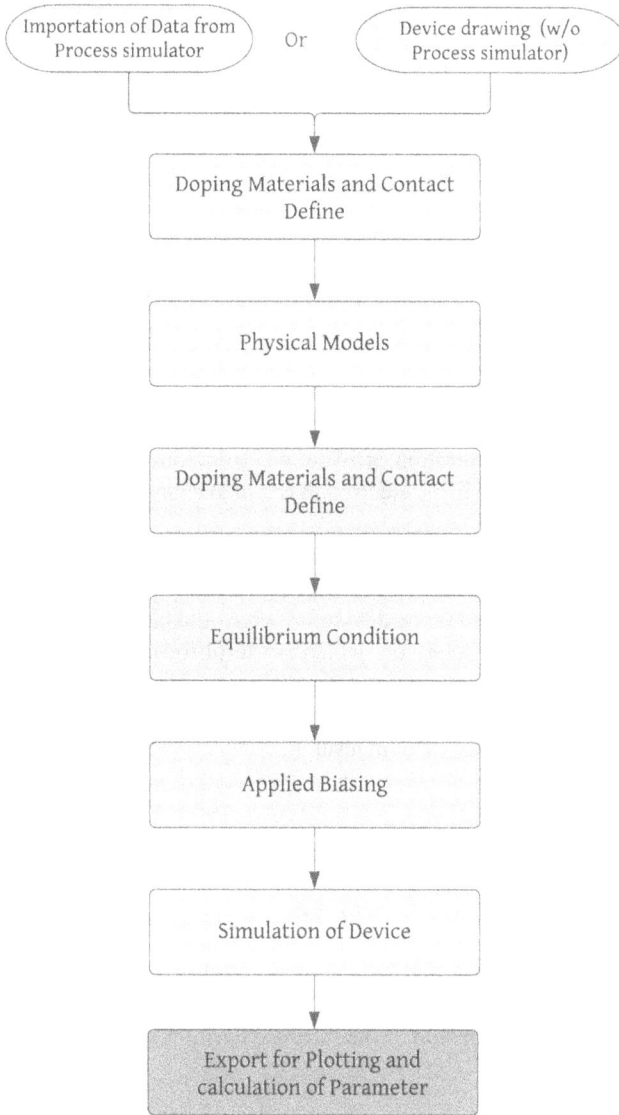

FIGURE 12.1 Flowchart of the technology computer-aided design (TCAD) simulation process.

12.2 THE DRIFT-DIFFUSION MODEL

TCAD device simulation uses the drift-diffusion model to simulate the behavior of charges present in semiconductors. These models are based on the drift-diffusion equations, which describe the momentum of charges present in an electric field. The electron–hole equations along with concentration equations can be modeled along with an electric field. These equations are used to calculate the current density, which is then used to determine how devices will behave under different conditions. Drift-diffusion models can be used to simulate the behavior of transistors, diodes, and other semiconductor devices. They can also be used to simulate the behavior of materials and structures, such as quantum wells and quantum dots. Poisson's equation and continuity equations can be solved by the drift-diffusion model with the help of partial differential equations (PDEs).

12.2.1 POISSON'S EQUATION

Equations that use PDEs, which describe the behavior of electric or gravitational fields in three-dimensional (3D) space, can be expressed as [7]:

$$\nabla.\nabla\varepsilon\Psi = -q\left(p - n + N_D^+ - N_A^-\right) \tag{12.1}$$

where p is the hole concentration; n is the electron concentration; N_D^+, N_A^- are the impurity concentrations; q is the electron magnitude charge; and Ψ denotes the electrostatic potential.

The effect of doping performance on the device can be understood with the help of the drift-diffusion model. The drift-diffusion model for electrons and holes predicts the effect of dopants present in the device, which would be helpful for generating the current–voltage (I-V) characteristic, which is useful for investigating how a device functions in a range of environmental conditions such as temperature and voltage dependency, and for optimizing the device's design for a particular application.

The relationship between vacuum level Ψ, conduction band E_C, and valence band E_V is:

$$E_C = -q\Psi - \chi - \nabla E_C \tag{12.2}$$

$$E_V = E_C - E_g - \nabla E_V \tag{12.3}$$

where ∇E_C, ∇E_V are the shifted bandgaps because of heavy doping or produced mechanical strain; χ is the electron affinity; and E_g is the bandgap.

Furthermore, the relationship of the intrinsic Fermi potential to the vacuum level is:

$$\Psi = \Psi_{intrinsic} - \frac{\chi}{q} - \frac{E_g}{2q} - \frac{K_bT}{2q}\ln\left(\frac{N_C}{N_V}\right) \tag{12.4}$$

The intrinsic Fermi level, an equilibrium state, is used as reference energy, which is set at zero eV.

12.2.2 CONTINUITY EQUATIONS

A continuity equation describes the conservation of charge in a semiconductor material. Mathematically, this can be represented as [8]:

$$\frac{\partial n}{\partial t} = \frac{1}{q}\nabla \cdot \vec{J_n} - (R - G) \tag{12.5}$$

$$\frac{\partial p}{\partial t} = -\frac{1}{q}\nabla \cdot \vec{J_p} - (R - G) \tag{12.6}$$

where $\vec{J_p}$, $\vec{J_n}$ are the hole and electron current densities; R is the recombination and generation rates; and G is electrons and holes.

12.2.3 DRIFT-DIFFUSION EQUATION OF CHARGES

Drift-diffusion current equations describe the charge densities flowing through a semiconductor device. The equations are given by [9]:

$$\vec{J_n} = q\mu_n n \vec{E_n} + qD_n \nabla n \tag{12.7}$$

$$\vec{J_p} = q\mu_p p \vec{E_p} - qD_p \nabla p \tag{12.8}$$

where μ_p, μ_n are the hole and electron mobility; and D_n, D_p, are diffusion constants for electrons and holes, respectively.

An effective electric field of holes and electrons can be represented by $\vec{E_p}$, $\vec{E_n}$, which is related to the band diagram.

$$\vec{E_n} = \frac{1}{q}\nabla E_c - \frac{K_b T}{q}\nabla\left(\ln(N_C) - \ln\left(T^{\frac{3}{2}}\right)\right) \tag{12.9}$$

$$\vec{E_p} = \frac{1}{q}\nabla E_V - \frac{K_b T}{q}\nabla\left(\ln(N_C) - \ln\left(T^{\frac{3}{2}}\right)\right) \tag{12.10}$$

The equation for the drift-diffusion model is generated by replacing the existing density expressions with those from the drift-diffusion model, combined with Poisson's equation.

$$\frac{\partial n}{\partial t} = \nabla \cdot \left(\mu_n n \vec{E_n} + \mu_n \frac{K_b T}{q}\nabla n\right) - (R - G) \tag{12.11}$$

$$\frac{\partial p}{\partial t} = \nabla \cdot \left(\mu_p p \vec{E_p} + \mu_p \frac{K_b T}{q}\nabla p\right) - (R - G) \tag{12.12}$$

This drift-diffusion model is suitable for bipolar junction transistors (BJTs) and metal–oxide–semiconductor field-effect transistor (MOSFET) simulations.

12.3 MOBILITY MODEL

TCAD device simulation uses a mobility model, which uses a mathematical representation of the motion of charges present in a semiconductor device [10]. It is a key component in the simulation of charge transport and device performance. The mobility model considers various physical effects, such as scattering due to phonons, impurities, and defects, as well as quantum effects like band structure, band-to-band tunneling, and Auger recombination. By accurately incorporating these effects, this model provides a more accurate prediction of device performance than simpler models that ignore these effects. In addition, the model is used to evaluate new materials and device structures for device performance. An analytical field model is the default field model for all materials; it is a temperature- and concentration-dependent empirical model, expressed as [11]:

$$\mu_o = \mu_{min} + \frac{\mu_{max}\left(\dfrac{T}{300k}\right)^v - \mu_{min}}{1+\left(\dfrac{T}{300k}\right)^\xi \left(\dfrac{N_{Total}}{N_{ref}}\right)^\alpha} \qquad (12.13)$$

where $N_{Total} = N_A + N_D$ is the total impurity concentration.

12.3.1 FERMI–DIRAC STATISTICS (FDS)

FDS is utilized to accurately model the thermodynamic properties of holes and electrons in a semiconductor by considering the effects of quantum mechanics. FDS is used to calculate the energy distributions of holes and electrons, which helps to decide the Fermi level of a given material [12, 13]. The Fermi level is used to calculate device characteristics like I-V characteristics and capacitance. FDS can also be used to calculate the chemical potential of a semiconductor material and the bandgap width. FDS is essential to accurately simulate the behavior of semiconductor devices used in device development. FDS is also used in the optimization of existing devices and device performance.

12.3.2 BOUNDARY CONDITIONS

TCAD device simulation is an application used in semiconductor device modeling, which is used to predict the electrostatic behavior of a semiconductor device. It combines numerical method simulation with device physics to analyze the behavior of semiconductor devices, including their structures, fabrication processes, and electrical properties.

Boundary conditions are the conditions that need to be satisfied at the device boundaries or edges. These conditions are used to determine the behavior of the device when it is subjected to an input [14]. In TCAD device simulation, boundary conditions include the type of contact, the type of device being modeled, the temperature, and the applied electric field. In addition, the boundary conditions can also include other environmental conditions such as the presence of radiation, light, and active dopants.

The boundary conditions in TCAD device simulation can be used to model a variety of scenarios, includes device performance in different conditions, the design of a device for a specific application, and the analysis of a device for process optimization. For example, boundary conditions can be used to model the performance of a device under different temperatures. It might be implemented to enhance a device's design for a specific task. Similarly, boundary conditions are used to analyze a device's behavior, which can be used to optimize the device's fabrication process.

Boundary conditions have a significant role in TCAD device simulation, because they are used to ensure that the simulated device behaves as expected in a given application. By setting the appropriate boundary conditions, engineers can design and analyze devices that are optimized for their intended applications.

- *Ohmic contact*: This is a type of electrical contact that has low resistance, which allows an efficient flow of charges in the semiconductor material. In an ohmic contact, the energy band diagram shows a flat energy level across the metal–semiconductor interface, indicating that there is no energy barrier present for electrons to move from one material to the other. Figure 12.2(a) shows the energy band diagram, and Figure 12.2(b) shows the ohmic contact. The conduction band and Fermi level of the metal are aligned, indicating that there is no energy barrier present for electrons to pass through the metal into the semiconductor. Similarly, the valence band and Fermi level of the metal are aligned, indicating that there is no energy barrier present for holes to pass through the metal into the semiconductor.

  In a semiconductor device simulation software, such as TCAD, there is a region at which the interface of semiconductor and metal electric current can be determined by the applied voltage. It is a region of low resistance, typically less than 10 ohms, which is used to connect a semiconductor device with an external circuit [15, 16]. The ohmic contact is important for optimization of the performance of the device. It is used to reduce electrical losses through the contact and to ensure that the charge carriers are transferred efficiently from the semiconductor to the external circuit.

$$\phi_n^s = \phi_p^s = -E_{fn}^s = -E_{fp}^s = V_{applied} \qquad (12.14)$$

  The application of ohmic contact in TCAD simulations includes the analysis of device performance and electrical characteristics. For example, ohmic contact can be used to analyze the I-V characteristics of a device, such as the impact of contact resistance on the device's output.

  It can also be used to study the effects of contact material and contact geometry on device performance. Dirichlet boundary conditions are used to implement ohmic contact.

- *Schottky contact*: This is a contact in which there is a semiconductor junction between a metal and a semiconductor material that exhibits very low conduction and low contact resistance. It is commonly used in semiconductor

(a)   Metal   N-type Semiconductor

(b)   Ohmic contact band alignment

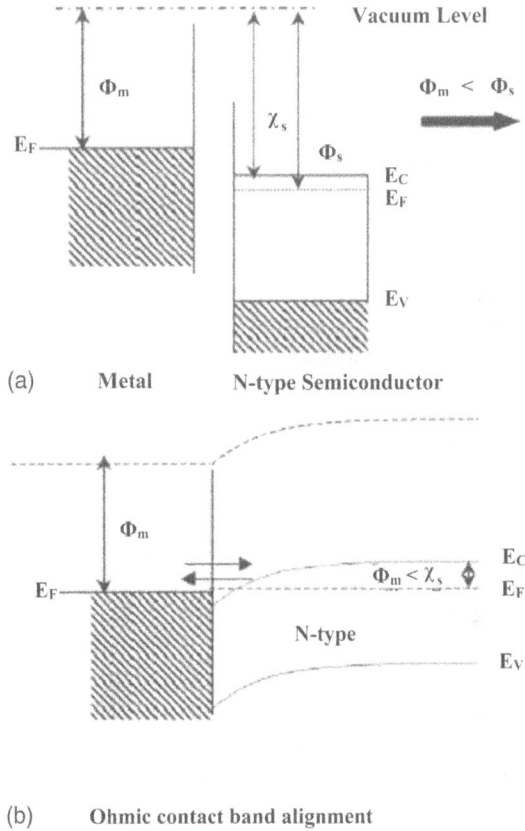

FIGURE 12.2   (a) Ohmic contact and (b) band formation diagram.

devices such as transistors and diodes [17, 18]. TCAD device simulation for Schottky contacts can be modeled with the electrical and electronic properties of the device, like I-V characteristics, capacitance, and transconductance. Schottky contacts have many applications in the electronics industry, including power rectifiers, logic gates, signal processing, and signal conditioning. In these applications, the Schottky contact for a metal and a semiconductor provides a low-resistive path between them, allowing higher current flow and lower voltage drop. It also reduces the risk of device failure due to excessive current. Schottky contacts are also used in solar cell applications to improve efficiency by allowing for the collection of more photons. The surface potential for a Schottky contact is expressed as:

$$V_s = \chi - \chi_{ref} - \phi_b + V_{applied} \qquad (12.15)$$

Figure 12.3 shows (a) an energy band diagram of a Schottky contact as zero potential, (b) an energy band diagram of a Schottky contact as positive potential, (c) an energy band diagram of a Schottky contact as negative

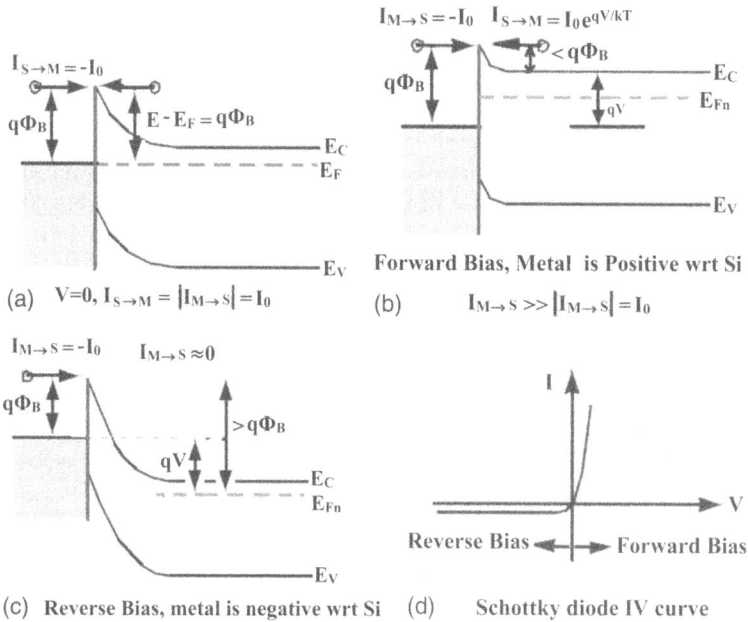

FIGURE 12.3 A Schottky contact band diagram for a metal and a semiconductor: (a) Forward bias with a lower metal work function, (b) forward bias with a higher metal work function, (c) reverse bias, and (d) Schottky diode I-V characteristics.

potential, and (d) the I-V characteristics of a Schottky diode. In the energy band diagram above (Figure 12.3(c)), the Fermi level (E_F) of the metal is located within the bandgap of the semiconductor, and is typically higher than the conduction band edge (E_C) of the semiconductor. The conduction band edge of the semiconductor is shifted upward due to the presence of the metal, creating a barrier to electron flow.

In a TCAD simulation, the Schottky contact can be modeled using parameters such as the Schottky barrier height, series resistance, contact resistance, and saturation current. These parameters are used to accurately model the electrical characteristics of the device, including its I-V characteristics, capacitance, and transconductance. The Schottky contact can also use the simulated device behavior under different operating conditions, like voltage bias and temperature.

• *Neumann boundaries*: TCAD device simulation is used in the semiconductor industry to model the behavior of devices like transistors, diodes, and other ICs. The simulation process involves solving the device's PDEs to predict its behavior under various conditions. By imposing certain boundary conditions on the PDEs, the behavior of the device can be accurately simulated. One type of boundary condition that is commonly used in TCAD device simulations is the Neumann boundary condition [19, 20]. This boundary condition involves the specification of the derivative of the solution of the PDE at the boundaries of the device model. This condition is important, since it is closely related to the conservation of charge and ensures that the total charge present in the system should be conserved. The Neumann boundary condition can be used to simulate the behavior of various devices like diodes, transistors, and ICs. For example, it can be used to simulate the behavior of a transistor when it is connected to a power supply. By specifying the derivative of the solution of the PDE at the boundaries of the device model, the behavior of the transistor can be accurately simulated. The Neumann boundary condition can also be used to simulate the behavior of an IC when it is connected to a power supply. By specifying the derivative of the solution of the PDE at the boundaries of the device model, the behavior of the IC can be accurately simulated, allowing engineers to design more efficient and reliable ICs. In addition, the Neumann boundary condition can be used to simulate the behavior of a circuit when it is connected to a signal source. By specifying the derivative of the solution of the PDE at the boundaries of the device model, the behavior of the circuit can be accurately simulated, allowing engineers to design more efficient and reliable circuits. Overall, the Neumann boundary condition is an essential tool for accurately simulating the behavior of electronic devices. By specifying the derivative of the solution of the PDE at the boundaries of the device model, engineers can accurately simulate the behavior of transistors, diodes, ICs, and other electronic devices.

• *Lumped elements*: Lumped elements in TCAD device simulation are used to model electrical circuit elements like the inductor, capacitor, and resistor. These elements are linked together to create a circuit model, which can then be used to predict and analyze the behavior of a device. For example, a circuit model is used to predict the behavior of the operating characteristics for a solar cell [21]. See Figure 12.4.

The application of lumped elements in TCAD simulation is to help device designers understand how their device will work under different operating conditions. With lumped elements, designers can rapidly and precisely determine the behavior of the device under various operating conditions. This information can then be used to make design decisions that will improve device performance. Lumped elements are also used for optimizing and designing the device to maximize its efficiency.

12.3.3 MESH ISSUES

Mesh issues in TCAD device simulation can arise from several sources. Poor mesh resolution or incorrect mesh settings can lead to inaccurate or incomplete results. Other issues can arise from incorrect material parameters or boundary conditions. In addition, the mesh may be too coarse or too fine, leading to inaccurate results or convergence problems. Finally, the mesh may be limited in its ability to resolve complex geometries or physics [22]. All of these issues can be addressed through careful mesh selection, optimization, and refinement. In terms of applications, mesh issues can lead to incorrect or incomplete predictions of device performance and

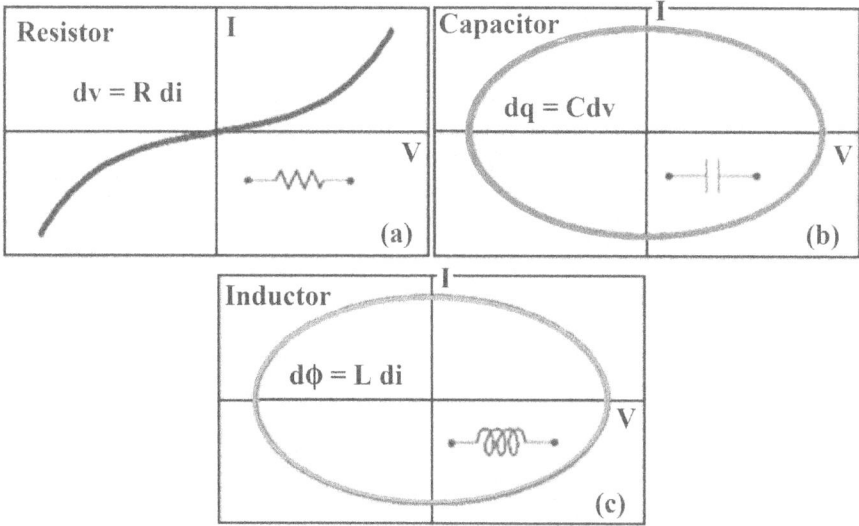

FIGURE 12.4 Lumped-element graph. (a) Resistor symbol and I-V curve, (b) capacitor symbol and I-V curve, and (c) inductor symbol and I-V curve.

reliability. For example, an inadequate mesh may lead to inaccurate predictions of carrier mobility or junction leakage. Mesh issues can also lead to problems with numerical stability and convergence, which can lead to incorrect results. In addition, mesh issues may lead to erroneous predictions of device stress and fatigue, which can lead to device failure. Finally, mesh issues can lead to inaccurate predictions of device lifetime and yield, which can lead to high manufacturing costs and reduced profitability.

12.3.4 PHYSICAL MODELS

A physical model in TCAD device simulation is a mathematical representation of a physical system or process. It is used to understand the device behavior with the help of simulating the physical properties and interactions of the components. Applications of physical models in TCAD device simulation include the following:

- *Design optimization*: Physical models are used to optimize the designed device by accounting for factors such as temperature, stress, and strain [23].
- *Electrical characterization*: Physical models can be used to characterize device performance by accounting for factors like I-V characteristics and capacitance–voltage (CV) characteristics.
- *Reliability analysis*: Physical models can be used to predict the reliability of a device by accounting for factors such as fatigue, wear, and degradation.
- *Process optimization*: Physical models can be used to optimize the fabrication process by accounting for factors such as doping concentration and etch rate.

12.3.5 CARRIER STATISTICS

Carrier statistics in TCAD device simulation comprise a powerful set of tools used to model the behavior of carriers (electrons and holes) in semiconductor devices. This suite of tools allows for the accurate characterization and simulation of the electrical characteristics of devices, such as transistors, diodes, and other electronic components. The carrier statistics suite of tools is developed using advanced numerical techniques, such as Monte Carlo methods, to simulate the behavior of carriers in a device at the microscopic level. This suite of tools is used to analyze the performance of semiconductor devices and to optimize their design [24]. The carrier statistics suite of tools has been used in various applications, such as the design of high-performance transistors, the study of optoelectronic devices, and the characterization of quantum well devices. It is also used in the development of advanced power electronics, such as power converters and inverters. This suite of tools has also been used to study effects like temperature and device operation for the desired performance of devices, as well as in the study of device reliability.

12.3.6 INCOMPLETE IONIZATION OF IMPURITIES

Incomplete ionization of impurities in TCAD device simulation is a big issue that needs to be modified. It affects the accuracy of simulated results, as the charge distribution and mobility of the carriers are significantly altered by the presence of ions [12]. This issue can be addressed by applying a self-consistent approach to the simulation, which considers the effects of the ionized impurities. This approach is applied to various applications, like predicting the performance of the solar cells, designing MOSFETs, and analyzing the behavior of nanoelectronics. The variables f_D and f_A are occupancy coefficients that can be expressed by [14]:

$$f_D = \frac{1}{1 + g_d^{-1}\exp\left[\dfrac{E_D - E_{fn}}{KT}\right]} \tag{12.16}$$

$$f_A = \frac{1}{1 + g_a\exp\left[\dfrac{E_A - E_{fp}}{KT}\right]} \tag{12.17}$$

where f_D, f_A are occupancy coefficients used to determine the degree of ionization; subscripted A, D show the donor and acceptor atoms; and g_d, g_a are degeneracy levels.

12.3.7 HEAVY DOPING EFFECT

Heavy doping is used in TCAD device simulations to improve device performance. It is used to increase carrier mobility and reduce the device's ON-state resistance. Heavy doping can also be used to reduce the threshold voltage and increase the breakdown voltage of the device. Heavy doping also has several applications in complementary metal–oxide–semiconductor (CMOS) technology, such as reducing the junction capacitance and improving the immunity to hot-carrier effects [25]. Heavy doping also improves device robustness and reliability. In addition, heavy doping can be used to improve the performance of nanoscale transistors and circuits. The ionized energy, which is a function of concentration, can be obtained by [26]:

$$E_D = E_{D0} \left[1 - \left(\frac{N_D}{N_{cri}} \right)^{\frac{1}{3}} \right] \tag{12.18}$$

12.3.8 SHOCKLEY–READ–HALL (SRH) AND AUGER RECOMBINATION

SRH recombination is an important phenomenon observed in semiconductor devices. It occurs when an electron and hole move away from each other in opposite directions, and as a result a new electron is created at the location of the hole process known as SRH recombination. SRH recombination is useful in various aspects of semiconductor device simulation, such as the modeling of minority carrier lifetime, carrier mobility, and device I-V characteristics. In a TCAD device simulation, SRH recombination is typically modeled using the drift-diffusion equation, which describes the motion of holes and electrons in a semiconductor device [26, 27]. The SRH recombination process is then included in the equation as a source term, which accounts for recombination of charge carriers. These improved predictions of device characteristics enable device designers to achieve better performance from their devices.

A process in which electrons and holes are recombined and result in the emission of photons is known as the "Auger recombination process." Semiconductor device simulation is a useful process, and it is often used to model the performance of photodetectors and photovoltaic devices. Auger recombination can be simulated using TCAD device simulation software. In this simulation, the Auger recombination rate can be varied in order to study different device behaviors. For example, the Auger recombination rate can be increased to study the effect of increased recombination on device performance. Auger recombination has several important applications in photodetectors and photovoltaic devices. For example, it can be used to model the effect of increased recombination performance of the device, as well as to study the temperature variation effects on device performance. In addition, Auger recombination can be used to simulate the effect of optical feedback performance, and the impact of the surface recombination performance of the device.

12.3.9 IMPACT IONIZATION COEFFICIENT

This is a physical phenomenon in which the impact of an energetic electron with a semiconductor material can cause the generation of electron–hole pairs, resulting in an increase of current flow. This phenomenon is commonly observed in semiconductor devices such as bipolar transistors and MOSFETs. The impact ionization coefficient (IIC) is a measure of the relative likelihood of this effect occurring in a given material. It is the ratio of current generated due to impact ionization to current flow from the device. TCAD simulation is used to model semiconductor devices. It is used to understand the impact ionization behavior of a device under various operating conditions [28, 29]. The IIC can be used in TCAD simulations to accurately predict the performance of a device under varying conditions. In particular, it can be used to estimate the breakdown voltage of a device, which is an important parameter in determining its ability to withstand large currents. Furthermore, the IIC can also be used in conjunction with other device parameters to optimize the device design for better performance. For example, it can be used to optimize the doping profile of a device, therefore reducing the leakage current and improving device efficiency. Overall, the IIC is an important parameter in semiconductor device simulation, and its use can lead to improved device performance.

12.3.10 BARAFF MODEL

The Baraff model is a physically based device-level simulation model for semiconductor devices that was developed in the mid-1980s. It is based on the drift-diffusion equation with a non-parabolic band structure. The model is widely used in TCAD device simulations and widely accepted in the semiconductor industry. The Baraff model is used to simulate various semiconductor device properties, such as electrical characteristics, capacitance, and breakdown voltages. It is also used to simulate the effects of device scaling, process variation, and the performance of high-speed devices. Moreover, it is used to predict how new devices will behave under different operating conditions [30, 31]. The Baraff model is a powerful tool to understand device physics after a prediction of the device behavior. It is a versatile tool that can be used for a variety of device simulations, from simple to complex.

$$\alpha = \frac{e^{g(r,x)}}{\lambda} \tag{12.19}$$

The Baraff model can be used for device optimization, for designing high-performance devices, and for predicting the behavior of future devices. It has been widely used in TCAD device simulations for decades and is highly regarded for its accuracy and reliability. It has been used to develop advanced device models and to optimize device performance. It has also been employed in the study of the physics behind novel device designs and in the forecasting of future device behavior. High-performance device design and process variation effect simulation are two other applications of the concept..

12.3.11 FULOP'S APPROXIMATION

Fulop's approximation is a numerical method used in TCAD device simulation. It is a technique used to solve the PDEs that are used to simulate the physical behavior of devices like BJTs and field-effect transistors (FETs). Fulop's approximation is an iterative method that uses a combination of numerical integration and linear interpolation to solve PDEs [32, 33]. It is a powerful tool for accurately simulating the

behavior of devices, as it can take into account the effects of nonlinearity, temperature, and device geometry.

$$\alpha_F = 1.8 \times 10^{-35} E^7 \tag{12.20}$$

Fulop's approximation has been used in various applications, including the device
development model used for device simulation in semiconductor technology. This model
is used to accurately model the I-V characteristics of the device, allowing for more accurate simulations of their performance. In addition, it is used to simulate the device under
different operating conditions, such as extreme temperatures and device geometries.
This can be useful for designing more energy-efficient devices and for optimizing their
performance. Fulop's approximation is also useful for simulating device failure mechanisms. By accurately modeling the behavior of devices, it can be used to predict how they
will fail under certain conditions. This can be used to identify potential failure modes
and design strategies to prevent them. In addition, Fulop's approximation can be used to
simulate the effects of aging on devices, allowing for more accurate predictions of the
lifetime of a device. Overall, Fulop's approximation is a powerful tool for accurately
simulating the behavior of devices. It can be used to develop device models, design more
efficient devices, and simulate device failure mechanisms. It is an invaluable tool for
TCAD device simulation, and its applications are only growing.

12.3.12 OKUTO–CROWELL MODEL

The Okuto–Crowell model is a 2D device simulation model used for analyzing the
electrical behavior of thin-film transistors (TFTs). It is commonly used in TCAD
device simulations. The Okuto–Crowell model depends on drift-diffusion equations,
which consider the effects of electron mobility, current density, and electric field for
the electrical parameters for TFTs [34, 35]. It is used to model the electrical behavior
of TFTs under a wide range of operating conditions.

$$\alpha(F) = a.\left[1 + c\left(T - T_0\right)\right].E^\gamma.e^{\left\{\frac{b\left[1 + d\left(T - T_0\right)\right]}{F}\right\}^\delta} \tag{12.21}$$

The model is especially useful for simulating TFTs with high mobility, such as
those based on amorphous silicon or organic materials. The model can be used to
analyze the threshold voltage, subthreshold swing, transfer characteristics, and other
electrical properties of TFTs. The Okuto–Crowell model is also used to analyze
the effects of device-designing parameters like the channel doping profile, channel
length, gate oxide thickness, and electrical parameters of the TFTs. This model can
be used to predict the performance of TFTs in various applications, including displays, radiofrequency circuits, and logic circuits.

12.3.13 LECKNER MODEL

The Leckner model is a physics-based model used to simulate the device behavior
for charge carriers present in the semiconductor device. This model is based on the

Boltzmann transport equation, which describes the motion of charge particles in a semiconductor. The Leckner model is used to simulate the behavior of semiconductor devices [36, 37]. This model is used to study the electrical and optical properties of a device, the effect of doping concentration, and the effect of temperature. In addition, it is useful for studying the effects of device geometry, such as the impact of gate length on device characteristics. The Leckner model is an important tool for device designers and can help them optimize the performance of their designs.

$$\alpha_n = \frac{\gamma b_n}{Z} . e^{-\frac{\gamma b_n}{E}}$$

$$\alpha_p = \frac{\gamma a_p}{Z} . e^{-\frac{\gamma b_p}{E}}$$

where α_n, α_p are IICs for electrons and holes, respectively.

$$Z = 1 + \frac{\gamma b_n}{Z} . e^{-\frac{\gamma b_n}{E}} + \frac{\gamma a_p}{Z} . e^{-\frac{\gamma b_p}{E}} \qquad (12.22)$$

$$\gamma = \frac{\tanh\left(\frac{h\omega_{op}}{2kT_0}\right)}{\tanh\left(\frac{h\omega_{op}}{2kT}\right)} \qquad (12.23)$$

12.3.14 MEAN FREE PATH (MFP) MODEL

The MFP model is a widely used model in TCAD device simulations used for the transport of electrons and holes in devices. The MFP calculates the distance of electrons and holes before their collision with another particle [38]. The mobility of electrons and holes can be calculated using the MFP model, which is used to understand the performance of devices. The MFP model is used to understand the behavior of semiconductor devices like transistors, diodes, solar cells, and microelectromechanical (MEMS) devices. It can be used to simulate the effects of temperature, doping, and electric fields. The MFP model is also used to predict the electrical behavior of the device, like I-V characteristics. This model is used to vary simulation conditions like different bias, temperatures, electric fields, and device doping profiles. It is also used to design and optimize the performance of the device. For example, the MFP model can be used to design devices with improved mobility, which can lead to improved device performance.

$$E_C = \frac{E_g}{\lambda}$$

$$\lambda = \lambda_0 \tanh\left(\frac{E_p}{2kT}\right) \qquad (12.24)$$

where λ is the MFP for holes and electrons.

12.4 DESIGN AND SIMULATION OF A SINGLE-GATE MOSFET USING THE COGENDA TCAD SIMULATOR

TCAD is used to simulate the behavior of semiconductor devices such as diodes, transistors, and ICs. COGENDA TCAD software is a powerful tool for designing and simulating MOSFET devices [39, 40]. COGENDA TCAD Simulator allows engineers to model and simulate the performance of semiconductor devices under different conditions, such as temperature, electric field, and material composition. This helps to optimize the design and performance of semiconductor devices before they are manufactured, reducing development time and costs. The simulator includes a range of physics models, including quantum mechanics, electrostatics, and thermo-dynamics. It also has simulation capabilities for both 2D and 3D structures, which makes this tool versatile for a wide range of applications. Overall, COGENDA TCAD Simulator is a valuable tool for semiconductor device design and optimization, allowing engineers to simulate and test the performance of devices before they are manufactured. Here are the steps to use this software for MOSFET device designing and simulation.

12.4.1 SIMULATION SOFTWARE SET UP

Launch the COGENDA TCAD software and create a new project, then select Device Drawing.

12.4.2 SELECTION OF DEVICE GEOMETRY

Use the built-in editor to create the MOSFET device structure. This involves defining the gate, source, drain, and other elements, as shown in Figure 12.5, with the

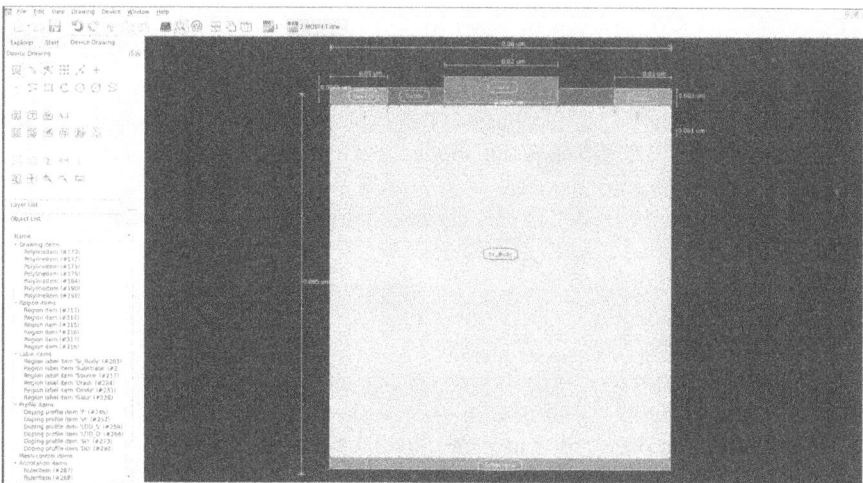

FIGURE 12.5 Selection of device geometry.

TABLE 12.1
Device Geometry

Parameter	Conventional MOSFET
Gate length, L_g (nm)	20
Source/drain length, $L_{S/D}$ (nm)	10
Gate oxide thickness, T_{ox} (nm)	0.5
Channel doping N_{ch} (cm^{-3})	1×10^{16}
Source/drain doping, $N_{S/D}$ (cm^{-3})	1×10^{19}
Gate metal work function, ϕ_M (eV)	4.5 (N Poly Si)

selection of device geometry with a doping profile. Selection of gate oxide thickness, which is a thin layer of insulator material, is used to separate the gate and body parts of the MOSFET. The thickness of the oxide effect is the capacitance that turns on the switching and power consumption of the device.

A thinner oxide leads to higher capacitance and causes faster switching of the device, but it also increases the risks of gate leakage and breakdown. The channel length is the distance measure from the source to the drain regions of the MOSFET. It determines the resistance of the channel and the current-carrying capacity of the MOSFET. The effect of the channel length has a significant impact on device performance; for example, short-channel leads low on resistance cause high current density. Table 12.1 shows the device geometry profile for the proposed device structure.

12.4.3 PROCESS PARAMETERS

Define the process parameters for your MOSFET device; this includes specifying the doping profile, gate oxide thickness, and channel length. For the doping profile, MOSFETs are typically fabricated on a silicon substrate, which is doped to create either a p-type or n-type semiconductor. The doping profile of the substrate affects the V_{th} of the MOSFET, which is the voltage at which the device starts to conduct. A higher doping concentration leads to a lower V_{th}, which means that the MOSFET turns on at a lower voltage. The doping concentration of the proposed device is shown in Figure 12.6.

12.4.4 MESHING

Meshing is a technique used in finite element analysis to break down complex geometries into smaller, simpler elements. In the context of MOSFET modeling, meshing is used to divide the device structure into smaller parts for simulation. The process of meshing involves dividing the MOSFET structure into small triangular or quadrilateral elements.

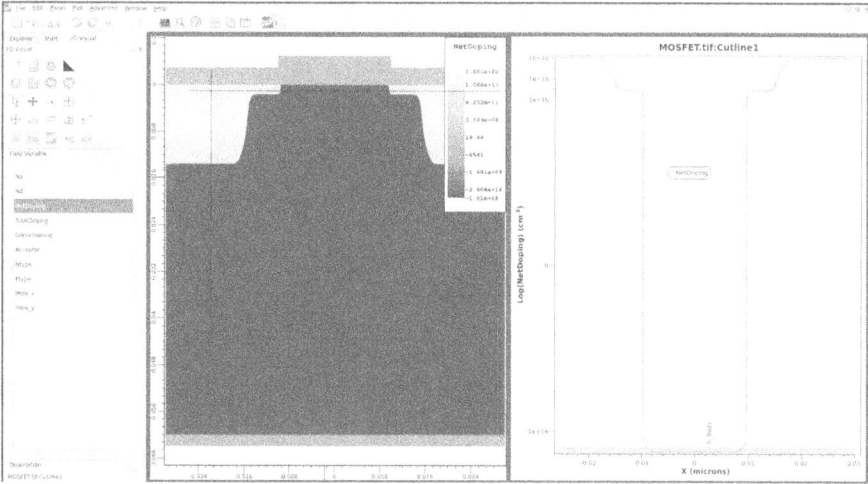

FIGURE 12.6 Doping concentration of the device.

The size and shape of the elements are chosen to ensure accurate simulation results while keeping the computational cost reasonable. Typically, the smaller the size of the element, the more accurate the simulation, but a simulation takes a long time. The meshing process is critical to accurately simulate MOSFET behavior because it affects the accuracy of the resulting numerical solution. A poorly meshed MOSFET model can lead to inaccurate results, while a well-meshed model can provide a more accurate representation of the device behavior. Do the meshing and refine the mesh according to the device structure, as shown in Figure 12.7.

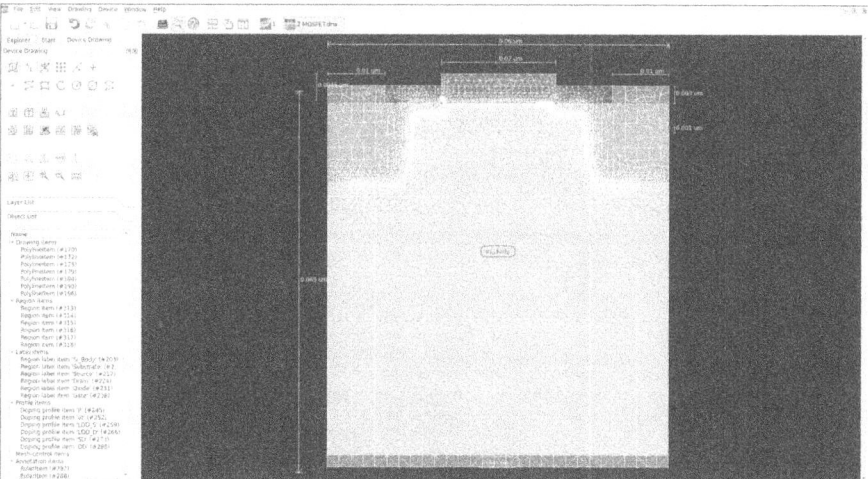

FIGURE 12.7 Meshing profile of the designed device.

12.4.5 SIMULATION PARAMETER SETUP

Set up the simulation environment. This includes defining the simulation temperature, bias conditions, and other parameters. Figure 12.8 shows the device simulation setup.

- First, let's define the simulation temperature at which the simulation environment can analyze the behavior of the MOSFET. Typically, MOSFET simulations are performed at room temperature, which is 300 K (Kelvin).
- Next, let's define the range of applied voltage bias for the MOSFET, which is the conditions at which the device will achieve the desired performance.
- The source is typically grounded (0 V), so the bias conditions for the MOSFET can be specified using two parameters, V_{DS} and V_{GS}.
- To set up the simulation environment, you will need to specify values for V_{DS} and V_{GS}. The specific values will depend on the operating point you want to simulate.
- For example, if you want to simulate the MOSFET in the saturation region, you might set V_{DS} to a fixed value (e.g., 5 V) and vary V_{GS} from 0 V to a value that puts the MOSFET in saturation. Overall, the simulation environment for a conventional MOSFET would typically include the following:

  Temperature: 300 K

  MOSFET terminals: Source (S), drain (D), and gate (G)

  Source voltage: 0 V

  Drain voltage: A fixed value (0.5 V)

  Gate voltage: Varied to achieve the desired operating point (0–1 V)

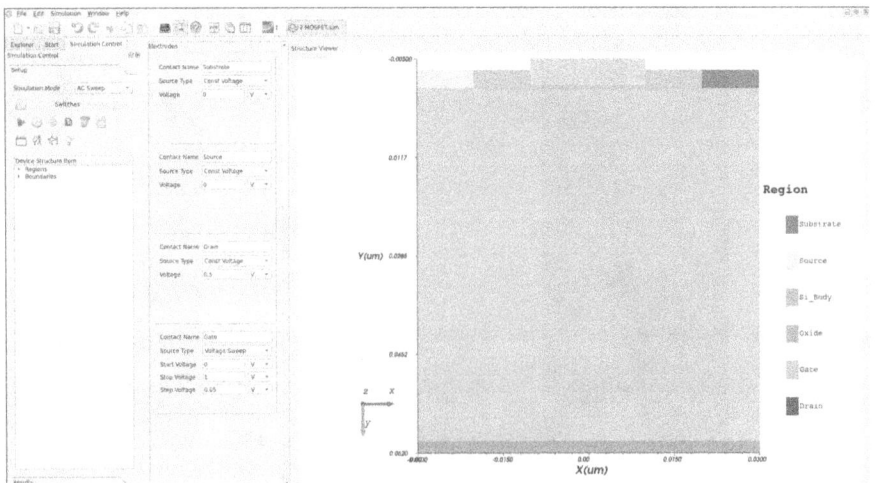

FIGURE 12.8 Simulation setup for the designed device.

12.4.6 START SIMULATION

Run the simulation to obtain I-V curves, CV curves, and other important performance metrics.

12.4.7 ANALYSIS OF RESULTS

Analyze the simulation results to determine if the device meets your design specifications. If not, adjust the device structure or simulation parameters, and repeat the simulation.

12.4.8 CHECKING RESULTS MEETS DESIGN SPECIFICATION OR NOT

The proposed device is an n-channel MOSFET; n-typed doping has been done for the source and drain regions, which means they contain an excess of negatively charged electrons. When a positive voltage is applied to the gate, it repels some of these electrons away from the surface, creating a depletion region. As the gate voltage becomes more positive, the depletion region gets wider and the channel becomes narrower, which reduces the flow of current. Figure 12.9 shows the electron concentration in the proposed device.

12.4.9 SIMULATED MESHING PROFILE WITH ELECTRONS

The flow of electrons in a MOSFET is controlled by the electric field created by the applied potential to the gate. By changing the gate voltage, the conductivity of the channel between the source and drain can be controlled, allowing the MOSFET to function as a switch or amplifier. Figure 12.10 shows the potential profile for the proposed device.

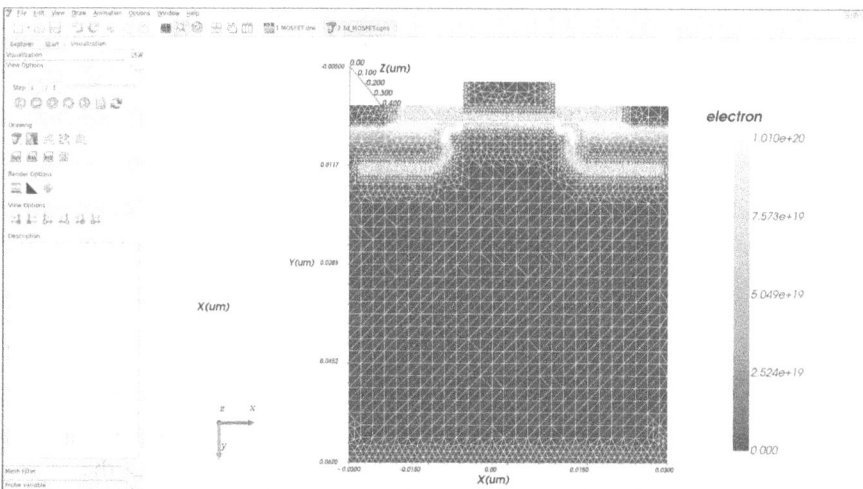

FIGURE 12.9 Charge flow diagram of the designed device.

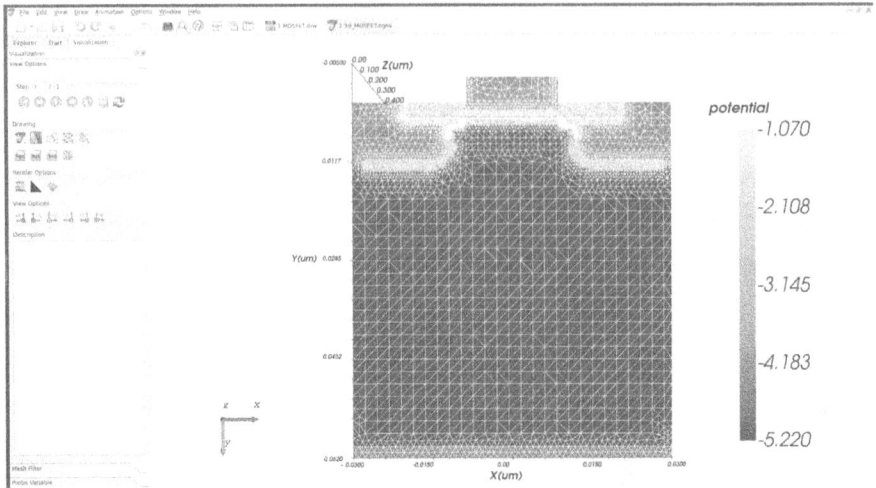

FIGURE 12.10   Potential diagram of the proposed device.

12.4.10 ELECTRON FLOW

The shape of the I-V curve depends on the specific MOSFET and its operating conditions, such as the gate-source voltage and the device dimensions. The I-V curve is an important characteristic of a MOSFET and is used to determine the device's operating region and its performance in various applications. When a MOSFET is operated with a fixed drain-source voltage (V_{ds}) of 0.5 V and a variable gate-source voltage (V_{gs}) between 0 and 1 V, the resulting I-V curve can be explained as follows:

- At very low gate-source voltages (V_{gs} < threshold voltage V_t), the MOSFET is in the cutoff region, meaning that there is no current flow between the drain and source terminals.
- As the V_{gs} is increased beyond the V_{th}, the MOSFET enters the linear region, where the drain current (I_{ds}) increases linearly with increasing V_{gs}. In this region, the MOSFET behaves like a resistor with a resistance that depends on the V_{gs}.
- As the gate-source voltage is further increased, the MOSFET enters the saturation region, where the drain current reaches a maximum value and remains relatively constant, even if the V_{gs} is increased. In this region, the MOSFET behaves like a current source. Figure 12.11 shows the I-V characteristic curve for the proposed device.

12.4.11 CURRENT VOLTAGE (I–V) CHARACTERISTICS USING APPLIED POTENTIAL

The gate capacitance of a MOSFET varies with the applied gate-source voltage and drain-source voltage, and it is strongly dependent on the dimensions of the device, such as channel length, channel width, and gate oxide thickness. The gate capacitance curve of a MOSFET represents the variation of the gate capacitance with

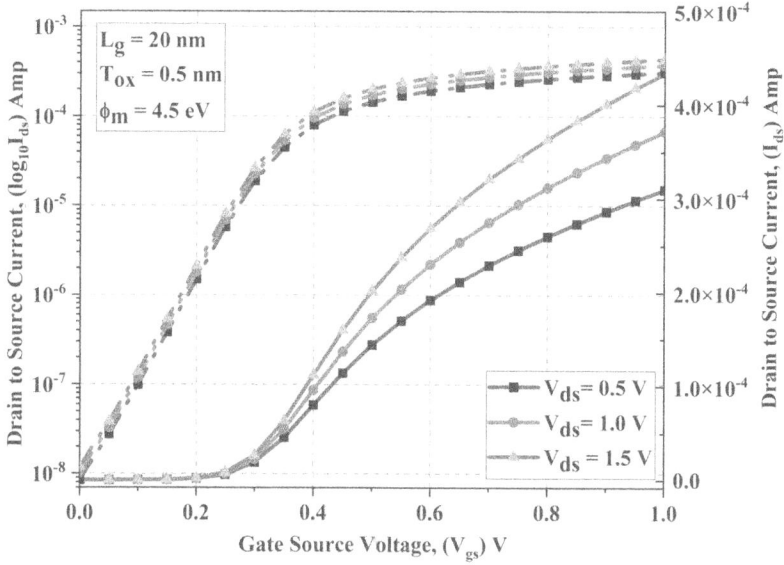

FIGURE 12.11   I-V characteristic curve for the proposed device.

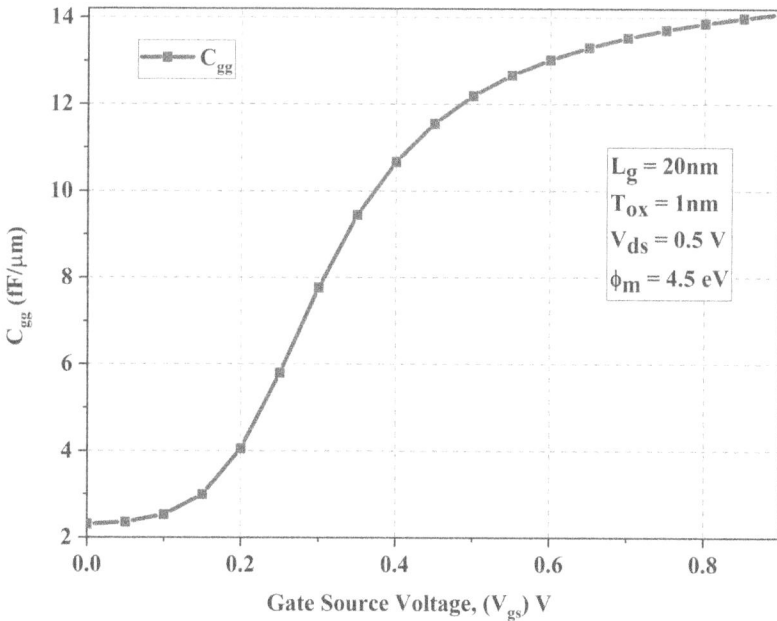

FIGURE 12.12   Capacitance curve of the proposed device.

TABLE 12.2
Calculated Parameters for the Proposed Device

Parameter	Conventional MOSFET
Drain-to-source current, I_{DS} (Amp)	3.12×10^{-4}
$I_{OFF}/I_{Leakage}$ (Amp)	10^{-8}
Switching ratio (I_{ON}/I_{OFF})	10^4
Gate oxide capacitance, C_{gg} (fF)	14.02

respect to the gate-source voltage (V_{gs}). At a fixed drain-source voltage (V_{ds}), which in this case is 0.5 V, the gate capacitance curve can be obtained by measuring the gate-source capacitance (C_{gs}) at different values of V_{gs} ranging from 0 V to 1 V. Initially, when V_{gs} is 0 V, the MOSFET is in the cutoff region, and the gate-source capacitance is at its minimum value, known as the "zero-bias capacitance" (C_{gs0}). As V_{gs} is increased, the MOSFET enters the triode or linear region, and the gate-source capacitance starts increasing due to the formation of an inversion layer in the channel. As the gate voltage continues to increase, the MOSFET enters the saturation region, where the channel is fully inverted, and the gate-source capacitance reaches its maximum value, known as the "saturation capacitance" (C_{gs_sat}). At this point, the gate-source voltage is high enough to fully deplete the channel under the oxide layer. Figure 12.12 shows the CV curve of the proposed device.

12.4.12 GATE CAPACITANCE CHARACTERISTICS

Analyze the simulation results to determine if the device meets your design specifications. If not, adjust the device structure or simulation parameters, and repeat the simulation. Table 12.2 shows the simulated results of the proposed device in terms of ON-current I_{ON}, OFF-current I_{OFF}, switching ratio I_{ON}/I_{OFF}, and gate capacitance C_{gg}.

Once you are satisfied with the simulation results, export the device structure and simulation data for use in further analysis or device fabrication.

12.5 CONCLUSION

A conventional silicon MOSFET has been designed using the COGENDA TCAD simulator tool for understanding the device simulation process, and each step of the simulation process has been described. TCAD-based simulation of MOSFETs is an effective tool for predicting the behavior of the device in terms of both electrical performance and reliability. It provides a detailed understanding of the device physics, enabling engineers to design better MOSFETs for specific applications. It also helps to optimize the device performance, reduce leakage current, and improve device reliability. TCAD-based simulations are becoming increasingly important in the design and fabrication of modern semiconductor devices, and are likely to become even more important in the future.

REFERENCES

1. Gerrer, L., Brown, A.R., Millar, C., Hussin, R., Amoroso, S.M., Cheng, B., Reid, D., Alexander, C., Fried, D., Hargrove, M. and Greiner, K., 2015. Accurate simulation of transistor-level variability for the purposes of TCAD-based device-technology cooptimization. IEEE Transactions on Electron Devices, 62(6), pp. 1739–1745.
2. Maiti, C.K., 2017. Introducing Technology Computer-Aided Design (TCAD): Fundamentals, Simulations, and Applications. CRC Press.
3. Saha, S.K., 1999. Managing technology CAD for competitive advantage: An efficient approach for integrated circuit fabrication technology development. IEEE Transactions on Engineering Management, 46(2), pp. 221–229.
4. Pati, G.K., 2014. Two dimensional analytical threshold voltage modeling of dual material gate S-SOI MOSFET (Doctoral dissertation).
5. Yuan, J.S. and Liou, J.J., 1998. Semiconductor Device Physics and Simulation. Springer Science & Business Media.
6. Goldsman, N., Henrickson, L. and Frey, J., 1991. A physics-based analytical/numerical solution to the Boltzmann transport equation for use in device simulation. Solid-State Electronics, 34(4), pp. 389–396.
7. Glynn, P.W. and Meyn, S.P., 1996. A Lyapunov bound for solutions of the poisson equation. The Annals of Probability, 24, pp. 916–931.
8. Ambrosio, L. and Crippa, G., 2014. Continuity equations and ODE flows with non-smooth velocity. Proceedings of the Royal Society of Edinburgh Section A: Mathematics, 144(6), pp. 1191–1244.
9. Boris, J.P. and Book, D.L., 1976. Solution of continuity equations by the method of flux-corrected transport. Controlled Fusion, 16, pp. 85–129.
10. Vasileska, D., 2006. Drift-diffusion model: Introduction. online materials to nanohub.org. resource link: https://nanohub.org/resources/1545/download/ddmodel_introductory_part_word.pdf.
11. Würfel, U., Cuevas, A. and Würfel, P., 2014. Charge carrier separation in solar cells. IEEE Journal of Photovoltaics, 5(1), pp. 461–469.
12. Hellings, G., Eneman, G., Krom, R., De Jaeger, B., Mitard, J., De Keersgieter, A., Hoffmann, T., Meuris, M. and De Meyer, K., 2010. Electrical TCAD simulations of a germanium pMOSFET technology. IEEE Transactions on Electron Devices, 57(10), pp. 2539–2546.
13. Coppolelli, B., 2022. Modeling of Tunnel-FETs: accurate calibration of numerical and semi-analytical models (Doctoral dissertation, Politecnico di Torino).
14. Altermatt, P.P., Schumacher, J.O., Cuevas, A., Kerr, M.J., Glunz, S.W., King, R.R., Heiser, G. and Schenk, A., 2002. Numerical modeling of highly doped Si: P emitters based on Fermi–Dirac statistics and self-consistent material parameters. Journal of Applied Physics, 92(6), pp. 3187–3197.
15. Jin, Y., Zeng, C., Ma, L. and Barlage, D., 2007. Analytical threshold voltage model with TCAD simulation verification for design and evaluation of tri-gate MOSFETs. Solid-State Electronics, 51(3), pp. 347–353.
16. Fan, Z., Mohammad, S.N., Kim, W., Aktas, Ö., Botchkarev, A.E. and Morkoç, H., 1996. Very low resistance multilayer ohmic contact to n-GaN. Applied Physics Letters, 68(12), pp. 1672–1674.
17. Matsuzawa, K., Uchida, K. and Nishiyama, A., 2000. A unified simulation of Schottky and ohmic contacts. IEEE Transactions on Electron Devices, 47(1), pp. 103–108.
18. Trew, R.J., Yan, J.B. and Mock, P.M., 1991. The potential of diamond and SiC electronic devices for microwave and millimeter-wave power applications. Proceedings of the IEEE, 79(5), pp. 598–620.

19. Lee, K., Shur, M., Fjeldly, T.A. and Ytterdal, T., 1993. Semiconductor Device Modeling for VLSI. ACM Digital Library.

20. Singh, A.P., Shankar, P.N., Baghel, R.K. and Tirkey, S., 2023, February. A review on graphene transistors. In 2023 IEEE International Students' Conference on Electrical, Electronics and Computer Science (SCEECS) (pp. 1–6). IEEE.

21. Chung, I.Y., Jang, H., Lee, J., Moon, H., Seo, S.M. and Kim, D.H., 2012. Simulation study on discrete charge effects of SiNW biosensors according to bound target position using a 3D TCAD simulator. Nanotechnology, 23(6), p. 065202.

22. Lee, J., Kim, K.W., Huh, Y., Bendix, P. and Kang, S.M., 2003. Chip-level charged-device modeling and simulation in CMOS integrated circuits. IEEE Transactions on Computer-Aided Design of Integrated Circuits and Systems, 22(1), pp. 67–81.

23. Shigyo, N., Tanimoto, H. and Enda, T., 2000. Mesh related problems in device simulation: Treatments of meshing noise and leakage current. Solid-State Electronics, 44(1), pp. 11–16.

24. Stanojevic, Z., Baumgartner, O., Mitterbauer, F., Demel, H., Kernstock, C., Karner, M., Eyert, V., France-Lanord, A., Saxe, P., Freeman, C. and Wimmer, E., 2015, December. Physical modeling-A new paradigm in device simulation. In 2015 IEEE International Electron Devices Meeting (IEDM) (pp. 5–1). IEEE.

25. Donato, N. and Udrea, F., 2018. Static and dynamic effects of the incomplete ionization in superjunction devices. IEEE Transactions on Electron Devices, 65(10), pp. 4469–4475.

26. Bankapalli, Y.S. and Wong, H.Y., 2019, September. TCAD augmented machine learning for semiconductor device failure troubleshooting and reverse engineering. In 2019 International Conference on Simulation of Semiconductor Processes and Devices (SISPAD) (pp. 1–4). IEEE.

27. Kuik, M., Wetzelaer, G.J.A., Nicolai, H.T., Craciun, N.I., De Leeuw, D.M. and Blom, P.W., 2014. 25th anniversary article: Charge transport and recombination in polymer light-emitting diodes. Advanced Materials, 26(4), pp. 512–531.

28. Min, B., Wagner, H., Dastgheib-Shirazi, A., Kimmerle, A., Kurz, H. and Altermatt, P.P., 2014. Heavily doped Si: P emitters of crystalline Si solar cells: Recombination due to phosphorus precipitation. Physica Status Solidi (RRL)–Rapid Research Letters, 8(8), pp. 680–684.

29. Beard, M.C., Midgett, A.G., Hanna, M.C., Luther, J.M., Hughes, B.K. and Nozik, A.J., 2010. Comparing multiple exciton generation in quantum dots to impact ionization in bulk semiconductors: Implications for enhancement of solar energy conversion. Nano Letters, 10(8), pp. 3019–3027.

30. Mayer, F., Le Royer, C., Le Carval, G., Clavelier, L. and Deleonibus, S., 2006. Static and dynamic TCAD analysis of IMOS performance: From the single device to the circuit. IEEE Transactions on Electron Devices, 53(8), pp. 1852–1857.

31. Li, S. and Li, S., 2011. 3D TCAD Simulation for Semiconductor Processes, Devices and Optoelectronics. Springer Science & Business Media.

32. Li, S., Fu, Y., Li, S. and Fu, Y., 2012. Advanced Theory of TCAD Device Simulation. 3D TCAD Simulation for Semiconductor Processes, Devices and Optoelectronics, pp. 41–80, Springer.

33. Reggiani, S., Barone, G., Gnani, E., Gnudi, A., Baccarani, G., Poli, S., Wise, R., Chuang, M.Y., Tian, W., Pendharkar, S. and Denison, M., 2014. Characterization and modeling of electrical stress degradation in STI-based integrated power devices. Solid-State Electronics, 102, pp. 25–41.

34. Hueting, R.J., Heringa, A., Boksteen, B.K., Dutta, S., Ferrara, A., Agarwal, V. and Annema, A.J., 2016. An improved analytical model for carrier multiplication near breakdown in diodes. IEEE Transactions on Electron Devices, 64(1), pp. 264–270.

35. Qiu, W.C., Hu, W.D., Chen, L., Lin, C., Cheng, X.A., Chen, X.S. and Lu, W., 2015. Dark current transport and avalanche mechanism in HgCdTe electron-avalanche photodiodes. IEEE Transactions on Electron Devices, 62(6), pp. 1926–1931.

36. Mandurrino, M., Cartiglia, N., Staiano, A., Arcidiacono, R., Obertino, M.M., Ferrero, M., Cenna, F., Sola, V., Boscardin, M., Patetnoster, G. and Ficorella, F., 2017, October. Numerical simulation of charge multiplication in ultra-fast silicon detectors (ufsd) and comparison with experimental data. In 2017 IEEE Nuclear Science Symposium and Medical Imaging Conference (NSS/MIC) (pp. 1–4). IEEE.

37. Strüder, L., Lutz, G., Lechner, P., Soltau, H., Holl, P., Conde, C.A.N., Kurakado, M., Galeazzi, M., Figueroa, Feliciano, E., Dąbrowski, W. and Gryboś, P., 2004. X-Ray Detectors. X-Ray Spectrometry: Recent Technological Advances, pp. 133–275, Wiley Online Library.

38. Drozdov, D.G., Prokopenko, N.N., Savchenko, E.M., Dukanov, P.A., Rodin, V.G. and Grushin, A.I., 2020. Technological and devices modeling of complementary JFETs over a wide temperature range. Microelectronics Journal, 105, p. 104911.

39. Martinie, S., Le Carval, G., Munteanu, D., Soliveres, S. and Autran, J.L., 2008. Impact of ballistic and quasi-ballistic transport on performances of double-gate MOSFET-based circuits. IEEE Transactions on Electron Devices, 55(9), pp. 2443–2453.

40. Singh, A.P., Baghel, R.K. and Tirkey, S., 2023, February. Enhanced low dimensional MOSFETs with variation of high K dielectric materials. In 2023 IEEE International Students' Conference on Electrical, Electronics and Computer Science (SCEECS) (pp. 1–5). IEEE.

Index

For Product Safety Concerns and Information please contact our EU
representative GPSR@taylorandfrancis.com
Taylor & Francis Verlag GmbH, Kaufingerstraße 24, 80331 München, Germany

www.ingramcontent.com/pod-product-compliance
Lightning Source LLC
Chambersburg PA
CBHW060337220326
41598CB00023B/2737

9 781032 493879